T0350478

CAMBRIDGE TRACTS IN MATHEMATICS

General Editors

B. BOLLOBÁS, W. FULTON, A. KATOK, F. KIRWAN,
P. SARNAK, B. SIMON, B.TOTARO

192 Normal Approximations with Malliavin Calculus

CAMBRIDGE TRACTS IN MATHEMATICS

GENERAL EDITORS

B. BOLLOBÁS, W. FULTON, A. KATOK, F. KIRWAN, P. SARNAK,

B. SIMON, B.TOTARO

A complete list of books in the series can be found at www.cambridge.org/mathematics.
Recent titles include the following:

Normal Approximations with Malliavin Calculus

From Stein's Method to Universality

IVAN NOURDIN
Université de Nancy I, France

GIOVANNI PECCATI
Université du Luxembourg

CAMBRIDGE
UNIVERSITY PRESS

University Printing House, Cambridge CB2 8BS, United Kingdom

Cambridge University Press is part of the University of Cambridge.

It furthers the University's mission by disseminating knowledge in the pursuit of education, learning and research at the highest international levels of excellence.

www.cambridge.org
Information on this title: www.cambridge.org/9781107017771

First published 2012

A catalogue record for this publication is available from the British Library

Library of Congress Cataloguing in Publication data
Nourdin, Ivan.
Normal approximations with Malliavin calculus : from Stein's method to universality / Ivan Nourdin, Giovanni Peccati.
p. cm. – (Cambridge tracts in mathematics ; 192)
Includes bibliographical references and index.
ISBN 978-1-107-01777-1 (hardback)
1. Approximation theory. 2. Malliavin calculus.
I. Peccati, Giovanni, 1975– II. Title.
QA221.N68 2012
519.2′3–dc23
2012010132

ISBN 978-1-107-01777-1 Hardback

Additional resources for this publication at
www.iecn.u-nancy.fr/~nourdin/steinmalliavin.htm

To Lili, Juliette and Delphine.
To Emma Elīza and Ieva.

Contents

Preface

This is a text about *probabilistic approximations*, which are mathematical statements providing estimates of the distance between the laws of two random objects. As the title suggests, we will be mainly interested in approximations involving one or more *normal* (equivalently called *Gaussian*) random elements. Normal approximations are naturally connected with central limit theorems (CLTs), i.e. convergence results displaying a Gaussian limit, and are one of the leading themes of the whole theory of probability.

The main thread of our text concerns the normal approximations, as well as the corresponding CLTs, associated with random variables that are functionals of a given Gaussian field, such as a (fractional) Brownian motion on the real line. In particular, a pivotal role will be played by the elements of the so-called *Gaussian Wiener chaos*. The concept of Wiener chaos generalizes to an infinite-dimensional setting the properties of the Hermite polynomials (which are the orthogonal polynomials associated with the one-dimensional Gaussian distribution), and is now a crucial object in several branches of theoretical and applied Gaussian analysis.

The cornerstone of our book is the combination of two probabilistic techniques, namely the *Malliavin calculus of variations* and *Stein's method* for probabilistic approximations.

The Malliavin calculus of variations is an infinite-dimensional differential calculus, whose operators act on functionals of general Gaussian processes. Initiated by Paul Malliavin (starting from the seminal paper [69], which focused on a probabilistic proof of Hörmander's 'sum of squares' theorem), this theory is based on a powerful use of infinite-dimensional integration by parts formulae. Although originally exploited for studying the regularity of the laws of Wiener functionals (such as the solutions of stochastic differential equations), the scope of its actual applications, ranging from density estimates to concentration inequalities, and from anticipative stochastic calculus to the computations of 'Greeks' in mathematical finance, continues to grow. For a classic presentation of this subject, the reader can consult the three texts by Malliavin [70], Nualart [98] and Janson [57]. Our book is the first monograph providing a self-contained introduction to Malliavin calculus from the specific standpoint of limit theorems and probabilistic approximations.

Stein's method can be roughly described as a collection of probabilistic techniques for assessing the distance between probability distributions by means of differential operators. This approach was originally developed by Charles Stein in the landmark paper [135], and then further refined in the monograph [136]. In recent years, Stein's method has become one of the most popular and powerful tools for computing explicit bounds in probabilistic limit theorems, with applications to fields as diverse as random matrices, random graphs, probability on groups and spin glasses (to name but a few). The treatise [22], by Chen, Goldstein and Shao, provides an exhaustive discussion of the theoretical foundations of Stein's method for normal approximations, as well as an overview of its many ramifications and applications (see also the two surveys by Chen and Shao [23] and Reinert [117]).

We shall show that the integration by parts formulae of Malliavin calculus can be fruitfully combined with the differential operators arising in Stein's method. This interaction will be exploited to produce a set of flexible and far-reaching tools, allowing general CLTs (as well as explicit rates of convergence) to be deduced for sequences of functionals of Gaussian fields. It should be noted that the theory developed in this book virtually replaces every technique previously used to establish CLTs for Gaussian-subordinated random variables, e.g. those based on moment/cumulant computations (see, for example, Peccati and Taqqu [110]).

As discussed at length in the text, the theoretical backbone of the present monograph originates from the content of five papers.

- Nualart and Peccati [101] give an exhaustive (and striking) characterization of CLTs inside a fixed Wiener chaos. This result, which we will later denote as the 'fourth-moment theorem', yields a drastic simplification of the classic *method of moments and cumulants*, and is one of the main topics discussed in the book.
- Peccati and Tudor [111] provide multidimensional extensions of the findings of [101]. In view of the *Wiener–Itô chaotic representation property* (see Chapter 2), the findings of [111] pave the way for CLTs involving general functionals of Gaussian fields (not necessarily living inside a fixed Wiener chaos).
- The paper by Nualart and Ortiz-Latorre [100] contains a crucial methodological breakthrough, linking CLTs on Wiener chaos to the

asymptotic behavior of Malliavin operators. In particular, a prominent role is played by the norms of the *Malliavin derivatives* of multiple Wiener–Itô integrals.

- Nourdin and Peccati [88] establish the above-mentioned connection between Malliavin calculus and Stein's method, thus providing substantial refinements of the findings of [100, 101, 111]. Along similar lines, the multivariate case is dealt with by Nourdin, Peccati and Réveillac [95].
- Nourdin, Peccati and Reinert [94] link the above results to the so-called *universality phenomenon*, according to which the asymptotic behavior of (correctly rescaled) large random systems does not depend on the distribution of their components. Universality results, also known as 'invariance principles', are almost ubiquitous in probability: distinguished examples are the classic central limit theorem and the circular and semicircular laws in random matrix theory. See Chapter 11 for further discussions.

The above-mentioned references have been the starting point of many developments and applications. These include density estimates, concentration inequalities, Berry–Esseen bounds for power variations of Gaussian-subordinated processes, normalization of Brownian local times, random polymers, random matrices, parametric estimation in fractional models and the study of polyspectra associated with stationary fields on homogeneous spaces. Several of these extensions and applications are explicitly described in our book. See the webpage

```
http://www.iecn.u-nancy.fr/~nourdin/steinmalliavin.htm
```

for a constantly updated reference list. See the monographs [74] and [110], respectively, for further applications to random fields on the sphere (motivated by cosmological data analysis), and for a discussion of the combinatorial structures associated with the Gaussian Wiener chaos.

The book is addressed to researchers and graduate students in probability and mathematical statistics, wishing to acquire a thorough knowledge of modern Gaussian analysis as used to develop asymptotic techniques related to normal approximations.

With very few exceptions (where precise references are given), every result stated in the book is proved (sometimes through a detailed exercise), and even the most basic elements of Malliavin calculus and Stein's method are motivated, defined and studied from scratch. Several proofs are new, and each chapter contains a set of exercises (with some hints!), as well as a number of bibliographic comments. Due to these features, the text is more or less self-contained, although our ideal reader should have attended a basic course in modern probability (corresponding, for example, to the books by Billingsley [13] and Chung [25]), and also have some knowledge of functional analysis (covering, for example, the content of chapters 5–6 in Dudley's book [32]). Some facts and definitions concerning operator theory are used – we find that a very readable reference in this respect is [47], by Hirsch and Lacombe.

Acknowledgements. We heartily thank Simon Campese, David Nualart and Mark Podolskij for a careful reading of some earlier drafts of the book. All remaining errors are, of course, their sole responsibility.

Ivan Nourdin, Nancy
Giovanni Peccati, Luxembourg

Introduction

Let $F = (F_n)_{n \geq 1}$ be a sequence of random variables, and assume that F satisfies a *central limit theorem* (CLT), that is, there exists a (non-zero) Gaussian random variable Z such that, as $n \to \infty$,

$$P(F_n \leq z) \to P(Z \leq z), \quad \text{for every } z \in \mathbb{R}$$

(here $(\Omega, \mathcal{F}, P)$ denotes the underlying probability space). A natural question is therefore the following: *for a fixed n, can we assess the distance between the laws of F_n and Z?* In other words, *is it possible to quantify the error one makes when replacing F_n with Z?* Answering questions of this type customarily requires us to produce uniform upper bounds of the type

$$\sup_{h \in \mathcal{H}} |E[h(F_n)] - E[h(Z)]| \leq \phi(n), \quad n \geq 1,$$

where $\mathcal{H}$ is a rich enough class of test functions, E denotes mathematical expectation, and $(\phi(n))_{n \geq 1}$ is a positive numerical sequence (sometimes called a *rate of convergence*) such that $\phi(n) \to 0$. Finding explicit rates of convergence can have an enormous impact on applications. For instance, if F_n is some statistical estimator with unknown distribution, then a small value of $\phi(n)$ implies that a Gaussian likelihood may be appropriate; if F describes the evolution of some random system exhibiting asymptotically Gaussian fluctuations, then a rate $\phi(n)$ rapidly converging to zero implies that one can safely consider such fluctuations to be Gaussian, up to some negligible error.

The aim of this monograph is to build an exhaustive theory, allowing the above questions to be answered whenever the sequence F is composed of sufficiently regular functionals of a (possibly infinite-dimensional) Gaussian field. As made clear by the title, our main tools will be the *Malliavin calculus of variations* and *Stein's method for normal approximations*. Both topics will be developed from first principles.

The book is organized as follows:

- **Chapter 1** deals with Malliavin operators in the special case where the underlying Gaussian space is one-dimensional. This chapter is meant as a 'smooth' introduction to Malliavin calculus, as well as to some more advanced topics discussed later in the book, such as variance expansions and second-order Poincaré inequalities. Several useful computations concerning one-dimensional Gaussian distributions and Hermite polynomials are also carefully developed.
- **Chapter 2** contains all the definitions and results on Malliavin calculus that are needed throughout the text. Specific attention is devoted to the derivative and divergence operators, and to the properties of the so-called Ornstein–Uhlenbeck semigroup. The notions of Wiener chaos, Wiener–Itô multiple integrals and chaotic decompositions are also introduced from scratch.
- **Chapter 3** introduces Stein's method for normal approximations in the one-dimensional case. General Stein's equations are studied in detail, and Stein-type bounds are obtained for the total variation, Kolmogorov and Wasserstein distances.
- **Chapter 4** discusses multidimensional Gaussian approximations. The main proofs in this chapter are based on the use of Malliavin operators.
- **Chapter 5** is arguably the most important of the book. Here we show how to explicitly combine Stein's method with Malliavin calculus. One of the main achievements is a complete characterization of CLTs on a fixed Wiener chaos, in terms of 'fourth-moment conditions'. Several examples are discussed in detail.
- **Chapter 6** extends the findings of Chapter 5 to the multidimensional case. In particular, the results of this chapter yield some useful characterizations of CLTs for vectors of chaotic random variables.
- **Chapter 7** contains a detailed application to the so-called *Breuer–Major CLTs*. These convergence results are one of the staples of asymptotic Gaussian analysis. They typically involve sequences of the form $F_n = n^{-1/2} \sum_{i=1}^{n} f(X_i)$, $n \geq 1$, where $(X_i)_{i \geq 1}$ is a stationary Gaussian sequence (for instance, given by the increments of a fractional Brownian motion) and f is some deterministic mapping. This framework is both very natural and very challenging, and is perfectly tailored to demonstrate the power of the techniques developed in the preceding chapters.
- In **Chapter 8** we provide some applications of Malliavin calculus to the recursive computations of cumulants of (possibly vector-valued) random elements. The results of this section may be seen as a simpler alternative

to the familiar moments/cumulants computations based on graphs and diagrams (see [110]).

– **Chapter 9** deals with the delicate issue of optimality of convergence rates. Some connections with Edgeworth expansions are also discussed.

– **Chapter 10** deals with some explicit formulae for the density of the laws of functionals of Gaussian fields. This chapter, which is mainly based on [96], provides expressions for densities that differ from those usually obtained via Malliavin calculus (see, for example, [98, chapter 1]).

– **Chapter 11** establishes an explicit connection between the previous material and the so-called 'universality phenomenon'. The results of this chapter are tightly connected with a truly remarkable paper by Mossel, O'Donnel and Oleszkiewicz (see [79]), providing an extension of the Lindeberg principle to the framework of polynomial functionals of sequences of independent random variables.

– The book concludes with five appendices. **Appendix A** deals with Gaussian random variables, cumulants and Edgeworth expansions. **Appendix B** focuses on Hilbert spaces and contractions. **Appendix C** collects some useful results about distances between probability measures. **Appendix D** is an introduction to fractional Brownian motion. Finally, **Appendix E** discusses some miscellaneous results from functional analysis.

1

Malliavin operators in the one-dimensional case

As anticipated in the Introduction, in order to develop the main tools for the normal approximations of the laws of random variables, we need to define and exploit a modicum of *Malliavin-type operators* – such as the *derivative*, *divergence* and *Ornstein–Uhlenbeck* operators. These objects act on random elements that are functionals of some Gaussian field, and will be fully described in Chapter 2. The aim of this chapter is to introduce the reader into the realm of Malliavin operators, by focusing on their one-dimensional counterparts. In particular, in what follows we are going to define derivative, divergence and Ornstein–Uhlenbeck operators acting on random variables of the type $F = f(N)$, where f is a deterministic function and $N \sim \mathcal{N}(0, 1)$ has a standard Gaussian distribution. As we shall see below, one-dimensional Malliavin operators basically coincide with familiar objects of functional analysis. As such, one can describe their properties without any major technical difficulties. Many computations detailed below are further applied in Chapter 3, where we provide a thorough discussion of Stein's method for one-dimensional normal approximations.

For the rest of this chapter, every random object is defined on an appropriate probability space $(\Omega, \mathscr{F}, P)$. The symbols 'E' and 'Var' denote, respectively, the expectation and the variance associated with P.

1.1 Derivative operators

Let us consider the probability space $(\mathbb{R}, \mathscr{B}(\mathbb{R}), \gamma)$, where γ stands for the standard Gaussian probability measure, that is,

$$\gamma(A) = \frac{1}{\sqrt{2\pi}} \int_A e^{-x^2/2} dx,$$

for every Borel set A. A random variable N with distribution γ is called *standard Gaussian*; equivalently, we write $N \sim \mathcal{N}(0, 1)$. We start with a simple (but crucial) statement.

Lemma 1.1.1 *Let $f : \mathbb{R} \to \mathbb{R}$ be an absolutely continuous function such that $f' \in L^1(\gamma)$. Then $x \mapsto xf(x) \in L^1(\gamma)$ and*

$$\int_{\mathbb{R}} xf(x)d\gamma(x) = \int_{\mathbb{R}} f'(x)d\gamma(x). \tag{1.1.1}$$

Proof Since $\int_{\mathbb{R}} |x| d\gamma(x) < \infty$, we can assume that $f(0) = 0$ without loss of generality. We first prove that the mapping $x \mapsto xf(x)$ is in $L^1(\gamma)$. Indeed,

$$\begin{aligned}
\int_{-\infty}^{\infty} |f(x)| \, |x| d\gamma(x) &= \frac{1}{\sqrt{2\pi}} \int_{-\infty}^{\infty} \left| \int_0^x f'(y)dy \right| |x| e^{-x^2/2} dx \\
&\leq \frac{1}{\sqrt{2\pi}} \int_{-\infty}^0 \left(\int_x^0 |f'(y)| dy \right) (-x) e^{-x^2/2} dx \\
&\quad + \frac{1}{\sqrt{2\pi}} \int_0^{\infty} \left(\int_0^x |f'(y)| dy \right) x e^{-x^2/2} dx \\
&= \int_{-\infty}^{\infty} |f'(y)| d\gamma(y) < \infty,
\end{aligned}$$

where the last equality follows from a standard application of the Fubini theorem. To show relation (1.1.1), one can apply once again the Fubini theorem and infer that

$$\begin{aligned}
\int_{-\infty}^{\infty} f(x)x d\gamma(x) &= \frac{1}{\sqrt{2\pi}} \int_{-\infty}^0 \left(\int_x^0 f'(y)dy \right) (-x) e^{-x^2/2} dx \\
&\quad + \frac{1}{\sqrt{2\pi}} \int_0^{\infty} \left(\int_0^x f'(y)dy \right) x e^{-x^2/2} dx = \int_{-\infty}^{\infty} f'(y) d\gamma(y).
\end{aligned}$$

$\square$

Remark 1.1.2 Due to the fact that the assumptions in Lemma 1.1.1 are minimal, we proved relation (1.1.1) by using a Fubini argument instead of a (slightly more natural) integration by parts. Observe that one cannot remove the 'absolutely continuous' assumption on f. For instance, if $f = \mathbf{1}_{[0,\infty)}$, then $\int_{-\infty}^{\infty} xf(x)d\gamma(x) = \frac{1}{\sqrt{2\pi}}$, whereas $\int_{-\infty}^{\infty} f'(x)d\gamma(x) = 0$.

We record a useful consequence of Lemma 1.1.1, consisting in a characterization of the moments of γ, which we denote by

$$m_n(\gamma) = \int_{\mathbb{R}} x^n d\gamma(x), \quad n \geq 0. \tag{1.1.2}$$

Corollary 1.1.3 *The sequence $(m_n(\gamma))_{n \geq 0}$ satisfies the induction relation*

$$m_{n+1}(\gamma) = n \times m_{n-1}(\gamma), \quad n \geq 0. \tag{1.1.3}$$

In particular, one has $m_n(\gamma) = 0$ if n is odd, and

$$m_n(\gamma) = n!/(2^{n/2}(n/2)!) = (n-1)!! = 1 \cdot 3 \cdot 5 \cdot \ldots \cdot (n-1) \quad \text{if } n \text{ is even.}$$

Proof To obtain the induction relation (1.1.3), just apply (1.1.1) to the function $f(x) = x^n$, $n \geq 0$. The explicit value of $m_n(\gamma)$ is again computed by an induction argument. $\qquad\square$

In what follows, we will denote by $\mathcal{S}$ the set of C^∞-functions $f : \mathbb{R} \to \mathbb{R}$ such that f and all its derivatives have at most polynomial growth. We call any element of $\mathcal{S}$ a *smooth* function.

Remark 1.1.4 The relevance of smooth functions is explained by the fact that the operators introduced below are all defined on domains that can be obtained as the closure of $\mathcal{S}$ with respect to an appropriate norm. We will see in the next chapter that an analogous role is played by the collection of the *smooth functionals* of a general Gaussian field. In the one-dimensional case, the reason for the success of this 'approximation procedure' is nested in the following statement.

Proposition 1.1.5 *The monomials $\{x^n : n = 0, 1, 2, \ldots\}$ generate a dense subspace of $L^q(\gamma)$ for every $q \in [1, \infty)$. In particular, for any $q \in [1, \infty)$ the space $\mathcal{S}$ is a dense subset of $L^q(\gamma)$.*

Proof Elementary Hahn–Banach theory (see Proposition E.1.3) implies that it is sufficient to show that, for every $\eta \in (1, \infty]$, if $g \in L^\eta(\gamma)$ is such that $\int_{\mathbb{R}} g(x) x^k d\gamma(x) = 0$ for every integer $k \geq 0$, then $g = 0$ almost everywhere. So, let $g \in L^\eta(\gamma)$ satisfy $\int_{\mathbb{R}} g(x) x^k d\gamma(x) = 0$ for every $k \geq 0$, and fix $t \in \mathbb{R}$. We have, for all $x \in \mathbb{R}$,

$$\left| g(x) e^{-\frac{x^2}{2}} \sum_{k=0}^{n} \frac{(itx)^k}{k!} \right| \leq |g(x)| e^{|tx| - \frac{x^2}{2}},$$

so that, by dominated convergence,[1] we have

$$\int_{\mathbb{R}} g(x) e^{itx} d\gamma(x) = \lim_{n \to \infty} \sum_{k=0}^{n} \frac{(it)^k}{k!} \int_{\mathbb{R}} g(x) x^k d\gamma(x) = 0.$$

---

[1] Indeed, by the Hölder inequality and by using the convention $\frac{\infty - 1}{\infty} = 1$ to deal with the case $\eta = \infty$, one has that

$$\int_{-\infty}^{\infty} |g(x)| e^{|tx| - \frac{x^2}{2}} dx = \sqrt{2\pi} \int_{-\infty}^{\infty} |g(x)| e^{|tx|} d\gamma(x) \leq \sqrt{2\pi} \|g\|_{L^\eta(\gamma)} \|e^{|t \cdot|}\|_{L^{(\eta-1)/\eta}(\gamma)} < \infty.$$

We have therefore proved that $\int_{\mathbb{R}} g(x)\exp(itx)d\gamma(x) = 0$ for every $t \in \mathbb{R}$, from which it follows immediately (by injectivity of the Fourier transform) that $g = 0$ almost everywhere. □

Fix $f \in S$; for every $p = 1, 2, \ldots$, we write $f^{(p)}$ or, equivalently, $D^p f$ to indicate the pth derivative of f. Note that the mapping $f \mapsto D^p f$ is an operator from S into itself. We now prove that this operator is closable.

Lemma 1.1.6 *The operator $D^p : S \subset L^q(\gamma) \to L^q(\gamma)$ is closable for every $q \in [1, \infty)$ and every integer $p \geq 1$.*

Proof We only consider the case $q > 1$; due to the duality $L^1(\gamma)/L^\infty(\gamma)$, the case $q = 1$ requires some specific argument and is left to the reader. Let (f_n) be a sequence of S such that: (i) f_n converges to zero in $L^q(\gamma)$; (ii) $f_n^{(p)}$ converge to some η in $L^q(\gamma)$. We have to prove that η is equal to zero. Let $g \in S$, and define $\delta^p g \in S$ iteratively by $\delta^r g = \delta^1 \delta^{r-1} g, r = 2, \ldots, p$, where $\delta^1 g(x) = \delta g(x) = xg(x) - g'(x)$ (note that this notation is consistent with the content of Section 1.2, where the operator δ will be fully characterized). We have, using Lemma 1.1.1 several times,

$$\int_{\mathbb{R}} \eta(x)g(x)d\gamma(x) = \lim_{n\to\infty} \int_{\mathbb{R}} f_n^{(p)}(x)g(x)d\gamma(x)$$

$$= \lim_{n\to\infty} \int_{\mathbb{R}} f_n^{(p-1)}(x)\big(xg(x) - g'(x)\big)d\gamma(x)$$

$$= \lim_{n\to\infty} \int_{\mathbb{R}} f_n^{(p-1)}(x)\delta g(x)d\gamma(x)$$

$$= \ldots$$

$$= \lim_{n\to\infty} \int_{\mathbb{R}} f_n(x)\delta^p g(x)d\gamma(x).$$

Hence, since $f_n \to 0$ in $L^q(\gamma)$ and $\delta^p g$ belongs to $S \subset L^{\frac{q}{q-1}}(\gamma)$, we deduce, by applying the Hölder inequality, that $\int_{\mathbb{R}} \eta(x)g(x)d\gamma(x) = 0$. Since it is true for any $g \in S$, we deduce from Proposition 1.1.5 that $\eta = 0$ almost everywhere, and the proof of the lemma is complete. □

Fix $q \in [1, \infty)$ and an integer $p \geq 1$. We set $\mathbb{D}^{p,q}$ to be the closure of S with respect to the norm

$$\|f\|_{\mathbb{D}^{p,q}} = \left(\int_{\mathbb{R}} |f(x)|^q d\gamma(x) + \int_{\mathbb{R}} |f'(x)|^q d\gamma(x) + \ldots \right.$$
$$\left. + \int_{\mathbb{R}} |f^{(p)}(x)|^q d\gamma(x) \right)^{1/q}.$$

In other words, a function f is an element of $\mathbb{D}^{p,q}$ if and only if there exists a sequence $(f_n)_{n\geq1} \subset \mathcal{S}$ such that (as $n \to \infty$): (i) f_n converges to f in $L^q(\gamma)$; and (ii) for every $j = 1, \ldots, p$, $f_n^{(j)}$ is a Cauchy sequence in $L^q(\gamma)$. For such an f, one defines

$$f^{(j)} = D^j f = \lim_{n\to\infty} D^j f_n = \lim_{n\to\infty} f_n^{(j)}, \tag{1.1.4}$$

where $j = 1, \ldots, p$, and the limit is in the sense of $L^q(\gamma)$. Observe that

$$\mathbb{D}^{p,q+\epsilon} \subset \mathbb{D}^{p+m,q}, \quad \forall m \geq 0, \forall \epsilon \geq 0. \tag{1.1.5}$$

We write $\mathbb{D}^{\infty,q} = \bigcap_{p\geq1} \mathbb{D}^{p,q}$.

Remark 1.1.7 Equivalently, $\mathbb{D}^{p,q}$ is the Banach space of all functions in $L^q(\gamma)$ whose derivatives up to the order p in the sense of distributions also belong to $L^q(\gamma)$ – see, for example, Meyers and Serrin [78].

Definition 1.1.8 For $p = 1, 2, \ldots$, the mapping

$$D^p : \mathbb{D}^{p,q} \to L^q(\gamma) : f \mapsto D^p f, \tag{1.1.6}$$

as defined in (1.1.4), is called the pth **derivative operator** (associated with the $L^q(\gamma)$ norm). Note that, for every $q \neq q'$, the operators $D^p : \mathbb{D}^{p,q} \to L^q(\gamma)$ and $D^p : \mathbb{D}^{p,q'} \to L^{q'}(\gamma)$ coincide when acting on the intersection $\mathbb{D}^{p,q} \cap \mathbb{D}^{p,q'}$. When $p = 1$, we will often write D instead of D^1.

Since $L^2(\gamma)$ is a Hilbert space, the case $q = 2$ is very important,. In the next section, we characterize the adjoint of the operator $D^p : \mathbb{D}^{p,2} \to L^2(\gamma)$.

1.2 Divergences

Definition 1.2.1 We denote by Dom δ^p the subset of $L^2(\gamma)$ composed of those functions g such that there exists $c > 0$ satisfying the property that, for all $f \in \mathcal{S}$ (or, equivalently, for all $f \in \mathbb{D}^{p,2}$),

$$\left| \int_{\mathbb{R}} f^{(p)}(x)g(x)d\gamma(x) \right| \leq c \sqrt{\int_{\mathbb{R}} f^2(x)d\gamma(x)}. \tag{1.2.1}$$

Fix $g \in \text{Dom}\,\delta^p$. Since condition (1.2.1) holds, the linear operator $f \mapsto \int_{\mathbb{R}} f^{(p)}(x)g(x)d\gamma(x)$ is continuous from $\mathcal{S}$, equipped with the $L^2(\gamma)$-norm, into $\mathbb{R}$. Thus, we can extend this operator to a linear operator from $L^2(\gamma)$ into $\mathbb{R}$. By the Riesz representation theorem, there exists a unique element in $L^2(\gamma)$, denoted by $\delta^p g$, such that $\int_{\mathbb{R}} f^{(p)}(x)g(x)d\gamma(x) = \int_{\mathbb{R}} f(x)\delta^p g(x)d\gamma(x)$ for all $f \in \mathcal{S}$.

Definition 1.2.2 Fix an integer $p \geq 1$. The pth **divergence operator** δ^p is defined as follows. If $g \in \text{Dom}\,\delta^p$, then $\delta^p g$ is the unique element of $L^2(\gamma)$ characterized by the following duality formula: for all $f \in \mathcal{S}$ (or, equivalently, for all $f \in \mathbb{D}^{p,2}$),

$$\int_{\mathbb{R}} f^{(p)}(x)g(x)d\gamma(x) = \int_{\mathbb{R}} f(x)\delta^p g(x)d\gamma(x). \tag{1.2.2}$$

When $p = 1$, we shall often write δ instead of δ^1.

Remark 1.2.3 Taking f to be equal to a constant in (1.2.2), we deduce that, for every $p \geq 1$ and every $g \in \text{Dom}\,\delta^p$,

$$\int_{\mathbb{R}} \delta^p g(x)d\gamma(x) = 0. \tag{1.2.3}$$

Notice that the operator δ^p is closed (being the adjoint of D^p). Also,

$$\delta^p g = \delta(\delta^{p-1}g) = \delta^{p-1}(\delta g) \tag{1.2.4}$$

for every $g \in \text{Dom}\,\delta^p$. In particular, the first equality in (1.2.4) implies that, if $g \in \text{Dom}\,\delta^p$, then $\delta^{p-1}g \in \text{Dom}\,\delta$, whereas from the second equality we infer that, if $g \in \text{Dom}\,\delta^p$, then $\delta g \in \text{Dom}\,\delta^{p-1}$.

Exercise 1.2.4 Prove the two equalities in (1.2.4).

For every $f, g \in \mathcal{S}$, we can write, by virtue of Lemma 1.1.1,

$$\int_{\mathbb{R}} f'(x)g(x)d\gamma(x) = \int_{\mathbb{R}} xf(x)g(x)d\gamma(x) - \int_{\mathbb{R}} f(x)g'(x)d\gamma(x). \tag{1.2.5}$$

Relation (1.2.5) implies that $\mathcal{S} \subset \text{Dom}\,\delta$ and, for $g \in \mathcal{S}$, that $\delta g(x) = xg(x) - g'(x)$. By approximation, we deduce that $\mathbb{D}^{1,2} \subset \text{Dom}\,\delta$, and also that the previous formula for δg continues to hold when $g \in \mathbb{D}^{1,2}$, that is, $\delta g = G - Dg$ for every $g \in \mathbb{D}^{1,2}$, where $G(x) = xg(x)$. More generally, we can prove that $\mathbb{D}^{p,2} \subset \text{Dom}\,\delta^p$ for any $p \geq 1$.

1.3 Ornstein–Uhlenbeck operators

Definition 1.3.1 The **Ornstein–Uhlenbeck semigroup**, written $(P_t)_{t\geq 0}$, is defined as follows. For $f \in \mathcal{S}$ and $t \geq 0$,

$$P_t f(x) = \int_{\mathbb{R}} f(e^{-t}x + \sqrt{1 - e^{-2t}}\,y)d\gamma(y), \quad x \in \mathbb{R}. \tag{1.3.1}$$

The semigroup characterization is proved in Proposition 1.3.3. An explicit connection with Ornstein–Uhlenbeck stochastic processes is provided in Exercise 1.7.4.

Plainly, $P_0 f(x) = f(x)$. By using the fact that f is an element of S and by dominated convergence, it is immediate that $P_\infty f(x) := \lim_{t \to \infty} P_t f(x) = \int_{\mathbb{R}} f(y) d\gamma(y)$. On the other hand, by applying the Jensen inequality to the right-hand side of (1.3.1), we infer that, for every $q \in [1, \infty)$,

$$
\int_{\mathbb{R}} |P_t f(x)|^q d\gamma(x) = \int_{\mathbb{R}} \left| \int_{\mathbb{R}} f(e^{-t}x + \sqrt{1 - e^{-2t}}y) d\gamma(y) \right|^q d\gamma(x)
$$

$$
\leq \int_{\mathbb{R}^2} |f(e^{-t}x + \sqrt{1 - e^{-2t}}y)|^q d\gamma(x) d\gamma(y)
$$

$$
= \int_{\mathbb{R}} |f(x)|^q d\gamma(x). \tag{1.3.2}
$$

The last equality in (1.3.2) follows from the well-known (and easily checked using the characteristic function) fact that, if N, N' are two independent standard Gaussian random variables, then $e^{-t}N + \sqrt{1 - e^{-2t}}N'$ is also standard Gaussian.

The relations displayed in (1.3.2), together with Proposition 1.1.5, show that the expression on the right-hand side of (1.3.1) is indeed well defined for $f \in L^q(\gamma)$, $q \geq 1$. Moreover, a contraction property holds.

Proposition 1.3.2   *For every $t \geq 0$ and every $q \in [1, \infty)$, P_t extends to a linear contraction operator on $L^q(\gamma)$.*

As anticipated, the fundamental property of the class $(P_t)_{t \geq 0}$ is that it is a semigroup of operators.

Proposition 1.3.3   *For any $s, t \geq 0$, we have $P_t P_s = P_{t+s}$ on $L^1(\gamma)$.*

Proof   For all $f \in L^1(\gamma)$, we can write

$$
P_t P_s f(x) = \int_{\mathbb{R}^2} f(e^{-s-t}x + e^{-s}\sqrt{1 - e^{-2t}}y + \sqrt{1 - e^{-2s}}z) d\gamma(y) d\gamma(z)
$$

$$
= \int_{\mathbb{R}} f(e^{-s-t}x + \sqrt{1 - e^{-2s-2t}}y) d\gamma(y) = P_{t+s} f(x),
$$

where the second inequality follows from the easily verified fact that, if N, N' are two independent standard Gaussian random variables, then $e^{-s}\sqrt{1 - e^{-2t}}N + \sqrt{1 - e^{-2s}}N'$ and $\sqrt{1 - e^{-2(s+t)}}N$ have the same law. $\square$

The following result shows that P_t and D can be interchanged on $\mathbb{D}^{1,2}$.

Proposition 1.3.4   *Let $f \in \mathbb{D}^{1,2}$ and $t \geq 0$. Then $P_t f \in \mathbb{D}^{1,2}$ and $DP_t f = e^{-t} P_t Df$.*

Proof   Suppose that $f \in S$. Then, for any $x \in \mathbb{R}$,

$$DP_t f(x) = e^{-t} \int_{\mathbb{R}} f'(e^{-t}x + \sqrt{1 - e^{-2t}}y)d\gamma(y) = e^{-t}P_t f'(x) = e^{-t}P_t Df(x).$$

The case of a general f follows from an approximation argument, as well as from the contraction property stated in Proposition 1.3.2.   □

Now denote by $L = \frac{d}{dt}\big|_{t=0}P_t$ the infinitesimal generator of $(P_t)_{t \geq 0}$ on $L^2(\gamma)$, and by Dom L its domain.

Remark 1.3.5   We recall that Dom L is defined as the collection of those $f \in L^2(\gamma)$ such that the expression $\frac{P_h f - f}{h}$ converges in $L^2(\gamma)$, as h goes to zero.

On S one has that, for any $t \geq 0$,

$$\frac{d}{dt}P_t = \lim_{h \to 0} \frac{P_{t+h} - P_t}{h} = \lim_{h \to 0} P_t \frac{P_h - Id}{h} = P_t \lim_{h \to 0} \frac{P_h - Id}{h}$$

$$= P_t \frac{d}{dh}\bigg|_{h=0} P_h = P_t L,$$

and, similarly, $\frac{d}{dt}P_t = LP_t$. On the other hand, for $f \in S$ and $x \in \mathbb{R}$, we can write, by differentiating with respect to t in (1.3.1) (note that the interchanging of differentiation and integration is justified by the fact that f is smooth),

$$\frac{d}{dt}P_t f(x) = -xe^{-t} \int_{\mathbb{R}} f'(e^{-t}x + \sqrt{1 - e^{-2t}}y)d\gamma(y) + \frac{e^{-2t}}{\sqrt{1 - e^{-2t}}}$$

$$\int_{\mathbb{R}} f'(e^{-t}x + \sqrt{1 - e^{-2t}}y)yd\gamma(y)$$

$$= -xe^{-t} \int_{\mathbb{R}} f'(e^{-t}x + \sqrt{1 - e^{-2t}}y)d\gamma(y) + e^{-2t}$$

$$\int_{\mathbb{R}} f''(e^{-t}x + \sqrt{1 - e^{-2t}}y)d\gamma(y),$$

where we used Lemma 1.1.1 to get the last inequality. In particular, by specializing the previous calculations to $t = 0$ we infer that

$$Lf(x) = -xf'(x) + f''(x). \tag{1.3.3}$$

This fact is reformulated in the next statement.

Proposition 1.3.6   *For any $f \in S$, we have $Lf = -\delta Df$.*

The seemingly innocuous Proposition 1.3.6 is indeed quite powerful. As an illustration, we now use it in order to prove an important result about concentration of Gaussian random variables, known as *Poincaré inequality*.

Proposition 1.3.7 (Poincaré inequality)   *Let $N \sim \mathcal{N}(0, 1)$ and $f \in \mathbb{D}^{1,2}$.*
Then

$$\text{Var}[f(N)] \leq E[f'^2(N)]. \tag{1.3.4}$$

Proof   By an approximation argument, we may assume without loss of generality that $f \in \mathcal{S}$. Since we are now dealing with a function in $\mathcal{S}$, we can freely interchange derivatives and integrals, and write

$$\text{Var}[f(N)] = E[f(N)(f(N) - E[f(N)])] = E[f(N)(P_0 f(N) - P_\infty f(N))]$$
$$= -\int_0^\infty E[f(N)\frac{d}{dt} P_t f(N)]dt$$
$$= \int_0^\infty E[f(N) \delta D P_t f(N)]dt \quad (\text{using } \frac{d}{dt}P_t = LP_t = -\delta D P_t)$$
$$= \int_0^\infty E[f'(N) D P_t f(N)]dt \quad (\text{by the duality formula (1.2.2)})$$
$$= \int_0^\infty e^{-t} E[f'(N) P_t f'(N)]dt \quad (\text{by Proposition 1.3.4})$$
$$\leq \int_0^\infty e^{-t} \sqrt{E[f'^2(N)]}\sqrt{E[(P_t f')^2(N)]}dt \quad (\text{by Cauchy–Schwarz})$$
$$\leq E[f'^2(N)] \int_0^\infty e^{-t}dt \quad (\text{by Proposition 1.3.2})$$
$$= E[f'^2(N)],$$

yielding the desired conclusion.   □

By using their definitions, one can immediately prove that δ and D enjoy the following 'Heisenberg commutativity relationship':

Proposition 1.3.8   *For every $f \in \mathcal{S}$, $(D\delta - \delta D)f = f$.*

Exercise 1.3.9   Combine Proposition 1.3.8 with an induction argument to prove that, for any integer $p \geq 2$,

$$(D\delta^p - \delta^p D)f = p\delta^{p-1}f \quad \text{for all } f \in \mathcal{S}. \tag{1.3.5}$$

See also Proposition 2.6.1.

In the subsequent sections, we present three applications of the theory developed above. In Section 1.4 we define and characterize an orthogonal basis of $L^2(\gamma)$, known as the class of *Hermite polynomials*. Section 1.5 deals with decompositions of variances. Section 1.6 provides some basic examples of normal approximations for the law of random variables of the type $F = f(N)$.

1.4 First application: Hermite polynomials

Definition 1.4.1 Let $p \geq 0$ be an integer. We define the pth **Hermite polynomial** as $H_0 = 1$ and $H_p = \delta^p 1$, $p \geq 1$. Here, 1 is a shorthand notation for the function that is identically one, which is of course an element of Dom δ^p for every p. For instance, $H_1(x) = x$, $H_2(x) = x^2 - 1$, $H_3(x) = x^3 - 3x$, and so on. We shall also use the convention that $H_{-1}(x) = 0$.

The main properties of Hermite polynomials are gathered together in the next statement (which is one of the staples of the entire book):

Proposition 1.4.2 (i) *For any $p \geq 0$, we have $H_p' = pH_{p-1}$, $LH_p = -pH_p$ and $P_t H_p = e^{-pt} H_p$, $t \geq 0$.*

(ii) *For any $p \geq 0$, $H_{p+1}(x) = xH_p(x) - pH_{p-1}(x)$.*

(iii) *For any $p, q \geq 0$,*

$$\int_{\mathbb{R}} H_p(x) H_q(x) d\gamma(x) = \begin{cases} p! & \text{if } p = q \\ 0 & \text{otherwise.} \end{cases}$$

(iv) *The family $\left\{ \frac{1}{\sqrt{p!}} H_p : p \geq 0 \right\}$ is an orthonormal basis of $L^2(\gamma)$.*

(v) *If $f \in \mathbb{D}^{\infty,2}$ then $f = \sum_{p=0}^{\infty} \frac{1}{p!} \left(\int_{\mathbb{R}} f^{(p)}(x) d\gamma(x) \right) H_p$ in $L^2(\gamma)$.*

(vi) *For all $c \in \mathbb{R}$, we have $e^{cx - c^2/2} = \sum_{p=0}^{\infty} \frac{c^p}{p!} H_p(x)$ in $L^2(\gamma)$.*

(vii) *(Rodrigues's formula) For any $p \geq 1$, $H_p(x) = (-1)^p e^{x^2/2} \frac{d^p}{dx^p} e^{-x^2/2}$.*

(viii) *For every $p \geq 0$ and every real x, $H_p(-x) = (-1)^p H_p(x)$.*

Proof (i) By the definition of H_p, we have $H_p' = D\delta^p 1$. Hence, by applying the result of Exercise 1.3.9, we get $H_p' = p\delta^{p-1} 1 + \delta^p D1 = pH_{p-1}$. We deduce that $LH_p = -\delta DH_p = -\delta H_p' = -p\delta H_{p-1} = -pH_p$. Fix $x \in \mathbb{R}$, and define $y_x : \mathbb{R}_+ \to \mathbb{R}$ by $y_x(t) = P_t H_p(x)$. We have $y_x(0) = P_0 H_p(x) = H_p(x)$ and, for $t > 0$,

$$y_x'(t) = \frac{d}{dt}\bigg|_{t=0} P_t H_p(x) = P_t L H_p(x) = -p P_t H_p(x) = -p y_x(t).$$

Hence $y_x(t) = e^{-pt} H_p(x)$, that is, $P_t H_p = e^{-pt} H_p$.

(ii) We can take $p \geq 1$. We have that $H_{p+1} = \delta^{p+1} 1 = \delta\delta^p 1 = \delta H_p$. It follows that, by the definition of δ, $H_{p+1}(x) = xH_p(x) - H_p'(x)$. Since $H_p'(x) = pH_{p-1}(x)$ by part (i), we deduce the conclusion.

(iii) The case $p > q = 0$ is a direct consequence of (1.2.3). If $p \geq q \geq 1$, one can write

$$
\int_{\mathbb{R}} H_p(x)H_q(x)d\gamma(x) = \int_{\mathbb{R}} H_p(x)\delta^q 1(x)d\gamma(x)
$$
$$
= \int_{\mathbb{R}} H_p'(x)\delta^{q-1}1(x)d\gamma(x) \quad \text{(by the duality}
$$
$$
\text{formula (1.2.2))}
$$
$$
= p \int_{\mathbb{R}} H_{p-1}(x)H_{q-1}(x)d\gamma(x) \quad \text{(by (i))}.
$$

Hence, the desired conclusion is proved by induction.

(iv) By the previous point, the family $\left\{ \frac{1}{\sqrt{p!}} H_p : p \geq 0 \right\}$ is orthonormal in $L^2(\gamma)$. On the other hand, it is simple to prove (e.g. by induction) that, for any $p \geq 0$, the polynomial H_p has degree p. Hence, the claim in this part is equivalent to saying that the monomials $\{x^p : p = 0, 1, 2, \ldots\}$ generate a dense subspace of $L^2(\gamma)$. The conclusion is now obtained by using Proposition 1.1.5 in the case $q = 2$.

(v) By part (iv), for any $f \in L^2(\gamma)$,

$$
f = \sum_{p=0}^{\infty} \frac{1}{p!} \left(\int_{\mathbb{R}} f(x)H_p(x)d\gamma(x) \right) H_p.
$$

If $f \in \mathbb{D}^{\infty,2}$ then, by applying (1.2.2) repeatedly, we can write

$$
\int_{\mathbb{R}} f(x)H_p(x)d\gamma(x) = \int_{\mathbb{R}} f(x)\delta^p 1(x)d\gamma(x) = \int_{\mathbb{R}} f^{(p)}(x)d\gamma(x).
$$

Hence, in this case, we also have

$$
f = \sum_{p=0}^{\infty} \frac{1}{p!} \left(\int_{\mathbb{R}} f^{(p)}(x)d\gamma(x) \right) H_p,
$$

as required.

(vi) If we choose $f(x) = e^{cx}$ in the previous identity for f, we get

$$
e^{cx} = \sum_{p=0}^{\infty} \frac{c^p}{p!} \left(\frac{1}{\sqrt{2\pi}} \int_{\mathbb{R}} e^{cx-x^2/2}dx \right) H_p(x) = e^{c^2/2} \sum_{p=0}^{\infty} \frac{c^p}{p!} H_p(x),
$$

which is the desired formula.

(vii) We have

$$e^{cx-c^2/2} = e^{x^2/2}e^{-(x-c)^2/2} = e^{x^2/2}\sum_{p=0}^{\infty}\frac{c^p}{p!}\times\left.\frac{d^p}{dc^p}\right|_{c=0}e^{-(x-c)^2/2}$$

$$= e^{x^2/2}\sum_{p=0}^{\infty}\frac{(-1)^p c^p}{p!}\frac{d^p}{dx^p}e^{-x^2/2}.$$

By comparing with the formula in (vi), we deduce the conclusion.

(viii) Since $H_0(x) = 1$ and $H_1(x) = x$, the conclusion is trivially true for $p = 0, 1$. Using part (ii), we deduce the desired result by an induction argument. $\qquad\square$

1.5 Second application: variance expansions

We will use the previous results in order to write two (infinite) series representations of the variance of $f(N)$, whenever $N \sim \mathcal{N}(0, 1)$ and $f : \mathbb{R} \to \mathbb{R}$ is sufficiently regular.

Proposition 1.5.1   *Let $N \sim \mathcal{N}(0, 1)$ and $f \in \mathbb{D}^{\infty,2}$. Then*

$$\mathrm{Var}[f(N)] = \sum_{n=1}^{\infty}\frac{1}{n!}E[f^{(n)}(N)]^2. \qquad (1.5.1)$$

If, moreover, $E[f^{(n)}(N)^2]/n! \to 0$ as $n \to \infty$ and $f \in \mathcal{S}$, we also have

$$\mathrm{Var}[f(N)] = \sum_{n=1}^{\infty}\frac{(-1)^{n+1}}{n!}E[f^{(n)}(N)^2]. \qquad (1.5.2)$$

(The convergence of the infinite series in (1.5.1) and (1.5.2) is part of the conclusion.)

Proof   The proof of (1.5.1) is easy: it suffices indeed to compute the $L^2(\gamma)$-norm on both sides of the formula appearing in part (v) of Proposition 1.4.2 – see also part (iv) therein.

For the proof of (1.5.2), let us consider the application

$$t \mapsto g(t) = E[(P_{\log(1/\sqrt{t})}f)^2(N)], \quad 0 < t \le 1.$$

We have $g(1) = E[f^2(N)]$. Moreover, g can be extended to a continuous function on the interval $[0, 1]$ by setting $g(0) = E[f(N)]^2$. In particular, $\mathrm{Var}[f(N)] = g(1) - g(0)$. For $t \in (0, 1)$, let us compute $g'(t)$ (note that we can interchange derivatives and expectations, due to the assumptions on f):

$$g'(t) = -\frac{1}{t} E[P_{\log(1/\sqrt{t})} f(N) \times L P_{\log(1/\sqrt{t})} f(N)]$$

$$= \frac{1}{t} E[P_{\log(1/\sqrt{t})} f(N) \times \delta D P_{\log(1/\sqrt{t})} f(N)] \quad \text{(by Proposition 1.3.6)}$$

$$= \frac{1}{t} E[(D P_{\log(1/\sqrt{t})} f)^2 (N)] \quad \text{(by the duality formula (1.2.2))}$$

$$= E[(P_{\log(1/\sqrt{t})} f')^2 (N)] \quad \text{(by Proposition 1.3.4)}.$$

By induction, we easily infer that, for all $n \geq 1$ and $t \in (0, 1)$,

$$g^{(n)}(t) = E[(P_{\log(1/\sqrt{t})} f^{(n)})^2 (N)],$$

and, in particular, $g^{(n)}$ can be extended to a continuous function on $[0, 1]$ by setting

$$g^{(n)}(0) = E[f^{(n)}(N)]^2 \quad \text{and} \quad g^{(n)}(1) = E[f^{(n)}(N)^2].$$

Taylor's formula yields, for any integer $m \geq 1$,

$$\left| g(0) - g(1) + \sum_{n=1}^{m} \frac{(-1)^{n+1}}{n!} g^{(n)}(1) \right| = \frac{1}{m!} \int_0^1 g^{(m+1)}(t) t^m \, dt.$$

By Proposition 1.3.2, we have that, for any $t \in (0, 1)$, $0 \leq g^{(m+1)}(t) \leq E[f^{(m+1)}(N)^2]$. Therefore,

$$0 \leq \frac{1}{m!} \int_0^1 g^{(m+1)}(t) t^m \, dt \leq \frac{E[f^{(m+1)}(N)^2]}{(m+1)!} \to 0 \quad \text{as } m \to \infty,$$

and the desired formula (1.5.2) follows. To finish, let us stress that we could recover (1.5.1) by using this time the expansion $g(1) = g(0) + \sum_{n=1}^{\infty} \frac{g^{(n)}(0)}{n!}$. (Such a series representation is valid because g is absolutely monotone; see, for example, Feller [38, p. 233].) $\qquad\square$

1.6 Third application: second-order Poincaré inequalities

We will now take a first step towards the combination of Stein's method and Malliavin calculus. Let F and N denote two integrable random variables defined on the probability space $(\Omega, \mathscr{F}, P)$. The difference between the laws of F and N can be assessed by means of the so-called *Wasserstein distance*:

$$d_{\mathrm{W}}(F, N) = \sup_{h \in \mathrm{Lip}(1)} \left| E[h(F)] - E[h(N)] \right|.$$

Here, $\mathrm{Lip}(K)$ stands for the set of functions $h : \mathbb{R} \to \mathbb{R}$ that are Lipschitz with constant $K > 0$, that is, satisfying $|h(x) - h(y)| \leq K|x - y|$ for all $x, y \in \mathbb{R}$.

When $N \sim \mathcal{N}(0, 1)$, there exists a remarkable result by Stein (which we will prove and discuss in full detail in Section 3.5) which states that

$$d_W(F, N) \leq \sup_{\phi \in \mathcal{C}^1 \cap \text{Lip}(\sqrt{2/\pi})} \left| E[F\phi(F)] - E[\phi'(F)] \right|. \tag{1.6.1}$$

In anticipation of the more general analysis that we will perform later on, we shall now study the case of F having the specific form $F = f(N)$, for $N \sim \mathcal{N}(0, 1)$ and $f : \mathbb{R} \to \mathbb{R}$ sufficiently regular. In particular, we shall prove a so-called *second-order Poincaré inequality*. The rationale linking Proposition 1.6.1 and the 'first-order' Poincaré inequality stated in Proposition 1.3.7 goes as follows. Formula (1.3.4) implies that a random variable of the type $f(N)$ is concentrated around its mean whenever f' is small, while inequality (1.6.2) roughly states that, whenever f'' is small compared to f', then $f(N)$ has approximately a Gaussian distribution.

Proposition 1.6.1 (Second-order Poincaré inequality) *Let $N \sim \mathcal{N}(0, 1)$ and $f \in \mathbb{D}^{2,4}$. Assume also that $E[f(N)] = 0$ and $E[f^2(N)] = 1$. Then*

$$d_W(f(N), N) \leq \frac{3}{\sqrt{2\pi}} \left(E\left[f''^4(N) \right] \right)^{1/4} \left(E\left[f'^4(N) \right] \right)^{1/4}. \tag{1.6.2}$$

Proof Assume first that $f \in \mathcal{S}$. By using the smoothness of f in order to interchange derivatives and integrals and by reasoning as in Proposition 1.3.7, we can write, for any $\mathcal{C}^1$ function $\phi : \mathbb{R} \to \mathbb{R}$ with bounded derivative:

$$E[f(N)\phi(f(N))] = E[(P_0 f(N) - P_\infty f(N))\phi(f(N))]$$
$$= -\int_0^\infty E\left[\frac{d}{dt} P_t f(N)\phi(f(N)) \right] dt$$
$$= \int_0^\infty E[\delta D P_t f(N)\phi(f(N))] dt$$
$$= \int_0^\infty e^{-t} E[P_t f'(N)\phi'(f(N)) f'(N)] dt$$
$$= E\left[\phi'(f(N)) f'(N) \int_0^\infty e^{-t} P_t f'(N) dt \right].$$

In particular, for $\phi(x) = x$, we obtain

$$1 = E[f^2(N)] = E\left[f'(N) \int_0^\infty e^{-t} P_t f'(N) dt \right].$$

Therefore, for any $\phi \in \mathcal{C}^1 \cap \mathrm{Lip}(\sqrt{2/\pi})$ and $f \in \mathcal{S}$,

$$\left| E[f(N)\phi(f(N))] - E[\phi'(f(N))] \right|$$

$$= \left| E\left[\phi'(f(N)) \left(f'(N) \int_0^\infty e^{-t} P_t f'(N)dt - 1 \right) \right] \right|$$

$$\leq \sqrt{\frac{2}{\pi}} E\left| f'(N) \int_0^\infty e^{-t} P_t f'(N)dt - 1 \right|$$

$$\leq \sqrt{\frac{2}{\pi}} \sqrt{\mathrm{Var}\left[f'(N) \int_0^\infty e^{-t} P_t f'(N)dt \right]}. \tag{1.6.3}$$

To go further, we apply the Poincaré inequality (1.3.4) and then the triangle inequality to deduce that

$$\sqrt{\mathrm{Var}\left[f'(N) \int_0^\infty e^{-t} P_t f'(N)dt \right]} \leq \sqrt{E\left[\left(f' \int_0^\infty e^{-t} P_t f' dt \right)'^2 (N) \right]}$$

$$\leq \sqrt{E\left[f''^2(N) \left(\int_0^\infty e^{-t} P_t f' dt \right)^2 (N) \right]}$$

$$+ \sqrt{E\left[f'^2(N) \left(\int_0^\infty e^{-2t} P_t f'' dt \right)^2 (N) \right]}. \tag{1.6.4}$$

Applying now the Cauchy–Schwarz inequality, Jensen inequality and the contraction property (see Proposition 1.3.2) for $p = 4$ yields

$$E\left[f''^2(N) \left(\int_0^\infty e^{-t} P_t f' dt \right)^2 (N) \right] \leq \sqrt{E\left[f''^4(N) \right]} \sqrt{E\left[f'^4(N) \right]}.$$

Similarly,

$$E\left[f'^2(N) \left(\int_0^\infty e^{-2t} P_t f'' dt \right)^2 (N) \right] \leq \frac{1}{4} \sqrt{E\left[f''^4(N) \right]} \sqrt{E\left[f'^4(N) \right]}.$$

By exploiting these two inequalities in order to bound the right-hand side of (1.6.4) and by virtue of (1.6.3), we obtain that

$$\left| E[f(N)\phi(f(N))] - E[\phi'(f(N))] \right| \leq \frac{3}{\sqrt{2\pi}} \left(E\left[f''^4(N) \right] \right)^{1/4} \left(E\left[f'^4(N) \right] \right)^{1/4}, \tag{1.6.5}$$

for any $\phi \in \mathcal{C}^1 \cap \mathrm{Lip}(\sqrt{2/\pi})$ and $f \in \mathcal{S}$. By approximation, inequality (1.6.5) continues to hold when f is in $\mathbb{D}^{2,4}$. Finally, the desired inequality (1.6.2) follows by plugging (1.6.5) into Stein's bound (1.6.1). $\qquad\square$

Note that Proposition 1.6.1 will be significantly generalized in Theorem 5.3.3; see also Exercise 5.3.4.

1.7 Exercises

1.7.1 Let $p, q \geq 1$ be two integers. Show the product formula for Hermite polynomials:

$$H_p H_q = \sum_{r=0}^{p \wedge q} r! \binom{p}{r} \binom{q}{r} H_{p+q-2r}.$$

(Hint: Use formula (v) of Proposition 1.4.2.)

1.7.2 Let $p \geq 1$ be an integer.

1. Show that $H_p(0) = 0$ if p is odd and $H_p(0) = \frac{(-1)^{p/2}}{2^{p/2}(p/2)!}$ if p is even.

   (Hint: Use Proposition 1.4.2(vi).)

2. Deduce that:

$$H_p(x) = \sum_{k=0}^{\lfloor p/2 \rfloor} \frac{p!(-1)^k}{k!(p-2k)!2^k} x^{p-2k}.$$

   (Hint: Use $H_p' = p H_{p-1}$ and proceed by induction.)

1.7.3 Let $f \in S$ and $x \in \mathbb{R}$. Show that

$$\int_0^\infty e^{-2t} P_t f''(x) dt - x \int_0^\infty e^{-t} P_t f'(x) dt = f(x) - \int_{\mathbb{R}} f(y) d\gamma(y).$$

(Hint: Use $\frac{d}{dt} P_t = L P_t = \ldots$)

1.7.4 (**Ornstein–Uhlenbeck processes**) Let $B = (B_t)_{t \geq 0}$ be a Brownian motion. Consider the linear stochastic differential equation

$$dX_t^x = \sqrt{2} dB_t - X_t^x dt, \quad t \geq 0, \quad X_0^x = x \in \mathbb{R}. \tag{1.7.1}$$

The solution of (1.7.1) is called an **Ornstein–Uhlenbeck** stochastic process.

1. For any $x \in \mathbb{R}$, show that the solution to (1.7.1) is given by

$$X_t^x = e^{-t} x + \sqrt{2} \int_0^t e^{-(t-s)} dB_s, \quad t \geq 0.$$

2. If $f \in S$ and $x \in \mathbb{R}$, show that $P_t f(x) = E[f(X_t^x)]$.

1.7.5   Let $N \sim \mathcal{N}(0, 1)$ and $f \in \mathcal{S}$. If $\frac{1}{2^n n!} E[f^{(n)}(N)^2] \to 0$ as $n \to \infty$, show that

$$\mathrm{Var}[f(N)] = \sum_{n=1}^{\infty} \frac{1}{2^n n!} \left(E[f^{(n)}(N)]^2 + (-1)^{n+1} E[f^{(n)}(N)^2] \right).$$

(Hint: Expand $g(1/2)$ around 0 and 1, with g the function introduced in the proof of Proposition 1.5.1.)

1.7.6   Let $f, g \in \mathbb{D}^{1,2}$, and let $N, \widetilde{N} \sim \mathcal{N}(0, 1)$ be jointly Gaussian, with covariance ρ.

1. Prove that $\mathrm{Cov}[f(N), g(\widetilde{N})] = E\big[f(N) \big(P_{\ln(1/\rho)} g(N) - P_\infty g(N) \big) \big]$.
   (Hint: Use the fact that $(N, \widetilde{N})$ and $(N, \rho N + \sqrt{1 - \rho^2}\hat{N})$ have the same law whenever $N, \hat{N} \sim \mathcal{N}(0, 1)$ are independent and $\rho = \mathrm{Cov}[N, \widetilde{N}]$.)
2. Deduce that $\big| \mathrm{Cov}[f(N), g(\widetilde{N})] \big| \leq |\mathrm{Cov}[N, \widetilde{N}]| \sqrt{E[f'^2(N)]} \sqrt{E[g'^2(N)]}$.

1.7.7   For $\varepsilon > 0$, let $p_\varepsilon(x) = \frac{1}{\sqrt{2\pi\varepsilon}} e^{-\frac{x^2}{2\varepsilon}}$ be the heat kernel with variance ε.

1. Show that $\int_{\mathbb{R}} p_\varepsilon(x - u) d\gamma(x) = p_{1+\varepsilon}(u)$ for all $u \in \mathbb{R}$.
2. For any $n \geq 0$ and $u \in \mathbb{R}$, show that $p_\varepsilon^{(n)}(u) = (-1)^n \varepsilon^{-n/2} p_\varepsilon(u) H_n(u/\sqrt{\varepsilon})$.
   (Hint: Use formula (vi) of Proposition 1.4.2.)
3. Deduce, for any $n \geq 0$ and $u \in \mathbb{R}$, that

$$\int_{\mathbb{R}} p_\varepsilon^{(n)}(x - u) d\gamma(x) = (1 + \varepsilon)^{-n/2} p_{1+\varepsilon}(u) H_n(u/\sqrt{1 + \varepsilon}).$$

4. Finally, prove the following identity in $L^2(\gamma)$:

$$p_\varepsilon = \frac{1}{\sqrt{2\pi(1 + \varepsilon)}} \sum_{n=0}^{\infty} \frac{(-1)^n}{n!(2n)!2^n (1 + \varepsilon)^n} H_{2n}.$$

(Hint: Use Proposition 1.4.2(v).)

1.8 Bibliographic comments

Sections 1.1–1.4 are strongly inspired by the first chapter in Malliavin's monograph [70]. The Poincaré inequality (1.3.4) was first proved by Nash in [80], and then rediscovered by Chernoff in [24] (both proofs use Hermite polynomials). A general, infinite-dimensional version of Poincaré inequality is proved by Houdré and Pérez-Abreu in [51]. The variance expansions presented in Section 1.5 are one-dimensional versions of the results proved by Houdré and Kagan [50]. An excellent reference for Stein's method is [22], by Chen, Goldstein and Shao. The reader is also referred to Stein's original paper [135] and monograph [136] – see Chapter 3 for more details. The concept of a second-order Poincaré inequality such as (1.6.2) first appeared in Chatterjee [20], in connection with normal fluctuations of eigenvalues of Gausssian-subordinated random matrices. Infinite-dimensional second-order Poincaré inequalities are proved in Nourdin *et al.* [93], building on the findings by Nourdin and Peccati [88]. In particular, [88] was the first reference to point out an explicit connection between Stein's method and Malliavin calculus.

2

Malliavin operators and isonormal Gaussian processes

We now present the general Malliavin operators that will be needed for the rest of this book. These operators act on spaces of random elements that are functionals of some possibly infinite-dimensional Gaussian field, such as a Brownian motion or a stationary Gaussian sequence. We will see that the properties of infinite-dimensional Malliavin operators are analogous to those of the classic objects introduced in Chapter 1.

As demonstrated below, a natural way to encode the properties of a given Gaussian field is to embed it in a so-called *isonormal Gaussian process*, that is, in a collection of jointly centered Gaussian random variables whose covariance structure reproduces the inner product of a real separable Hilbert space. We shall discuss these notions in detail. All the necessary notation and definitions concerning Hilbert spaces are gathered together in Appendix B.

2.1 Isonormal Gaussian processes

Fix a real separable Hilbert space $\mathfrak{H}$, with inner product $\langle \cdot, \cdot \rangle_{\mathfrak{H}}$ and norm $\langle \cdot, \cdot \rangle_{\mathfrak{H}}^{1/2} = \| \cdot \|_{\mathfrak{H}}$. We write $X = \{X(h) : h \in \mathfrak{H}\}$ to indicate an *isonormal Gaussian process* over $\mathfrak{H}$. This means that X is a centered Gaussian family (that is, a collection of jointly Gaussian random variables), defined on some probability space $(\Omega, \mathscr{F}, P)$ and such that $E\left[X(g)X(h)\right] = \langle g, h \rangle_{\mathfrak{H}}$ for every $g, h \in \mathfrak{H}$. Of course, if X' is another isonormal Gaussian process over $\mathfrak{H}$, then X and X' have the same law (since they are centered Gaussian families with the same covariance structure). For the rest of the book (when no further specification appears) we shall assume that $\mathscr{F}$ is generated by X. To simplify the notation, we write $L^2(\Omega)$ instead of $L^2(\Omega, \mathscr{F}, P)$.

Proposition 2.1.1 *Given a real separable Hilbert space $\mathfrak{H}$, there exists an isonormal Gaussian process over $\mathfrak{H}$.*

Proof Let $\mathfrak{H}$ be a real separable Hilbert space, let $\{e_i : i \geq 1\}$ be an orthonormal basis of $\mathfrak{H}$, and let $\{Z_i : i \geq 1\}$ be a sequence of independent and identically distributed (i.i.d.) $\mathcal{N}(0,1)$ random variables defined on some probability space $(\Omega, \mathscr{F}, P)$. By orthonormality, for every $h \in \mathfrak{H}$ one has that $\sum_{i=1}^{\infty} \langle h, e_i \rangle_{\mathfrak{H}}^2 = \|h\|_{\mathfrak{H}}^2 < \infty$. From this fact we deduce that, as $n \to \infty$, the sequence $\sum_{i=1}^{n} \langle h, e_i \rangle_{\mathfrak{H}} Z_i$ converges in $L^2(\Omega)$ and almost surely (see Remark 2.1.2) to some random variable, which we denote by $X(h) = \sum_{i=1}^{\infty} \langle h, e_i \rangle_{\mathfrak{H}} Z_i$, $h \in \mathfrak{H}$. By construction, the random variables $X(h)$ are jointly Gaussian. Moreover, since the Z_i are independent, centered and have unit variance, the $X(h)$s are centered and such that

$$E[X(h)X(h')] = \sum_{i=1}^{\infty} \langle h, e_i \rangle_{\mathfrak{H}} \langle h', e_i \rangle_{\mathfrak{H}} = \langle h, h' \rangle_{\mathfrak{H}},$$

where we have used Parseval's identity. This entails that $X = \{X(h) : h \in \mathfrak{H}\}$ is an isonormal Gaussian process over $\mathfrak{H}$. $\qquad\square$

Remark 2.1.2 Using the notation of the previous proof, the fact that the partial sums

$$\sum_{i=1}^{n} \langle h, e_i \rangle_{\mathfrak{H}} Z_i, \quad n \geq 1,$$

converge almost surely can be seen as a consequence of the so-called 'Lévy equivalence theorem'. According to this well-known result, for every sequence $X_1, X_2, \ldots$ of independent random variables, the following three conditions are equivalent: (i) $\sum_{i=1}^{\infty} X_i$ converges almost surely; (ii) $\sum_{i=1}^{\infty} X_i$ converges in probability; (iii) $\sum_{i=1}^{\infty} X_i$ converges in law. See, for example, [32, Theorem 9.7.1] for a proof.

As shown in the next five examples, the notion of the isonormal Gaussian process provides a convenient way to encode the structure of many remarkable Gaussian families.

Example 2.1.3 (Euclidean spaces) Fix an integer $d \geq 1$, set $\mathfrak{H} = \mathbb{R}^d$ and let $(e_1, \ldots, e_d)$ be the canonical orthonormal basis of $\mathbb{R}^d$ (with respect to the usual Euclidean inner product). Let $(Z_1, \ldots, Z_d)$ be a Gaussian vector whose components are i.i.d. $\mathcal{N}(0,1)$. For every $h = \sum_{j=1}^{d} c_j e_j$ (where the c_j are real and uniquely defined), set $X(h) = \sum_{j=1}^{d} c_j Z_j$, and define $X = \{X(h) : h \in \mathbb{R}^d\}$. Then X is an isonormal Gaussian process over $\mathbb{R}^d$ endowed with its canonical inner product.

Example 2.1.4 (Gaussian measures)   Consider the triple $(A, \mathscr{A}, \mu)$ where A is a Polish space (that is, A is metric, separable and complete), $\mathscr{A}$ is the associated Borel σ-field and the measure μ is positive, σ-finite and non-atomic. In this case, the real Hilbert space $L^2(A, \mathscr{A}, \mu)$ is separable. A (real) *Gaussian random measure* over $(A, \mathscr{A})$, with control μ, is a centered Gaussian family of the type

$$G = \{G(B) : B \in \mathscr{A}, \mu(B) < \infty\},$$

such that, for every $B, C \in \mathscr{A}$ of finite μ-measure, $E[G(B)G(C)] = \mu(B \cap C)$. Now consider the Hilbert space $\mathfrak{H} = L^2(A, \mathscr{A}, \mu)$, with inner product

$$\langle g, h \rangle_{\mathfrak{H}} = \int_A g(a)h(a)\mu(da).$$

For every $h \in \mathfrak{H}$, define $X(h) = \int_A h(a)G(da)$ to be the Wiener–Itô integral of h with respect to G. (Recall that $\int_A h(a)G(da)$ is defined as $\int_A h(a)G(da) = \sum_{i=1}^n a_i G(A_i)$ when h belongs to the set $\mathscr{E}$ of elementary functions of the form $h = \sum_{i=1}^n a_i \mathbf{1}_{A_i}$ with $a_i \in \mathbb{R}$ and $A_i \in \mathscr{A}$ such that $\mu(A_i) < \infty$, $i = 1, \ldots, n$; for general $h \in L^2(A, \mathscr{A}, \mu)$, $\int_A h(a)G(da)$ is defined as $\lim_{n \to \infty} \int_A h_n(a)G(da))$ in $L^2(\Omega)$, for any sequence $\{h_n\} \subset \mathscr{E}$ such that $\|h_n - h\|_{L^2(A, \mathscr{A}, \mu)} \to 0$.) Then $X = \{X(h) : h \in L^2(A, \mathscr{A}, \mu)\}$ defines a centered Gaussian family with covariance given by $E[X(g)X(h)] = \langle g, h \rangle_{\mathfrak{H}}$, thus yielding that X is an isonormal Gaussian process over $L^2(A, \mathscr{A}, \mu)$. For instance, by setting $A = [0, +\infty)$ and μ to be the Lebesgue measure, one obtains that the process $W_t = G([0, t))$, $t \geq 0$, is a standard Brownian motion starting from zero (of course, in order to meet the usual definition of a Brownian motion, one has also to select a continuous version of W), and X coincides with the $L^2(\Omega)$-closed linear Gaussian space generated by W.

Example 2.1.5 (Isonormal processes derived from covariances)   Let $Y = \{Y_t : t \geq 0\}$ be a real-valued centered Gaussian process indexed by the positive axis, and set $R(s, t) = E[Y_s Y_t]$ to be the covariance function of Y. One can embed Y into some isonormal Gaussian process as follows: (i) define $\mathscr{E}$ as the collection of all finite linear combinations of indicator functions of the type $\mathbf{1}_{[0,t]}$, $t \geq 0$; (ii) define $\mathfrak{H} = \mathfrak{H}_R$ to be the Hilbert space given by the closure of $\mathscr{E}$ with respect to the inner product

$$\langle f, h \rangle_R := \sum_{i,j} a_i c_j R(s_i, t_j),$$

where $f = \sum_i a_i \mathbf{1}_{[0,s_i]}$ and $h = \sum_j c_j \mathbf{1}_{[0,t_j]}$ are two generic elements of $\mathscr{E}$; (iii) for $h = \sum_j c_j \mathbf{1}_{[0,t_j]} \in \mathscr{E}$, set $X(h) = \sum_j c_j Y_{t_j}$; (iv) for $h \in \mathfrak{H}_R$, set $X(h)$ to be the $L^2(\Omega)$ limit of any sequence of the type $\{X(h_n)\}$, where $\{h_n\} \subset \mathcal{E}$ converges to h in $\mathfrak{H}_R$. Note that such a sequence $\{h_n\}$ necessarily exists and may not be unique (however, the definition of $X(h)$ does not depend on the choice of the sequence $\{h_n\}$). Then, by construction, the Gaussian space $\{X(h) : h \in \mathfrak{H}_R\}$ is an isonormal Gaussian process over $\mathfrak{H}_R$.

Example 2.1.6 (Even functions and symmetric measures) Other examples of isonormal Gaussian processes are given by objects of the type

$$X_\beta = \left\{ X_\beta(\psi) : \psi \in \mathfrak{H}_{E,\beta} \right\},$$

where β is a real non-atomic symmetric measure on $(-\pi, \pi]$ (that is, $d\beta(x) = d\beta(-x)$), and

$$\mathfrak{H} = \mathfrak{H}_{E,\beta} = L_E^2((-\pi, \pi], d\beta) \tag{2.1.1}$$

stands for the collection of all *real* linear combinations of complex-valued *even* functions that are square-integrable with respect to β (recall that a function ψ is even if $\overline{\psi(x)} = \psi(-x)$). The class $\mathfrak{H}_{E,\beta}$ is a real Hilbert space, endowed with the inner product

$$\langle \psi_1, \psi_2 \rangle_\beta = \int_{-\pi}^{\pi} \psi_1(x) \, \psi_2(-x) \, d\beta(x) \in \mathbb{R}. \tag{2.1.2}$$

This type of construction is used in the spectral theory of time series.

Example 2.1.7 (Gaussian free fields) Let $d \geq 2$ and let D be a domain in $\mathbb{R}^d$. Denote by $H_s(D)$ the space of real-valued continuously differentiable functions on $\mathbb{R}^d$ that are supported on a compact subset of D (note that this implies that the first derivatives of the elements of $H_s(D)$ are square-integrable with respect to the Lebesgue measure). Write $\mathfrak{H} = H(D)$ in order to denote the real Hilbert space obtained as the closure of $H_s(D)$ with respect to the inner product $\langle f, g \rangle = \int_{\mathbb{R}^d} \nabla f(x) \cdot \nabla g(x) dx$, where ∇ is the gradient. An isonormal Gaussian process of the type $X = \{X(h) : h \in H(D)\}$ is called a *Gaussian free field* (GFF).

Remark 2.1.8 When $X = \{X(h) : h \in \mathfrak{H}\}$ is an isonormal Gaussian process, we stress that the mapping $h \mapsto X(h)$ is linear. Indeed, for any $\lambda, \mu \in \mathbb{R}$ and $h, g \in \mathfrak{H}$, it is immediately verified (by expanding the square) that $E\left[\left(X(\lambda h + \mu g) - \lambda X(h) - \mu X(g)\right)^2\right] = 0$.

Remark 2.1.9 An isonormal Gaussian process is simply an isomorphism between a centered $L^2(\Omega)$-closed linear Gaussian space and a real separable Hilbert space $\mathfrak{H}$. Now, fix a generic centered $L^2(\Omega)$-closed linear Gaussian space, say $\mathcal{G}$. Since $\mathcal{G}$ is itself a real separable Hilbert space (with respect to the usual $L^2(\Omega)$ inner product) it follows that $\mathcal{G}$ can always be (trivially) represented as an isonormal Gaussian process, by setting $\mathfrak{H} = \mathcal{G}$. Plainly, the subtlety in the use of isonormal Gaussian processes is that one has to select an isomorphism that is well adapted to the specific problem at hand.

For the rest of the chapter, we fix an isonormal Gaussian process $X = \{X(h) : h \in \mathfrak{H}\}$.

2.2 Wiener chaos

The notion of *Wiener chaos* plays a role analogous to that of the Hermite polynomials $\{H_n : n \geq 0\}$ for the one-dimensional Gaussian distribution (recall Definition 1.4.1). The following property of Hermite polynomials is useful for our discussion.

Proposition 2.2.1 *Let* $Z, Y \sim \mathcal{N}(0, 1)$ *be jointly Gaussian. Then, for all* $n, m \geq 0$:

$$E[H_n(Z)H_m(Y)] = \begin{cases} n!\big(E[ZY]\big)^n & \text{if } n = m \\ 0 & \text{otherwise.} \end{cases}$$

Remark 2.2.2 In the previous statement, we used the convention that $0^0 = 1$ to deal with the case $E[ZY] = m = n = 0$.

Proof Set $\rho = E[ZY]$, and assume for the moment that $\rho > 0$. Let $N, \widetilde{N} \sim \mathcal{N}(0, 1)$ be independent, and observe that (Z, Y) and $(N, \rho N + \sqrt{1 - \rho^2}\widetilde{N})$ have the same law. Hence, for $n, m \geq 0$,

$$E[H_n(Z)H_m(Y)] = E[H_n(N)P_{\ln(1/\rho)}H_m(N)] \quad \text{(with } P_t \text{ defined by (1.3.1))}$$
$$= \rho^m E[H_n(N)H_m(N)] \quad \text{(by Proposition 1.4.2(i))}$$
$$= \begin{cases} n!\rho^n & \text{if } n = m \\ 0 & \text{otherwise} \end{cases} \quad \text{(by Proposition 1.4.2(iii)).}$$

If $\rho = 0$, the conclusion continues to hold, since in this case Z and Y are independent, and consequently

$$E[H_n(Z)H_m(Y)] = E[H_n(Z)] \times E[H_m(Y)] = \begin{cases} 1 & \text{if } n = m = 0 \\ 0 & \text{otherwise.} \end{cases}$$

Finally, if $\rho < 0$ we can proceed as in the first part of the proof to deduce that

$$E[H_n(Z)H_m(Y)] = E[H_n(N)P_{\ln(1/|\rho|)}H_m(-N)]$$
$$= |\rho|^m E[H_n(N)H_m(-N)]$$
$$= \begin{cases} n!(-1)^n|\rho|^n = n!\rho^n & \text{if } n = m \\ 0 & \text{otherwise,} \end{cases}$$

where we have used Proposition 1.4.2(viii) to infer that $H_m(-N) = (-1)^m$ $H_m(N)$. $\qquad\qquad\qquad\qquad\qquad\qquad\qquad\qquad\qquad\qquad\qquad\qquad\square$

Definition 2.2.3 For each $n \geq 0$, we write $\mathscr{H}_n$ to denote the closed linear subspace of $L^2(\Omega)$ generated by the random variables of type $H_n(X(h))$, $h \in \mathfrak{H}$, $\|h\|_{\mathfrak{H}} = 1$. The space $\mathscr{H}_n$ is called the nth **Wiener chaos** of X.

Plainly, $\mathscr{H}_0 = \mathbb{R}$ and $\mathscr{H}_1 = \{X(h) : h \in \mathfrak{H}\} = X$; see also Remark 2.1.8. By Proposition 2.2.1, if $n \neq m$ then $\mathscr{H}_n$ and $\mathscr{H}_m$ are orthogonal for the usual inner product of $L^2(\Omega)$. Therefore, the sum $\mathscr{H}_0 \oplus \mathscr{H}_1 \oplus \ldots = \bigoplus_{n=0}^{\infty} \mathscr{H}_n$ is direct in $L^2(\Omega)$. The next result shows that $\bigoplus_{n=0}^{\infty} \mathscr{H}_n$ coincides with $L^2(\Omega)$: this result is known as the *Wiener–Itô chaotic decomposition* of $L^2(\Omega)$.

Theorem 2.2.4 (i) *The linear space generated by the class* $\{H_n(X(h)) : n \geq 0, h \in \mathfrak{H}, \|h\|_{\mathfrak{H}} = 1\}$ *is dense in* $L^q(\Omega)$ *for every* $q \in [1, \infty)$.
(ii) *(Wiener–Itô chaos decomposition) One has that* $L^2(\Omega) = \bigoplus_{n=0}^{\infty} \mathscr{H}_n$. *This means that every random variable* $F \in L^2(\Omega)$ *admits a unique expansion of the type* $F = E[F] + \sum_{n=1}^{\infty} F_n$, *where* $F_n \in \mathscr{H}_n$ *and the series converges in* $L^2(\Omega)$.

Remark 2.2.5 For notational convenience, we shall sometimes use the symbols $\text{Proj}(F \mid \mathscr{H}_n)$ or $J_n(F)$, instead of F_n. See, in particular, Section 2.8.

Example 2.2.6 If $F = \varphi(X(h))$ with $h \in \mathfrak{H}$, $\|h\|_{\mathfrak{H}} = 1$, and $\varphi : \mathbb{R} \to \mathbb{R}$ is a Borel function such that $\frac{1}{\sqrt{2\pi}} \int_{\mathbb{R}} \varphi^2(x)e^{-x^2/2}dx < \infty$, then, according to Proposition 1.4.2, the (unique) Wiener–Itô decomposition of F is

$$F = \sum_{n=0}^{\infty} \frac{1}{n!}\left(\frac{1}{\sqrt{2\pi}}\int_{\mathbb{R}}\varphi(x)H_n(x)e^{-x^2/2}dx\right)H_n(X(h)).$$

Proof of Theorem 2.2.4 (i) An application of Proposition E.1.3 implies that it is sufficient to show that, for every $\eta \in (1, \infty]$, the only random variables $F \in L^\eta(\Omega)$ satisfying

$$E\big[F\,H_n(X(h))\big] = 0, \quad \forall n \geq 0, \forall h \in \mathfrak{H}, \|h\|_{\mathfrak{H}} = 1, \qquad (2.2.1)$$

are such that $F = 0$ a.s.-P. So, let $F \in L^\eta(\Omega)$ be such that (2.2.1) is satisfied. Since every monomial x^n can be expanded as a linear combination of the Hermite polynomials $H_r(x)$, $0 \leq r \leq n$, we get $E\big[F\,X(h)^n\big] = 0$ for every

$n \geq 0$ and every $h \in \mathfrak{H}$, $\|h\|_{\mathfrak{H}} = 1$. Therefore, by reasoning as in the proof of Proposition 1.1.5, one sees that $E\big[F\, e^{it X(h)}\big] = 0$ for all $t \in \mathbb{R}$ and $h \in \mathfrak{H}$ such that $\|h\|_{\mathfrak{H}} = 1$ or, equivalently, $E\big[F\, e^{i X(h)}\big] = 0$ for all $h \in \mathfrak{H}$. Now, let $(h_j)_{j \geq 1}$ be an orthonormal basis of $\mathfrak{H}$, and let us denote by $\mathscr{F}_m$ the σ-field generated by $X(h_j)$, $j \leq m$. The linearity of the map $h \mapsto X(h)$ implies

$$E\left[F\, e^{i \sum_{j=1}^{m} \lambda_j X(h_j)}\right] = 0, \quad \text{for every } m \geq 1 \text{ and } \lambda_1, \ldots, \lambda_m \in \mathbb{R},$$

which is the same as

$$E\left[E[F|\mathscr{F}_m] e^{i \sum_{j=1}^{m} \lambda_j X(h_j)}\right] = 0, \quad \text{for every } m \geq 1 \text{ and } \lambda_1, \ldots, \lambda_m \in \mathbb{R}.$$

$$(2.2.2)$$

Fix $m \geq 1$. Since $E[F|\mathscr{F}_m]$ is $\mathscr{F}_m$-measurable by definition, there exists a measurable function $\varphi : \mathbb{R}^m \to \mathbb{R}$ such that $E[F|\mathscr{F}_m] = \varphi\big(X(h_1), \ldots, X(h_m)\big)$. Combined with (2.2.2), this gives

$$\frac{1}{(2\pi)^{m/2}} \int_{\mathbb{R}^m} \varphi(x_1, \ldots, x_m)\, e^{-\frac{1}{2} \sum_{j=1}^{m} x_j^2}\, e^{i \sum_{j=1}^{m} \lambda_j x_j}\, dx_1 \ldots dx_m = 0,$$

for every $\lambda_1, \ldots, \lambda_m \in \mathbb{R}$. In particular, the Fourier transform of $x \mapsto \varphi(x_1, \ldots, x_m)\, e^{-\frac{1}{2} \sum_{j=1}^{m} x_j^2}$ is identically zero. As a consequence, $\varphi = 0$ almost everywhere, implying $E[F|\mathscr{F}_m] = 0$. Since the σ-fields $\mathscr{F}_m$, $m \geq 1$, generate $\mathscr{F} = \sigma\{X\}$, we deduce that $E(F|\mathscr{F}) = 0$. Since $F = E(F|\mathscr{F})$ by construction, the conclusion follows immediately.

(ii) This is a direct consequence of part (i) in the case $q = 2$ and of the fact that, according to Proposition 2.2.1, Wiener chaoses of different orders are orthogonal in $L^2(\Omega)$. □

2.3 The derivative operator

Let $\mathscr{S}$ denote the set of all random variables of the form

$$f\big(X(\phi_1), \ldots, X(\phi_m)\big), \quad (2.3.1)$$

where $m \geq 1$, $f : \mathbb{R}^m \to \mathbb{R}$ is a C^∞-function such that f and its partial derivatives have at most polynomial growth, and $\phi_i \in \mathfrak{H}$, $i = 1, \ldots, m$. A random variable belonging to $\mathscr{S}$ is said to be *smooth*.

Lemma 2.3.1 *The space $\mathscr{S}$ is dense in $L^q(\Omega)$ for every $q \geq 1$.*

Proof Observe that $\mathscr{S}$ contains the linear span of the class $\{H_n(X(h)) : n \geq 0, h \in \mathfrak{H}, \|h\|_{\mathfrak{H}} = 1\}$, so that the conclusion follows from Theorem 2.2.4. □

Definition 2.3.2 Let $F \in \mathscr{S}$ be given by (2.3.1), and $p \geq 1$ be an integer. The pth **Malliavin derivative** of F (with respect to X) is the element of $L^2(\Omega, \mathfrak{H}^{\odot p})$ (note the symmetric tensor product) defined by

$$D^p F = \sum_{i_1, \ldots, i_p = 1}^{m} \frac{\partial^p f}{\partial x_{i_1} \ldots \partial x_{i_p}} \left(X(\phi_1), \ldots, X(\phi_m) \right) \phi_{i_1} \otimes \ldots \otimes \phi_{i_p}.$$

The definition of $L^2(\Omega, \mathfrak{H}^{\odot p})$ is given in Section B.5 of Appendix B. When $p = 1$, we shall simply write 'the Malliavin derivative' instead of 'the first Malliavin derivative', and also write D instead of D^1.

Remark 2.3.3 1. In the definition of $D^p F$ the tensor products $\phi_{i_1} \otimes \ldots \otimes \phi_{i_p}$ are not necessarily symmetric. The symmetry of $D^p F$ is a consequence of the fact that the sum runs over all partial derivatives.
2. For $h \in \mathfrak{H}$ and for F as in (2.3.1), observe that, almost surely,

$$\langle DF, h \rangle_{\mathfrak{H}} = \lim_{\varepsilon \to 0} \frac{1}{\varepsilon} \left[f \left(X(\phi_1) + \varepsilon \langle \phi_1, h \rangle_{\mathfrak{H}}, \ldots, X(\phi_m) + \varepsilon \langle \phi_m, h \rangle_{\mathfrak{H}} \right) - F \right].$$

This shows that DF may be seen as a directional derivative. (Of course, an analogous interpretation holds for $D^p F$ as well.)

Proposition 2.3.4 *Let $q \in [1, \infty)$, and let $p \geq 1$ be an integer. Then the operator $D^p : \mathscr{S} \subset L^q(\Omega) \to L^q(\Omega, \mathfrak{H}^{\odot p})$ is closable.*

Proof We only consider the case $q > 1$; the proof for $q = 1$ (left to the reader) requires a specific argument, due to the duality $L^1(\Omega)/L^\infty(\Omega)$. We assume first that $p = 1$. Let $F, G \in \mathscr{S}$, and let $h \in \mathfrak{H}$ with $\|h\|_{\mathfrak{H}} = 1$. Notice that $FG \in \mathscr{S}$. Without loss of generality, we assume that FG has the form $f \left(X(\phi_1), \ldots, X(\phi_m) \right)$, with f a C^∞-function such that its partial derivatives have polynomial growth, $\phi_1 = h$ and $\phi_1, \ldots, \phi_m$ is an orthonormal system in $\mathfrak{H}$. Hence

$$E \left[\langle D(FG), h \rangle_{\mathfrak{H}} \right] = \frac{1}{(2\pi)^{m/2}} \int_{\mathbb{R}^m} \frac{\partial f}{\partial x_1} (x_1, \ldots, x_m) e^{-\frac{x_1^2 + \ldots + x_m^2}{2}} dx_1 \ldots dx_m$$

$$= \frac{1}{(2\pi)^{m/2}} \int_{\mathbb{R}^m} x_1 f(x_1, \ldots, x_m) e^{-\frac{x_1^2 + \ldots + x_m^2}{2}} dx_1 \ldots dx_m$$

(by integation by parts)

$$= E \left[X(h) FG \right].$$

It is immediate, using the definition of D on $\mathscr{S}$, that $D(FG) = FDG + GDF$. Therefore,

$$E \left[G \langle DF, h \rangle_{\mathfrak{H}} \right] = -E \left[F \langle DG, h \rangle_{\mathfrak{H}} \right] + E \left[X(h) FG \right]. \tag{2.3.2}$$

By linearity, (2.3.2) continues to hold for any $h \in \mathfrak{H}$ (i.e. not only when $\|h\|_{\mathfrak{H}} = 1$).

Now, let (F_n) be a sequence in $\mathscr{S}$ such that: (i) F_n converges to zero in $L^q(\Omega)$; (ii) DF_n converges to some η in $L^q(\Omega, \mathfrak{H})$. We have to prove that η is equal to 0 a.s.-P. Let $G \in \mathscr{S}$. We have, by (2.3.2):

$$
\begin{aligned}
E\big[G\langle \eta, h\rangle_{\mathfrak{H}}\big] &= \lim_{n \to \infty} E\big[G\langle DF_n, h\rangle_{\mathfrak{H}}\big] \\
&= \lim_{n \to \infty} -E\big[F_n\langle DG, h\rangle_{\mathfrak{H}}\big] + E\big[X(h)F_nG\big].
\end{aligned}
$$

Since $F_n \to 0$ in $L^q(\Omega)$ and $X(h)G$, $\langle DG, h\rangle_{\mathfrak{H}}$ belong to $L^{\frac{q}{q-1}}(\Omega)$, we deduce, using the Hölder inequality, that $E\big[G\langle \eta, h\rangle_{\mathfrak{H}}\big] = 0$. Because this is true for all $G \in \mathscr{S}$, it implies that, for every $h \in \mathfrak{H}$, $\langle \eta, h\rangle_{\mathfrak{H}} = 0$, a.s.-$P$. This last relation implies that, for every orthonormal basis $(e_i)_{i \geq 1}$ of $\mathfrak{H}$,

$$
P\big(\langle \eta, e_i\rangle_{\mathfrak{H}} = 0, \; \forall i\big) = 1,
$$

hence $P(\eta = 0) = 1$. The proof of the proposition in the case $p = 1$ is now complete. The proof for $p \geq 2$, which follows along the same lines, is left to the reader as a useful exercise. $\qquad\square$

Fix $q \in [1, \infty)$ and an integer $p \geq 1$, and let $\mathbb{D}^{p,q}$ denote the closure of $\mathscr{S}$ with respect to the norm

$$
\|F\|_{\mathbb{D}^{p,q}} = \Big(E[|F|^q] + E[\|DF\|^q_{\mathfrak{H}}] + \ldots + E[\|D^pF\|^q_{\mathfrak{H}^{\otimes p}}]\Big)^{1/q}.
$$

According to Proposition 2.3.4, for every $p \geq 1$ the operator D^p can be consistently extended to the set $\mathbb{D}^{p,q}$. We call $\mathbb{D}^{p,q}$ the *domain of D^p in $L^q(\Omega)$*. We also set $\mathbb{D}^{\infty,q} = \bigcap_{p \geq 1} \mathbb{D}^{p,q}$. We have the relation (already observed in the one-dimensional case)

$$
\mathbb{D}^{p,q+\epsilon} \subset \mathbb{D}^{p+m,q}, \quad \forall m \geq 0, \; \forall \epsilon \geq 0. \tag{2.3.3}
$$

Also, as a consequence of the closability of derivative operators, for every $q \neq q'$ the mappings $D^p : \mathbb{D}^{p,q} \to L^q(\Omega, \mathfrak{H}^{\odot p})$ and $D^p : \mathbb{D}^{p,q'} \to L^{q'}(\Omega, \mathfrak{H}^{\odot p})$ are *compatible*, that is, they coincide when acting on the intersection $\mathbb{D}^{p,q} \cap \mathbb{D}^{p,q'}$.

Exercise 2.3.5 Use Proposition 2.3.4 to prove that derivative operators are compatible.

Remark 2.3.6 For every $p \geq 1$, the space $\mathbb{D}^{p,2}$ is a Hilbert space with respect to the inner product

$$
\langle F, G\rangle_{\mathbb{D}^{p,2}} = E[FG] + \sum_{k=1}^{p} E\big[\langle D^kF, D^kG\rangle_{\mathfrak{H}^{\otimes k}}\big].
$$

The next statement is a useful *chain rule*. Thanks to the definition of D, this result is immediately shown for elements of $\mathscr{S}$ and then extended to $\mathbb{D}^{1,q}$ by an approximation argument (which we leave to the reader).

Proposition 2.3.7 (Chain rule) *Let $\varphi : \mathbb{R}^m \to \mathbb{R}$ be a continuously differentiable function with bounded partial derivatives. Suppose that $F = (F_1, \ldots, F_m)$ is a random vector whose components are elements of $\mathbb{D}^{1,q}$ for some $q \geq 1$. Then $\varphi(F) \in \mathbb{D}^{1,q}$ and*

$$D\varphi(F) = \sum_{i=1}^{m} \frac{\partial \varphi}{\partial x_i}(F) DF_i. \tag{2.3.4}$$

The conditions imposed on φ in Proposition 2.3.7 (that is, the partial derivatives of φ are bounded) are by no means optimal. For instance, by approximation, we can use the chain rule to prove that $D(X(h)^m) = mX(h)^{m-1}h$ for any $m \geq 1$ and $h \in \mathfrak{H}$, or that $De^{X(h)} = e^{X(h)}h$ for any $h \in \mathfrak{H}$. Moreover, the chain rule can sometimes be extended to the Lipschitz case as follows:

Proposition 2.3.8 (Extended chain rule) *Let $\varphi : \mathbb{R}^m \to \mathbb{R}$ be a Lipschitz function. Suppose that $F = (F_1, \ldots, F_m)$ is a random vector whose components are elements of $\mathbb{D}^{1,q}$ for some $q > 1$. If the law of F is absolutely continuous with respect to the Lebesgue measure on $\mathbb{R}^m$, then $\varphi(F) \in \mathbb{D}^{1,q}$ and (2.3.4) continues to hold (with $\frac{\partial \varphi}{\partial x_i}$ defined only a.e.).*

Proof It suffices to approximate φ by a sequence (φ_n) of sufficiently smooth functions, to apply Proposition 2.3.7 to each φ_n, and then to take the limit as $n \to \infty$. For a detailed proof, see [98, p. 29]. $\square$

Example 2.3.9 Let $\{h_i\}_{1 \leq i \leq m}$ be a collection of m linearly independent elements of $\mathfrak{H}$. Our aim is to use Proposition 2.3.8 in order to show that the Malliavin derivative of $F = \max_{1 \leq i \leq m} X(h_i)$ exists in $L^q(\Omega, \mathfrak{H})$ for all $q > 1$, and to compute it. First, we stress that the function $\max : \mathbb{R}^m \to \mathbb{R}$ is immediately shown (by induction on m) to be 1-Lipschitz with respect to the $\| \cdot \|_\infty$-norm, that is, it satisfies

$$\left| \max(y_1, \ldots, y_m) - \max(x_1, \ldots, x_m) \right| \leq \max \left(|y_1 - x_1|, \ldots, |y_m - x_m| \right)$$

for all $x_1, \ldots, x_m, y_1, \ldots, y_m \in \mathbb{R}$. Moreover, if Δ_i denotes the set $\{x = (x_1, \ldots, x_m) \in \mathbb{R}^m : \forall j, x_j \leq x_i\}$, we have

$$\frac{\partial}{\partial x_i} \max(x_1, \ldots, x_m) = \mathbf{1}_{\Delta_i}(x_1, \ldots, x_m) \quad \text{a.e.}$$

Set $I_0 = \text{argmax}_{1 \le i \le m} X(h_i)$, so that $F = X(h_{I_0})$. Since the Gaussian vector $\{X(h_i)\}_{1 \le i \le m}$ has a covariance matrix with rank m, we have that $P(X(h_i) = X(h_j)) = 0$ for $i \ne j$, so that I_0 is (almost surely) a well-defined random element of $\{1, \ldots, m\}$. Using Proposition 2.3.8, we deduce that $F \in \mathbb{D}^{1,q}$ and $DF = h_{I_0}$, because: (i) $X(h_i) \in \mathbb{D}^{1,q}$ for each i; (ii) the law of F is absolutely continuous with respect to the Lebesgue measure on $\mathbb{R}^m$ (indeed, see Proposition 2.1.11 in [98]); (iii) the function max is Lipschitz; (iv) $\frac{\partial}{\partial x_i} \max = \mathbf{1}_{\Delta_i}$ a.e. for each i.   □

The next exercise involves a 'Leibniz rule' for D^p.

Exercise 2.3.10   Fix an integer $p \ge 1$, and assume that F and G are in $\mathbb{D}^{p,4}$. Prove that $FG \in \mathbb{D}^{p,2}$ and that

$$D^p(FG) = \sum_{r=0}^{p} \binom{p}{r} D^r F \widetilde{\otimes} D^{p-r} G,$$

where $D^r F \widetilde{\otimes} D^{p-r} G$ denotes the symmetrization of the tensor product $D^r F \otimes D^{p-r} G$. The symmetrization $\widetilde{f}$ of any $f \in \mathfrak{H}^{\otimes p}$ is formally defined in formula (B.3.1) of Appendix B.

2.4 The Malliavin derivatives in Hilbert spaces

We consider an isonormal Gaussian process $X = \{X(h) : h \in \mathfrak{H}\}$, and let $\mathfrak{U}$ be another real separable Hilbert space. For $q \ge 1$, define the space $L^q(\Omega, \mathfrak{U}) = L^q(\Omega, \mathscr{F}, P; \mathfrak{U})$ as in Section B.5 of Appendix B (recall that $\mathscr{F} = \sigma\{X\}$). We adopt the following notational conventions:

– The space $\mathscr{S}_{\mathfrak{U}}$ is the collection of all smooth $\mathfrak{U}$-valued random elements of type $F = \sum_{j=1}^{n} F_j v_j$, where $F_j \in \mathscr{S}$ and $v_j \in \mathfrak{U}$.
– For every $p \ge 1$ we set $\mathscr{S}_p = \mathscr{S}_{\mathfrak{H}^{\otimes p}}$, that is, $\mathscr{S}_p$ is the subset of $L^1(\Omega, \mathfrak{H}^{\otimes p})$ composed of those random elements u of the form $u = \sum_{j=1}^{n} F_j h_j$, with $F_j \in \mathscr{S}$ and $h_j \in \mathfrak{H}^{\otimes p}$.

For $k \ge 1$, the kth Malliavin derivative of any $F \in \mathscr{S}_{\mathfrak{U}}$ is given by the $\mathfrak{H}^{\otimes k} \otimes \mathfrak{U}$-valued random element $D^k F = \sum_{j=1}^{n} D^k F_j \otimes v_j$. By following the same line of reasoning as in the previous section, one can prove the following three facts.

(i) For every $k \ge 1$, the operator D^k is closable from $\mathscr{S}_{\mathfrak{U}} \subset L^q(\Omega, \mathfrak{U})$ into $L^q(\Omega, \mathfrak{H}^{\otimes k} \otimes \mathfrak{U})$, for every $q \ge 1$.
(ii) By virtue of (i) above, one can extend the domain of D^k to the space $\mathbb{D}^{p,q}(\mathfrak{U})$, which is defined as the closure of $\mathscr{S}_{\mathfrak{U}}$ with respect to the norm

$$\|F\|_{\mathbb{D}^{p,q}(\mathfrak{U})} = \Big(E[\|F\|_{\mathfrak{U}}^q] + E[\|DF\|_{\mathfrak{H}\otimes\mathfrak{U}}^q] + \ldots + E[\|D^p F\|_{\mathfrak{H}^{\otimes p}\otimes\mathfrak{U}}^q] \Big)^{1/q}.$$

(iii) Relations analogous to (2.3.3) (as well as a compatibility property) hold.

Plainly, if $\mathfrak{U} = \mathbb{R}$ then $\mathbb{D}^{p,q}(\mathbb{R}) = \mathbb{D}^{p,q}$, as defined in the previous section. We sometimes write $\mathbb{D}^{0,q}(\mathfrak{U}) = L^q(\Omega, \mathfrak{U})$.

2.5 The divergence operator

Fix an integer $p \geq 1$. We will now define δ^p (the divergence operator of order p) as the adjoint of $D^p : \mathbb{D}^{p,2} \to L^2(\Omega, \mathfrak{H}^{\odot q})$. This is the exact analog of the operator δ^p introduced in Chapter 1.

Definition 2.5.1 Let $p \geq 1$ be an integer. We denote by $\mathrm{Dom}\,\delta^p$ the subset of $L^2(\Omega, \mathfrak{H}^{\otimes p})$ composed of those elements u such that that there exists a constant $c > 0$ satisfying

$$\big| E[\langle D^p F, u \rangle_{\mathfrak{H}^{\otimes p}}] \big| \leq c\sqrt{E[F^2]} \text{ for all } F \in \mathscr{S}$$

$$\text{(or, equivalently, for all } F \in \mathbb{D}^{p,2}). \quad (2.5.1)$$

Fix $u \in \mathrm{Dom}\,\delta^p$. Since condition (2.5.1) holds, the linear operator $F \mapsto E[\langle D^p F, u \rangle_{\mathfrak{H}^{\otimes p}}]$ is continuous from $\mathscr{S}$, equipped with the $L^2(\Omega)$-norm, into $\mathbb{R}$. Thus, we can extend it to a linear operator from $L^2(\Omega)$ into $\mathbb{R}$. By the Riesz representation theorem, there exists a unique element in $L^2(\Omega)$, denoted by $\delta^p(u)$, such that $E[\langle D^p F, u \rangle_{\mathfrak{H}^{\otimes p}}] = E[F \delta^p(u)]$ for all $F \in \mathscr{S}$. This fact leads to the next definition.

Definition 2.5.2 If $u \in \mathrm{Dom}\,\delta^p$, then $\delta^p(u)$ is the unique element of $L^2(\Omega)$ characterized by the following duality formula:

$$E[F \delta^p(u)] = E[\langle D^p F, u \rangle_{\mathfrak{H}^{\otimes p}}], \quad (2.5.2)$$

for all $F \in \mathscr{S}$. The operator $\delta^p : \mathrm{Dom}\,\delta^p \subset L^2(\Omega, \mathfrak{H}^{\otimes p}) \to L^2(\Omega)$ is called the **multiple divergence operator** of order p. When $p = 1$, we merely say 'the divergence operator', and write δ instead of δ^1. Also, we define δ^0 to be equal to the identity. Formula (2.5.2) is customarily called an **integration by parts formula**.

Notice that the operator δ^p is closed (as the adjoint of D^p). Moreover, by selecting F to be equal to a constant in (2.5.2), we deduce that, for every $u \in \mathrm{Dom}\,\delta^p$, $E[\delta^p(u)] = 0$.

Remark 2.5.3 By choosing G to be identically equal to one in (2.3.2), we get that $\mathfrak{H}$ is included in Dom δ, with $\delta(h) = X(h)$. Similarly, it is no more difficult to prove that $\mathfrak{H}^{\otimes p}$ is included in Dom δ^p for all $p \geq 1$; see also Theorem 2.5.5 below.

The following proposition shows that one can 'factor out' a scalar random variable in the divergence operator δ. This will be useful in many situations.

Proposition 2.5.4 *Let $F \in \mathbb{D}^{1,2}$ and $u \in$ Dom δ be such that the three expectations $E[F^2\|u\|^2_{\mathfrak{H}}]$, $E[F^2\delta(u)^2]$ and $E[\langle DF, u\rangle^2_{\mathfrak{H}}]$ are finite. Then $Fu \in$ Dom δ and*

$$\delta(Fu) = F\delta(u) - \langle DF, u\rangle_{\mathfrak{H}}.$$

For example, by using Proposition 2.5.4 we can readily check that $\delta(X(h)g) = X(h)X(g) - \langle h, g\rangle_{\mathfrak{H}}$ for all $g, h \in \mathfrak{H}$.

Proof of Proposition 2.5.4 For any $G \in \mathscr{S}$, we have

$$E[\langle DG, Fu\rangle_{\mathfrak{H}}] = E[F\langle DG, u\rangle_{\mathfrak{H}}] = E[\langle FDG, u\rangle_{\mathfrak{H}}]$$
$$= E[\langle u, D(FG) - GDF\rangle_{\mathfrak{H}}] = E[G(\delta(u)F - \langle DF, u\rangle_{\mathfrak{H}})].$$

Hence, using the assumptions, we get that $Fu \in$ Domδ with $\delta(Fu) = F\delta(u) - \langle DF, u\rangle_{\mathfrak{H}}$. $\qquad\square$

Recall the notation introduced in Section 2.4. For $p \geq 1$, it is not difficult to prove that $\mathscr{S}_p \subset$ Dom δ^p. For $u = \sum_{j=1}^n F_j h_j \in \mathscr{S}_1$, we have, using Proposition 2.5.4 (as well as Remark 2.5.3),

$$\delta(u) = \sum_{j=1}^n \delta(F_j h_j) = \sum_{j=1}^n \left(F_j X(h_j) - \langle DF_j, h_j\rangle_{\mathfrak{H}}\right). \tag{2.5.3}$$

In particular, we see that $\delta(u) \in \mathbb{D}^{1,2}$ with

$$D\delta(u) = \sum_{j=1}^n \left(X(h_j)DF_j + F_j h_j - \langle D^2 F_j, h_j\rangle_{\mathfrak{H}^{\odot 2}}\right).$$

On the other hand, it is immediate that $Du = \sum_{j=1}^n DF_j \otimes h_j$, and that $Du \in$ Dom δ. Thanks once again to Proposition 2.5.4 (and Remark 2.5.3), we can write

$$\delta(Du) = \sum_{j=1}^n \delta(DF_j \otimes h_j) = \sum_{j=1}^n \left(X(h_j)DF_j - \langle D^2 F_j, h_j\rangle_{\mathfrak{H}}\right). \tag{2.5.4}$$

Combining (2.5.3) and (2.5.4), we get the following Heisenberg commutativity property for δ and D (compare with Proposition 1.3.8):

$$D\delta(u) - \delta(Du) = u. \tag{2.5.5}$$

We conclude this section by stating without proof the following estimates, which are consequences of the so-called *Meyer inequalities* (see [98, Proposition 1.5.7]). We adopt the notation introduced in Section 2.4.

Theorem 2.5.5 (Meyer inequalities) *For any integers $k \geq p \geq 1$ and any $q \in [1, \infty)$, the operator δ^p is continuous from $\mathbb{D}^{k,q}(\mathfrak{H}^{\otimes p})$ to $\mathbb{D}^{k-p,q}$, that is, we have*

$$\|\delta^p(u)\|_{\mathbb{D}^{k-p,q}} \leq c_{k,p,q} \|u\|_{\mathbb{D}^{k,q}(\mathfrak{H}^{\otimes p})}$$

for all $u \in \mathbb{D}^{k,q}(\mathfrak{H}^{\otimes p})$, and some universal constant $c_{k,p,q} > 0$.

In particular, the last statement implies that $\mathbb{D}^{p,2}(\mathfrak{H}^{\otimes p}) \subset \mathrm{Dom}\,\delta^p$.

2.6 Some Hilbert space valued divergences

By following the same route as in the previous section, we could now define the adjoint operator of the generalized derivatives D^k discussed in Section 2.4. These operators would act on $\mathfrak{U} \otimes \mathfrak{H}^{\otimes k}$-valued random elements, where $\mathfrak{U}$ is an arbitrary Hilbert space. However, these general objects are not needed in this book, and we prefer to provide below a direct construction in the special case of a deterministic $u \in \mathfrak{U} \otimes \mathfrak{H}^{\otimes k}$.

Fix $k \geq 1$, let $\mathfrak{U}$ be a real separable Hilbert space and let $u \in \mathfrak{U} \otimes \mathfrak{H}^{\otimes k}$ have the form $u = \sum_{j=1}^{n} v_j \otimes h_j$, where $v_j \in \mathfrak{U}$ and $h_j \in \mathfrak{H}^{\otimes k}$. We set $\delta^k(u)$ to be the $\mathfrak{U}$-valued random element

$$\delta^k(u) = \sum_{j=1}^{n} v_j \delta^k(h_j), \tag{2.6.1}$$

where $\delta^k(h_j)$ has been defined in the previous section. Since vectors such as u are dense in $\mathfrak{U} \otimes \mathfrak{H}^{\otimes k}$, by using, for example, Theorem 2.5.5, we see that δ^k can be extended to a bounded operator from $\mathfrak{U} \otimes \mathfrak{H}^{\otimes k}$ into $L^2(\Omega, \mathfrak{U})$. Note that this construction allows a precise meaning to be given to the expression $\delta^k(f)$, where $f \in \mathfrak{H}^{\otimes p}$ and $p > k$. Indeed, since $\mathfrak{H}^{\otimes p} = \mathfrak{H}^{\otimes p-k} \otimes \mathfrak{H}^{\otimes k}$, we have that $\delta^k(f)$ is the element of $L^2(\Omega, \mathfrak{H}^{\otimes p-k})$, obtained by specializing the previous construction to the case $\mathfrak{U} = \mathfrak{H}^{\otimes p-k}$. Note that, for every $k = 1, \ldots, p-1$ and every $f \in \mathfrak{H}^{\otimes p}$, we also have that

$$\delta^p(f) = \delta^{p-k}(\delta^k(f)). \tag{2.6.2}$$

The following result partially generalizes relation (2.5.5) and Exercise 1.3.9:

Proposition 2.6.1   *Let $p \geq 1$ be an integer. For all $u \in \mathfrak{H}^{\otimes p}$, we have $\delta^p(u) \in \mathbb{D}^{1,2}$ and*

$$D\delta^p(u) = p\delta^{p-1}(u).$$

Proof   We proceed by induction. For $p = 1$, this is a direct consequence of (2.5.5). Now, assume that $D\delta^p(u) = p\delta^{p-1}(u)$, for some $p \geq 1$ and all $u \in \mathfrak{H}^{\otimes p}$. Then, for any $v \in \mathfrak{H}^{\otimes p+1}$, we have, again using (2.5.5),

$$D\delta^{p+1}(v) = D\delta(\delta^p v) = \delta D(\delta^p v) + \delta^p v$$
$$= p\delta(\delta^{p-1}(v)) + \delta^p v = (p+1)\delta^p(v),$$

which is the desired formula for $p + 1$.   □

2.7 Multiple integrals

2.7.1 Definition and first properties

Definition 2.7.1   Let $p \geq 1$ and $f \in \mathfrak{H}^{\odot p}$. The pth **multiple integral** of f (with respect to X) is defined by $I_p(f) = \delta^p(f)$.

We now state a fundamental hypercontractivity property for multiple integrals, which shows that, inside a fixed chaos, all the $L^q(\Omega)$-norms are equivalent.

Theorem 2.7.2 (Hypercontractivity)   *For every $q > 0$ and every $p \geq 1$, there exists a constant $0 < k(q, p) < \infty$ (depending only on q and p) such that*

$$E[|Y|^q]^{1/q} \leq k(q, p)E[Y^2]^{1/2} \tag{2.7.1}$$

for every random variable Y with the form of a pth multiple integral.

Proof   By definition, we have that $Y = \delta^p(f)$, where $f \in \mathfrak{H}^{\odot p}$. It follows from Theorem 2.5.5 in the case $u = f$ and $k = p$ (recall that $\mathbb{D}^{0,q} = L^q(\Omega)$ and observe that $\|f\|_{\mathbb{D}^{p,q}(\mathfrak{H}^{\otimes p})} = \|f\|_{\mathfrak{H}^{\otimes p}}$ because f is non-random), that

$$E[|Y|^q]^{1/q} \leq c_{0,p,q}\|f\|_{\mathfrak{H}^{\otimes p}} = \frac{c_{0,p,q}}{\sqrt{p!}}E[Y^2]^{1/2},$$

which is the desired conclusion.   □

Remark 2.7.3   A self-contained alternative proof of the previous statement will follow from the explicit estimates of Corollary 2.8.14, which are in turn based on the hypercontractivity properties of the Ornstein–Uhlenbeck semi-group. As an example (and for future use) we record here the following consequence of Corollary 2.8.14: if $Y = I_2(f)$, then for every $q > 2$,

$$E[|Y|^q]^{1/q} \leq (q-1) \times E[Y^2]^{1/2}, \tag{2.7.2}$$

that is, one can take $k(q, 2) = (q - 1)$ in (2.7.1).

Proposition 2.7.4 *Let $p \geq 1$ and $f \in \mathfrak{H}^{\odot p}$. For all $q \in [1, \infty)$, $I_p(f) \in \mathbb{D}^{\infty, q}$. Moreover, for all $r \geq 1$,*

$$D^r I_p(f) = \begin{cases} \frac{p!}{(p-r)!} I_{p-r}(f) & \text{if } r \leq p \\ 0 & \text{if } r > p. \end{cases}$$

Proof Using Proposition 2.6.1, we have that $I_p(f) \in \mathbb{D}^{1,2}$ and we can write

$$DI_p(f) = D\delta^p(f) = p\delta^{p-1}(f) = pI_{p-1}(f),$$

which is exactly the required formula in the case $r = 1$. By repeatedly applying this argument, we get that $I_p(f) \in \mathbb{D}^{\infty,2}$ and also the formula for a general $D^r I_p(f)$. Finally, it is a consequence of Theorem 2.5.5 that $I_p(f) \in \mathbb{D}^{\infty,q}$: indeed, for any $r \geq 1$, we have

$$\|I_p(f)\|_{\mathbb{D}^{r,q}} = \|\delta^p(f)\|_{\mathbb{D}^{r,q}} \leq c_{r,p,q} \|f\|_{\mathbb{D}^{r+p,q}(\mathfrak{H}^{\otimes p})} = c_{r,p,q} \|f\|_{\mathfrak{H}^{\otimes p}} < \infty. \qquad \square$$

Proposition 2.7.5 (Isometry property of integrals) *Fix integers $1 \leq q \leq p$, as well as $f \in \mathfrak{H}^{\odot p}$ and $g \in \mathfrak{H}^{\odot q}$. We have*

$$E[I_p(f)I_q(g)] = \begin{cases} p!\langle f, g \rangle_{\mathfrak{H}^{\otimes p}} & \text{if } p = q \\ 0 & \text{otherwise.} \end{cases} \tag{2.7.3}$$

In particular,

$$E[I_p(f)^2] = p!\|f\|_{\mathfrak{H}^{\otimes p}}^2. \tag{2.7.4}$$

Proof We proceed by duality:

$$E[I_p(f)I_q(g)] = E[\delta^p(f)I_q(g)] = E[\langle f, D^p I_q(g) \rangle_{\mathfrak{H}^{\otimes p}}]$$

$$= \begin{cases} p!\langle f, g \rangle_{\mathfrak{H}^{\otimes p}} & \text{if } p = q \\ 0 & \text{if } p > q, \end{cases}$$

the last equality being a consequence of Proposition 2.7.4. $\qquad \square$

Exercise 2.7.6 (Why are multiple integrals called 'integrals'?) In this exercise, we assume that $\mathfrak{H}$ is equal to $L^2(A, \mathscr{A}, \mu)$, where $(A, \mathscr{A})$ is a Polish space and μ is a σ-finite measure without atoms. When $B \in \mathscr{A}$ has finite

μ-measure, we write $X(B)$ instead of $X(\mathbf{1}_B)$ for simplicity. Observe that the application $B \mapsto X(B)$ can be thought as measure with values in the Hilbert space $L^2(\Omega)$. In particular, we have that, if $(B_n)_{n \geq 1}$ is a sequence of disjoint elements of $\mathscr{A}$ whose union has finite μ-measure, then

$$X\left(\bigcup_{n=1}^{\infty} B_n\right) = \sum_{n=1}^{\infty} X(B_n),$$

where the series converges in $L^2(\Omega)$; see also Example 2.1.4. Our aim is to prove that, in this case, multiple integrals are the limit in $L^2(\Omega)$ of multiple Riemann sums with respect to the vector-valued measure $B \mapsto X(B)$. Due to the fact that μ has no atoms, these sums can be chosen to be 'without diagonals'. Fix an integer $p \geq 1$, and denote by $\mathscr{E}_p$ the set of elementary functions of $\mathfrak{H}^{\odot p}$ of the form

$$f = \sum_{i_1,\ldots,i_p=1}^{n} a_{i_1\ldots i_p} \mathbf{1}_{A_{i_1}} \otimes \ldots \otimes \mathbf{1}_{A_{i_p}}, \tag{2.7.5}$$

where the A_i are some pairwise disjoint sets such that $\mu(A_i) < \infty$, and the $a_{i_1\ldots i_p}$ are a symmetric array of real numbers vanishing on diagonals (that is, $a_{i_1\ldots i_p} = a_{i_{\sigma(1)}\ldots i_{\sigma(p)}}$ for all $\sigma \in \mathfrak{S}_p$, and $a_{i_1\ldots i_p} = 0$ if any two of the indices $i_1,\ldots,i_p$ are equal). When f is given by (2.7.5), we set

$$\widetilde{I}_p(f) = \sum_{i_1,\ldots,i_p=1}^{n} a_{i_1\ldots i_p} X(A_{i_1})\ldots X(A_{i_p}).$$

1. Prove that $\widetilde{I}_p$ is a linear operator, and that the definition of $\widetilde{I}_p(f)$ does not depend on the particular representation of f.
2. Let $p, q \geq 1$ be integers, and let $f \in \mathscr{E}_p$ and $g \in \mathscr{E}_q$. Prove the following isometry property:

$$E\left[\widetilde{I}_p(f)\widetilde{I}_q(g)\right] = \begin{cases} p!\langle f, g\rangle_{\mathfrak{H}^{\otimes p}} & \text{if } p = q \\ 0 & \text{otherwise.} \end{cases} \tag{2.7.6}$$

3. Prove that $\mathscr{E}_p$ is dense in $\mathfrak{H}^{\odot p}$. (Hint: Use the fact that μ has no atoms.)
4. Deduce that $\widetilde{I}_p$ can be extended to a linear and continuous operator from $\mathfrak{H}^{\odot p}$ to $L^2(\Omega)$, still satisfying (2.7.6).
5. Prove that $\widetilde{I}_p : \mathfrak{H}^{\odot p} \to L^2(\Omega)$ coincides with $I_p : \mathfrak{H}^{\odot p} \to L^2(\Omega)$.

Now let $W = (W_t)_{t \in [0,T]}$ be a standard Brownian motion starting from zero. According to the discussion contained in Example 2.1.4, one can regard the Gaussian space generated by the paths of W as an isonormal Gaussian process over $\mathfrak{H} = L^2([0, T], dt) = L^2([0, T])$. As a consequence of the previous exercise, we see that if $f \in L^2([0, T]^p)$ is symmetric then

$$I_p(f) = \int_{[0,T]^p} f(t_1, \ldots, t_p) dW_{t_1} \ldots dW_{t_p},$$

where the right-hand side is the *multiple Wiener–Itô integral* of order p, of f with respect to W (as defined by Itô in [55]). In fact, taking into account that f is symmetric, we can also rewrite $I_p(f)$ as the following *iterated* adapted Itô stochastic integral:

$$I_p(f) = p! \int_0^T dW_{t_1} \int_0^{t_1} dW_{t_2} \ldots \int_0^{t_{p-1}} dW_{t_p} f(t_1, \ldots, t_p).$$

2.7.2 Multiple integrals as Hermite polynomials

Recall the definition of the Hermite polynomials $\{H_p : p \geq 0\}$ given in Definition 1.4.1.

Theorem 2.7.7 *Let $f \in \mathfrak{H}$ be such that $\|f\|_{\mathfrak{H}} = 1$. Then, for any integer $p \geq 1$, we have*

$$H_p\big(X(f)\big) = I_p(f^{\otimes p}). \tag{2.7.7}$$

As a consequence, the linear operator I_p provides an isometry from $\mathfrak{H}^{\odot p}$ (equipped with the modified norm $\frac{1}{\sqrt{p!}} \| \cdot \|_{\mathfrak{H}^{\otimes p}}$) onto the pth Wiener chaos $\mathscr{H}_p$ of X (equipped with the $L^2(\Omega)$-norm).

Proof First, we prove (2.7.7) by induction (on p). For $p = 1$, it is clear since $H_1(X(f)) = X(f) = \delta(f) = I_1(f)$. Assume that the property holds for $1, 2, \ldots, p$. We then have

$$\begin{aligned}
I_{p+1}&(f^{\otimes(p+1)}) \\
&= \delta(\delta^p(f^{\otimes(p)})f) = \delta(I_p(f^{\otimes p})f) \\
&= I_p(f^{\otimes p})\delta(f) - \langle DI_p(f^{\otimes p}), f \rangle_{\mathfrak{H}} \quad \text{(by Proposition 2.5.4)} \\
&= I_p(f^{\otimes p})X(f) - pI_{p-1}(f^{\otimes(p-1)})\|f\|_{\mathfrak{H}}^2 \quad \text{(by Proposition 2.7.4)} \\
&= H_p(X(f))X(f) - pH_{p-1}(X(f)) \quad \text{(by the induction property)} \\
&= H_{p+1}(X(f)) \quad \text{(by Proposition 1.4.2(ii))}.
\end{aligned}$$

Therefore, (2.7.7) is proved for all integers $p \geq 1$.

Let us now prove that I_p provides an isometry from $\mathfrak{H}^{\odot p}$ onto $\mathscr{H}_p$. For any $f \in \mathfrak{H}^{\odot p}$, we have, by (2.7.4), that $E[I_p(f)^2] = p!\|f\|_{\mathfrak{H}^{\otimes p}}^2 = \|f\|_{\mathfrak{H}^{\odot p}}^2$, so that the isometry property is deduced from the linearity of the operator $f \mapsto I_p(f)$. Hence, it remains to prove that I_p is onto, that is, that

$I_p(\mathfrak{H}^{\odot p}) := \{I_p(f) : f \in \mathfrak{H}^{\odot p}\} = \mathscr{H}_p$. Since $I_p(\mathfrak{H}^{\odot p})$ is a closed linear subspace of $L^2(\Omega)$, and by virtue of the isometry property, it is sufficient to show that there exists a class $U \subset \mathfrak{H}^{\odot p}$ such that: (i) the span of U is dense in $\mathfrak{H}^{\odot p}$; and (ii) the span of $\{I_p(f) : f \in U\}$ is dense in $\mathscr{H}_p$. By exploiting the first part of the statement one has that, for every $h \in \mathfrak{H}$ such that $\|h\|_{\mathfrak{H}} = 1$, $H_p(X(h)) = I_p(h^{\otimes p}) \in I_p(\mathfrak{H}^{\odot p})$. Since the span of $\{H_p(X(h)) : \|h\|_{\mathfrak{H}} = 1\}$ generates $\mathscr{H}_p$ by definition and since linear combinations of vectors of the type $h^{\otimes p}$, $\|h\|_{\mathfrak{H}} = 1$, are dense in $\mathfrak{H}^{\odot p}$, the conclusion is obtained by taking $U = \{h^{\otimes p} : \|h\|_{\mathfrak{H}} = 1\}$. □

The next statement provides a useful reformulation of the Wiener–Itô decomposition stated in Theorem 2.2.4(ii).

Corollary 2.7.8 (Chaos expansion and Stroock formula) *Every $F \in L^2(\Omega)$ can be expanded as*

$$F = E[F] + \sum_{p=1}^{\infty} I_p(f_p), \qquad (2.7.8)$$

for some unique collection of kernels $f_p \in \mathfrak{H}^{\odot p}$, $p \geq 1$. Moreover, if $F \in \mathbb{D}^{n,2}$ (for some $n \geq 1$) then $f_p = \frac{1}{p!} E[D^p F]$ for all $p \leq n$.

Proof The first part of the statement is a direct combination of Theorems 2.2.4 and 2.7.7. For the second part, using Proposition 2.7.4, we get

$$D^p F = \sum_{q=p}^{\infty} \frac{q!}{(q-p)!} I_{q-p}(f_q),$$

from which we deduce the relation $E[D^p F] = p! f_p$. □

We conclude this section with an important exercise, providing further characterizations of derivative operators.

Exercise 2.7.9   1. Let $F \in L^2(\Omega)$ be such that $F = E(F) + \sum_{p=1}^{\infty} I_p(f_p)$, with $f_p \in \mathfrak{H}^{\odot p}$. Use Proposition 2.7.4 to show that $F \in \mathbb{D}^{k,2}$ if and only if $\sum_{p=k}^{\infty} p^k p! \|f_p\|_{\mathfrak{H}^{\otimes p}}^2 < \infty$, and in this case

$$D^k F = \sum_{p=k}^{\infty} p(p-1)\ldots(p-k+1) I_{p-k}(f_p).$$

2. (Derivatives as stochastic processes) Assume that $\mathfrak{H} = L^2(A, \mathscr{A}, \mu)$, with μ non-atomic. Let $F \in L^2(\Omega)$ be such that $F = E(F) + \sum_{p=1}^{\infty} I_p(f_p)$, with $f_p \in \mathfrak{H}^{\odot p} = L_s^2(A^p, \mathscr{A}^p, \mu^p)$. Prove that, for every $k \geq 1$, the derivative $D^k F$ coincides with the stochastic process on A^k given by

$$(a_1, \dots, a_k) \mapsto D^k_{a_1, \dots, a_k} F$$

$$= \sum_{p=k}^{\infty} p(p-1) \dots (p-k+1) I_{p-k}(f_p(a_1, \dots, a_k, \cdot)),$$

where we write $I_{p-k}(f_p(a_1, \dots, a_k, \cdot))$ to indicate that only $p-k$ variables in the kernel f_p are integrated out (while the others act as free parameters).

2.7.3 The product formula

Let $p, q \geq 1$ be integers. Assume that $f \in \mathfrak{H}^{\odot p}$, $g \in \mathfrak{H}^{\odot q}$, and let $r \in \{0, \dots, p \wedge q\}$. We define the contraction $f \otimes_r g$, between f and g, according to Section B.4 in Appendix B (see, in particular, formula (B.4.4)). The symmetrization $f \tilde{\otimes}_r g$ is defined in formula (B.4.6).

We are now ready to state a crucial *product formula*, implying in particular that a product of two multiple integrals is indeed a finite sum of multiple integrals.

Theorem 2.7.10 (Product formula) *Let $p, q \geq 1$. If $f \in \mathfrak{H}^{\odot p}$ and $g \in \mathfrak{H}^{\odot q}$ then*

$$I_p(f)I_q(g) = \sum_{r=0}^{p \wedge q} r! \binom{p}{r}\binom{q}{r} I_{p+q-2r}(f \tilde{\otimes}_r g). \qquad (2.7.9)$$

Proof By Proposition 2.7.4, we have that $I_p(f)$ and $I_q(g)$ belong to $\mathbb{D}^{\infty,4}$, so that $I_p(f)I_q(g)$ belongs to $\mathbb{D}^{\infty,2}$; see Exercise 2.3.10. Hence, using Corollary 2.7.8, we can write

$$I_p(f)I_q(g) = \sum_{s=0}^{\infty} \frac{1}{s!} I_s(E[D^s(I_p(f)I_q(g))]).$$

Using the Leibniz rule for D (Exercise 2.3.10), we have

$$D^s(I_p(f)I_q(g)) = \sum_{k=0}^{s} \binom{s}{k} D^k(I_p(f)) \tilde{\otimes} D^{s-k}(I_q(g)).$$

Hence, by Proposition 2.7.4,

$$E[D^s(I_p(f)I_q(g))]$$

$$= \sum_{k=0 \vee (s-q)}^{s \wedge p} \binom{s}{k} \frac{p!}{(p-k)!} \frac{q!}{(q-s+k)!} E[I_{p-k}(f) \tilde{\otimes} I_{q-s+k}(g)].$$

By (2.7.3), the expectation $E[I_{p-k}(f) \tilde{\otimes} I_{q-s+k}(g)]$ is zero except if $p-k = q - s + k$. Moreover, the identity $p - k = q - s + k$ with

$0 \vee (s - q) \leq k \leq s \wedge p$ is equivalent to saying that s and $p + q$ have the same parity, that $k = (p - q + s)/2$, and that $|q - p| \leq s \leq p + q$. In this case, we have

$$E[I_{p-k}(f) \widetilde{\otimes} I_{q-s+k}(g)]$$

$$= \binom{s}{\frac{p-q+s}{2}} \frac{p!}{(\frac{p+q-s}{2})!} \frac{q!}{(\frac{p+q-s}{2})!} \binom{p+q-s}{2} ! \widetilde{\langle f, g \rangle}_{\mathfrak{H}^{\otimes(p+q-s)/2}}$$

$$= s! \binom{p+q-s}{2}! \binom{p}{\frac{p+q-s}{2}} \binom{q}{\frac{p+q-s}{2}} f \widetilde{\otimes}_{(p+q-s)/2} g.$$

Therefore,

$$I_p(f) I_q(g) = \sum_{s=|q-p|}^{p+q} \binom{p+q-s}{2}! \binom{p}{\frac{p+q-s}{2}} \binom{q}{\frac{p+q-s}{2}} I_s(f \widetilde{\otimes}_{(p+q-s)/2} g)$$

$$\times \mathbf{1}_{\{s \text{ and } p+q \text{ have same parity}\}}$$

$$= \sum_{r=0}^{p \wedge q} r! \binom{p}{r} \binom{q}{r} I_{p+q-2r}(f \widetilde{\otimes}_r g),$$

by setting $r = (p + q - s)/2$. The proof is complete. $\square$

2.7.4 Some properties of double integrals

In this section, we derive some further information on the elements of the second chaos of an isonormal process X, which are random variables of the type $F = I_2(f)$, with $f \in \mathfrak{H}^{\odot 2}$. Observe that, if $f = h \otimes h$, where $h \in \mathfrak{H}$ is such that $\|h\|_{\mathfrak{H}} = 1$, then $I_2(f) = X(h)^2 - 1 \overset{\text{Law}}{=} N^2 - 1$ by Theorem 2.7.10, where $N \sim \mathcal{N}(0, 1)$. One of the most effective ways of dealing with second chaos random variables is to associate with every kernel $f \in \mathfrak{H}^{\odot 2}$ the following two objects:

– The *Hilbert–Schmidt operator*

$$A_f : \mathfrak{H} \mapsto \mathfrak{H}; \quad g \mapsto f \otimes_1 g. \tag{2.7.10}$$

In other words, A_f transforms an element g of $\mathfrak{H}$ into the contraction $f \otimes_1 g \in \mathfrak{H}$. We write $\{\lambda_{f,j} : j \geq 1\}$ and $\{e_{f,j} : j \geq 1\}$, respectively, to indicate the eigenvalues of A_f and the corresponding eigenvectors (forming an orthonormal system in $\mathfrak{H}$).

– The sequence of auxiliary kernels

$$\left\{ f \otimes_1^{(p)} f : p \geq 1 \right\} \subset \mathfrak{H}^{\odot 2} \tag{2.7.11}$$

defined as follows: $f \otimes_1^{(1)} f = f$, and, for $p \geq 2$,

$$f \otimes_1^{(p)} f = \left(f \otimes_1^{(p-1)} f \right) \otimes_1 f. \tag{2.7.12}$$

In particular, $f \otimes_1^{(2)} f = f \otimes_1 f$.

Some useful relations between the objects introduced in (2.7.10) and (2.7.12) are explained in the next proposition. The proof relies on elementary functional analysis and is omitted (see, for example, Section 6.2 in [47]).

Proposition 2.7.11 *Let $f \in \mathfrak{H}^{\odot 2}$.*

1. *The series $\sum_{j=1}^{\infty} \lambda_{f,j}^p$ converges for every $p \geq 2$, and f admits the expansion*

$$f = \sum_{j=1}^{\infty} \lambda_{f,j} \; e_{f,j} \otimes e_{f,j}, \tag{2.7.13}$$

where the convergence takes place in $\mathfrak{H}^{\odot 2}$.

2. *For $p \geq 2$, one has the relations*

$$\mathrm{Tr}(A_f^p) = \langle f \otimes_1^{(p-1)} f, f \rangle_{\mathfrak{H}^{\otimes 2}} = \sum_{j=1}^{\infty} \lambda_{f,j}^p, \tag{2.7.14}$$

where $\mathrm{Tr}(A_f^p)$ stands for the trace of the pth power of A_f.

In the following statement we gather together some facts concerning the law of a real-valued random variable of type $I_2(f)$. We recall that, given a random variable F such that $E|F|^p < \infty, \forall p \geq 1$, the sequence of the *cumulants* of F, written $\{\kappa_p(F) : p \geq 1\}$, is defined by the relation (in the sense of *formal power series*):

$$\ln E[e^{i\mu F}] = \sum_{p=1}^{\infty} \frac{(i\mu)^p}{p!} \kappa_p(F), \quad \mu \in \mathbb{R} \tag{2.7.15}$$

(see Appendix A for details and references). For instance, κ_1 is the mean, and κ_2 is the variance. We will also use the content of the following exercise.

Exercise 2.7.12 Let F be a random variable such that $E[e^{t|F|}] < \infty$ for some $t > 0$. Show that $E|F|^r < \infty$ for every $r > 0$, and also that the law of F is determined by its moments, that is, if Y is another random variable such that $E[Y^n] = E[F^n]$ for every integer $n \geq 1$, then F and Y have necessarily the same law.

Proposition 2.7.13 *Let $F = I_2(f)$, with $f \in \mathfrak{H}^{\odot 2}$.*

1. *The following equality holds:*

$$F = \sum_{j=1}^{\infty} \lambda_{f,j} \left(N_j^2 - 1 \right), \tag{2.7.16}$$

where $(N_j)_{j\geq 1}$ is a sequence of i.i.d. $\mathcal{N}(0, 1)$ random variables, and the series converges in L^2 and almost surely.

2. For every $p \geq 2$,

$$\kappa_p(F) = 2^{p-1}(p-1)! \times \mathrm{Tr}(H_f^p) = 2^{p-1}(p-1)! \sum_{j=1}^{\infty} \lambda_{f,j}^p$$

$$= 2^{p-1}(p-1)! \times \langle f \otimes_1^{(p-1)} f, f \rangle_{\mathfrak{H}^{\otimes 2}}. \tag{2.7.17}$$

3. The law of the random variable F is determined by its moments or, equivalently, by its cumulants.

Proof. Relation (2.7.16) is an immediate consequence of (2.7.13), of the identity

$$I_2\left(e_{f,j} \otimes e_{f,j}\right) = I_1\left(e_{f,j}\right)^2 - 1,$$

as well as of the fact that the $\{e_{f,j}\}$ are orthonormal (implying that the sequence $\{I_1\left(e_{f,j}\right) : j \geq 1\}$ is i.i.d. $\mathcal{N}(0, 1)$).

To prove (2.7.17), first observe that (2.7.16) implies that

$$E\left[e^{i\mu F}\right] = \prod_{j=1}^{\infty} \frac{e^{-i\mu\lambda_{f,j}}}{\sqrt{1 - 2i\mu\lambda_{f,j}}}.$$

Thus, standard computations give

$$\ln E\left[e^{i\mu F}\right] = -i\mu \sum_{j=1}^{\infty} \lambda_{f,j} - \frac{1}{2}\sum_{j=1}^{\infty} \ln\left(1 - 2i\mu\lambda_{f,j}\right)$$

$$= \frac{1}{2}\sum_{p=2}^{\infty} \frac{(2i\mu)^p}{p} \sum_{j=1}^{\infty} \lambda_{f,j}^p = \sum_{p=2}^{\infty} 2^{p-1}\frac{(i\mu)^p}{p} \sum_{j=1}^{\infty} \lambda_{f,j}^p. \tag{2.7.18}$$

We can now identify the coefficients in the series (2.7.15) and (2.7.18), and so deduce that $\kappa_1(F) = E(F) = 0$ and

$$\frac{2^{p-1}}{p}\sum_{j=1}^{\infty} \lambda_{f,j}^p = \frac{\kappa_p(F)}{p!},$$

thus obtaining the desired conclusion, see also (2.7.14).

We are left with the proof of part 3. In view of Exercise 2.7.12, it is sufficient to show that there exists $t > 0$ such that $E[e^{t|F|}] < \infty$. To do so, we may assume without loss of generality that $E[F^2] = 1$, so that (2.7.2) implies that, for every $q > 2$,

$$E[|F|^q]^{1/q} \leq q - 1,$$

yielding that, for every $u > 0$, $P[|F| > u] \leq u^{-q}(q - 1)^q$. Choosing $q = q(u) = 1 + u/e$, the previous relation shows that $P[|F| > u] \leq e^{-u/e}$, for every $u > e$. By a Fubini argument,

$$E[e^{t|F|}] = 1 + t \int_0^\infty e^{tu} P[|F| > u] du,$$

hence $E[e^{t|F|}] < \infty$ for every $t \in [0, 1/e)$. The proof is concluded. $\qquad\square$

As a particular case, (2.7.17) gives immediately that, if $N \sim \mathcal{N}(0, 1)$, then

$$\kappa_p(N^2 - 1) = 2^{p-1}(p - 1)!, \quad p \geq 2. \tag{2.7.19}$$

Remark 2.7.14 1. One can prove that the laws of multiple integrals of order greater than 2 are, in general, not determined by their moments. In particular, if F belongs to a Wiener chaos of order at least 3, then $E[e^{t|F|}] = \infty$ for every $t > 0$. See Slud [133] or Janson [57, Chapter VI] for a complete picture.
2. In Chapter 8, we will extend (2.7.17) to the case $F = I_q(f)$, $f \in \mathfrak{H}^{\odot q}$, $q \geq 3$.

2.8 The Ornstein–Uhlenbeck semigroup

Recall from Theorem 2.2.4 that any $F \in L^2(\Omega)$ can be expanded as $F = E[F] + \sum_{p=1}^\infty J_p(F)$, where $J_p(F) = \text{Proj}(F|\mathscr{H}_p)$. See also Remark 2.2.5.

2.8.1 Definition and Mehler's formula

Definition 2.8.1 The **Ornstein–Uhlenbeck semigroup** $(P_t)_{t \geq 0}$ is defined, for all $t \geq 0$ and $F \in L^2(\Omega)$, by $P_t(F) = \sum_{p=0}^\infty e^{-pt} J_p(F) \in L^2(\Omega)$.

Note that the semigroup characterization of P_t is immediate from the definition: indeed, $P_{t+s}(F) = \sum_{p=0}^\infty e^{-pt} e^{-ps} J_p(F) = P_t(P_s(F))$. There is an alternative procedure to define P_t, much closer to the way we defined the one-dimensional Ornstein–Uhlenbeck semigroup in Chapter 1. Let $F \in L^1(\Omega)$, let X' be an independent copy of X, and assume that X and X' are defined on the product probability space $(\Omega \times \Omega', \mathscr{F} \otimes \mathscr{F}', P \times P')$. Since F is measurable with respect to X, we can write $F = \Psi_F(X)$ with $\Psi_F : \mathbb{R}^{\mathfrak{H}} \to \mathbb{R}$ a measurable mapping determined $P \circ X^{-1}$-a.s. As a consequence, for any $t \geq 0$ the random variable $\Psi_F(e^{-t}X + \sqrt{1 - e^{-2t}}X')$ is well defined $P \times P'$-a.s. (note, indeed, that $e^{-t}X + \sqrt{1 - e^{-2t}}X'$ and X have the same law). Reasoning exactly as in the proof of Proposition 1.3.3, we see that the collection of operators (with E' denoting the expectation with respect to P')

$$F \mapsto E'\big(\Psi_F(e^{-t}X + \sqrt{1 - e^{-2t}}X')\big), \quad t \geq 0,$$

which are well defined for $F \in L^1(\Omega)$, is indeed a semigroup. The next statement shows that this semigroup coincides with $(P_t)_{t \geq 0}$ on $L^2(\Omega)$.

Theorem 2.8.2 (Mehler's formula)   *For every $F \in L^2(\Omega)$ and every $t \geq 0$ we have*

$$P_t(F) = E'\big(\Psi_F(e^{-t}X + \sqrt{1 - e^{-2t}}X')\big), \tag{2.8.1}$$

where E' denotes the expectation with respect to P'.

Remark 2.8.3   Using Theorem 2.8.2, we can consistently extend the semigroup $(P_t)_{t \geq 0}$ to the space $L^1(\Omega)$, by setting $P_t(F)$ to be equal to the right-hand side of (2.8.1) for every $F \in L^1(\Omega)$.

Remark 2.8.4   An alternative way of expressing relation (2.8.1) (without specifying the form of the underlying probability space) is the following. Consider an independent copy X' of X, and assume that X, X' live on the same probability space $(\Omega, \mathscr{F}, P)$. Then

$$P_t(F) = E\big[\Psi_F(e^{-t}X + \sqrt{1 - e^{-2t}}X') | X\big]. \tag{2.8.2}$$

Proof of Theorem 2.8.2   The linear span of random variables F having the form $F = e^{X(h) - \frac{1}{2}}$, with $h \in \mathfrak{H}$ such that $\|h\|_{\mathfrak{H}} = 1$, is dense in $L^2(\Omega)$ (one can show this by using very similar arguments to those leading to the proof of Lemma 2.3.1). Therefore, it suffices to consider the case where F has this particular form. For $F = e^{X(h) - \frac{1}{2}}$, we have

$$E'\big(\Psi(e^{-t}X + \sqrt{1 - e^{-2t}}X')\big)$$

$$= E'\left(\exp\left(e^{-t}X(h) + \sqrt{1 - e^{-2t}}X'(h) - \frac{1}{2}\right)\right)$$

$$= \exp\left(e^{-t}X(h) - \frac{1}{2}e^{-2t}\right)$$

$$= \sum_{p=0}^{\infty} \frac{e^{-pt}}{p!} H_p(X(h)) \quad \text{(by Proposition 1.4.2(vi))}$$

$$= \sum_{p=0}^{\infty} \frac{e^{-pt}}{p!} I_p(h^{\otimes p}) \quad \text{(by (2.7.7))}.$$

2.8 The Ornstein–Uhlenbeck semigroup

On the other hand, still by Proposition 1.4.2(vi) and (2.7.7), we have

$$F = e^{X(h)-1/2} = \sum_{p=0}^{\infty} \frac{1}{p!} H_p(X(h)) = \sum_{p=0}^{\infty} \frac{1}{p!} I_p(h^{\otimes p}),$$

so that, by the definition of P_t,

$$P_t(F) = \sum_{p=0}^{\infty} \frac{e^{-pt}}{p!} I_p(h^{\otimes p}).$$

The desired conclusion follows. $\qquad\square$

Remark 2.8.5 It is possible to provide a further representation of $(P_t)_{t\geq 0}$ as the semigroup associated with a Markov process with values in $\mathbb{R}^{\mathfrak{H}}$. To do so, we consider an auxiliary isonormal Gaussian process B over $\widehat{\mathfrak{H}} = \mathfrak{H} \otimes L^2(\mathbb{R}, \mathcal{B}(\mathbb{R}), 2dx)$ (note the factor 2), where dx stands for the Lebesgue measure. Also, for $t \geq 0$, we denote by e_t the element of $L^2(\mathbb{R}, \mathcal{B}(\mathbb{R}), 2dx)$ given by the mapping $x \mapsto e^{-(t-x)}\mathbf{1}_{x<t}$. We define a process $X_t(h)$, on $\mathbb{R}_+ \times \mathfrak{H}$, as

$$X_t(h) = B(h \otimes e_t), \quad t \geq 0, \ h \in \mathfrak{H}.$$

One can check that: (i) for every fixed $h \in \mathfrak{H}$, the process $t \mapsto X_t(h)$ is a centered Gaussian process with covariance function $\mathbb{E}(X_t(h)X_s(h)) = \exp(-|t-s|)\|h\|_{\mathfrak{H}}^2$, that is, $t \mapsto X_t(h)$ is a multiple of a real-valued Ornstein–Uhlenbeck process starting from zero (see Exercise 1.7.4); (ii) for every fixed $t \geq 0$, the Gaussian family $X_t = \{X_t(h) : h \in \mathfrak{H}\}$ defines an isonormal Gaussian process over $\mathfrak{H}$ (that is, X_t has the same law as X). Now fix $F \in L^2(\Omega)$ and write $F = \Psi(X)$, where $\Psi : \mathbb{R}^{\mathfrak{H}} \to \mathbb{R}$ is the mapping considered at the beginning of this section. One can easily verify the following alternative representation of $(P_t)_{t\geq 0}$: for every $t, s \geq 0$,

$$E\left[\Psi(X_{t+s}) \mid X_u(h) : u \leq s, \ h \in \mathfrak{H}\right] = P_t\Psi(X_s).$$

In the previous formula, we wrote $P_t\Psi$ in order to indicate the mapping from $\mathbb{R}^{\mathfrak{H}}$ into $\mathbb{R}$ and defined $P \circ X^{-1}$ a.s., such that $P_t F = P_t\Psi(X)$.

From (2.8.1), it is easy to deduce a contraction property for P_t.

Proposition 2.8.6 *For every $t \geq 0$ and every $q \in [1, \infty)$, P_t (as defined in Remark 2.8.3) is a linear contraction operator on $L^q(\Omega)$.*

Proof We retain the notation of the beginning of this subsection. By applying the Jensen inequality to the right-hand side of (2.8.1), we get, for any $F \in L^q(\Omega)$, that

$$E[|P_t F|^q] = E[|E'[\Psi(e^{-t}X + \sqrt{1 - e^{-2t}}X')]|^q]$$
$$\leq E[E'[|\Psi(e^{-t}X + \sqrt{1 - e^{-2t}}X')|^q]] = E[|F|^q],$$

where the last equality holds because $e^{-t}X + \sqrt{1 - e^{-2t}}X'$ and X have the same law. $\qquad\square$

We will show in Section 2.8.3 that the Ornstein–Uhlenbeck semigroup is indeed hypercontractive.

2.8.2 The generator of the Ornstein–Uhlenbeck semigroup

Definition 2.8.7   We say that $F \in L^2(\Omega)$ belongs to DomL if

$$\sum_{p=1}^{\infty} p^2 E[J_p(F)^2] < \infty.$$

For such an F, we define $LF = -\sum_{p=1}^{\infty} p J_p(F)$.

It is easily shown that the operator L coincides with the *infinitesimal generator of the Ornstein–Uhlenbeck semigroup* $(P_t)_{t \geq 0}$, that is, $LF = \lim_{t \to 0} \frac{P_t F - F}{t}$ in $L^2(\Omega)$. The next result gives a crucial relationship between D, δ and L, which is the exact analog of Proposition 1.3.6.

Proposition 2.8.8   *Let $F \in L^2(\Omega)$. We have that $F \in \mathrm{Dom}L$ if and only if $F \in \mathbb{D}^{1,2}$ and $DF \in \mathrm{Dom}(\delta)$. In this case, $\delta(DF) = -LF$.*

Proof   When F has the form of a multiple integral, the proof of the formula $\delta(DF) = -LF$ is immediate. Indeed, if $F = I_p(f)$ for some $f \in \mathfrak{H}^{\odot p}$, we have $DI_p(f) = pI_{p-1}(f)$, implying $\delta(DI_p(f)) = pI_p(f) = -LI_p(f)$. For the general case, it suffices to use the chaotic representation of Corollary 2.7.8 and to proceed by approximation. $\qquad\square$

We also have the following property, which essentially says that L behaves as a second-order differential operator when it acts on smooth random variables (compare also with formula (1.3.3)).

Theorem 2.8.9   *Let $F \in \mathscr{S}$ be as in (2.3.1), that is, of the form $f(X(\phi_1), \dots, X(\phi_m))$ with $m \geq 1$, $f : \mathbb{R}^m \to \mathbb{R}$ a C^∞-function such that f and its partial derivatives have polynomial growth, and $\phi_i \in \mathfrak{H}$, $i = 1, \dots, m$. Then $F \in \mathrm{Dom}L$ and*

$$LF = \sum_{i,j=1}^{m} \frac{\partial^2 f}{\partial x_i \partial x_j}\big(X(\phi_1), \dots, X(\phi_m)\big)\langle h_i, h_j \rangle_{\mathfrak{H}}$$

$$- \sum_{i=1}^{m} \frac{\partial f}{\partial x_i}\big(X(\phi_1), \dots, X(\phi_m)\big)X(\phi_i).$$

Proof The fact that $F \in \mathrm{Dom}L$ is a consequence of Proposition 2.8.8. On the other hand,

$$LF = -\delta(DF) = -\delta \left(\sum_{i=1}^{m} \frac{\partial f}{\partial x_i} (X(\phi_1), \dots, X(\phi_m)) \phi_i \right)$$

$$= -\sum_{i=1}^{m} \delta \left(\frac{\partial f}{\partial x_i} (X(\phi_1), \dots, X(\phi_m)) \phi_i \right)$$

$$= -\sum_{i=1}^{m} \frac{\partial f}{\partial x_i} (X(\phi_1), \dots, X(\phi_m)) X(\phi_i)$$

$$+ \sum_{i,j=1}^{m} \frac{\partial^2 f}{\partial x_i \partial x_j} (X(\phi_1), \dots, X(\phi_m)) \langle \phi_i, \phi_j \rangle_{\mathfrak{H}},$$

the last equality coming from Proposition 2.5.4. $\qquad \square$

Definition 2.8.10 For any $F \in L^2(\Omega)$, we define $L^{-1}F = -\sum_{p=1}^{\infty} \frac{1}{p} J_p(F)$. The operator L^{-1} is called the **pseudo-inverse** of L.

The name of L^{-1} is justified by the following fact:

Proposition 2.8.11 *For any $F \in L^2(\Omega)$, we have $L^{-1}F \in \mathrm{Dom}L$ and $LL^{-1}F = F - E[F]$.*

Proof The fact that $L^{-1}F \in \mathrm{Dom}L$ is obvious (by the definition of $\mathrm{Dom}L$). On the other hand,

$$LL^{-1}F = L \left(-\sum_{p=1}^{\infty} \frac{1}{p} J_p(F) \right) = \sum_{p=1}^{\infty} J_p(F) = F - E[F]. \qquad \square$$

2.8.3 Hypercontractivity of the Ornstein–Uhlenbeck semigroup

The following statement provides the hypercontractive property of the Ornstein–Uhlenbeck semigroup. Recall that, according to Remark 2.8.3, the random variable $P_t F$ is well defined for every $F \in L^1(\Omega)$.

Theorem 2.8.12 (Nelson) *Let $F \in L^p(\Omega)$ for some $p > 1$. For any $t \geq 0$,*

$$E\left[|P_t F|^{1+e^{2t}(p-1)} \right]^{\frac{1}{1+e^{2t}(p-1)}} \leq E\left[|F|^p \right]^{\frac{1}{p}}. \tag{2.8.3}$$

Remark 2.8.13 When $t > 0$, we have $1 + e^{2t}(p-1) > p$ so that inequality (2.8.3) is strictly stronger than the conclusion of Proposition 2.8.6.

Proof of Theorem 2.8.12 Since smooth random variables are dense in every space $L^q(\Omega)$ ($q \geq 1$), and by virtue of the contraction property stated in Proposition 2.8.6, we can assume that F is an element of $\mathcal{S}$. Also, because of the positivity of P_t (that is, $F \geq 0$ implies $P_t F \geq 0$ – see (2.8.1)), we do assume without loss of generality that $F \geq 0$ and that F is not equal to zero almost surely. The proof of (2.8.3) consists of two steps.

Step 1. We shall prove that

$$E\big[F^p \log(F^p)\big] - E[F^p]\log\big(E[F^p]\big) \leq -\frac{p^2}{2(p-1)}E[F^{p-1}LF]. \quad (2.8.4)$$

First, observe that

$$
\begin{aligned}
E\big[F^2 &\log(F^2)\big] - E[F^2]\log\big(E[F^2]\big) \\
&= E\left[P_0(F^2)\log\big(P_0(F^2)\big)\right] - E\left[P_\infty(F^2)\log\big(P_\infty(F^2)\big)\right] \\
&= -\int_0^\infty E\left[\frac{d}{dt}P_t(F^2)\log\big(P_t(F^2)\big)\right]dt \\
&= -\int_0^\infty \left\{E\left[LP_t(F^2)\log\big(P_t(F^2)\big)\right] + \underbrace{E\left[LP_t(F^2)\right]}_{=0}\right\}dt \\
&= \int_0^\infty E\left[\delta\big(D(P_t(F^2))\big)\log\big(P_t(F^2)\big)\right]dt \\
&= \int_0^\infty E\left[\frac{\big\|D\big(P_t(F^2)\big)\big\|_{\mathfrak{H}}^2}{P_t(F^2)}\right]dt \\
&= \int_0^\infty e^{-2t} E\left[\frac{\big\|P_t\big(D(F^2)\big)\big\|_{\mathfrak{H}}^2}{P_t(F^2)}\right]dt \\
&= 4\int_0^\infty e^{-2t} E\left[\frac{\big\|P_t(FDF)\big\|_{\mathfrak{H}}^2}{P_t(F^2)}\right]dt.
\end{aligned}
$$

But

$$
\begin{aligned}
\|P_t(FDF)\|_{\mathfrak{H}} &\leq P_t\big(|F|\,\|DF\|_{\mathfrak{H}}\big) \quad \text{(by Jensen)} \\
&\leq \sqrt{P_t(F^2)}\sqrt{P_t\|DF\|_{\mathfrak{H}}^2} \quad \text{(by Cauchy–Schwarz)},
\end{aligned}
$$

so that

$$E\left[\frac{\|P_t(FDF)\|_{\mathfrak{H}}^2}{P_t(F^2)}\right] \leq E\big[P_t\|DF\|_{\mathfrak{H}}^2\big] = E\big[\|DF\|_{\mathfrak{H}}^2\big].$$

Thus,

$$E[F^2 \log(F^2)] - E[F^2] \log\left(E[F^2]\right) \le 2E[\|DF\|_{\mathfrak{H}}^2]. \qquad (2.8.5)$$

Hence, by considering $F^{p/2}$ instead of F in the previous identity, we get

$$E[F^p \log(F^p)] - E[F^p] \log\left(E[F^p]\right)$$

$$\le 2\,E[\|D(F^{p/2})\|_{\mathfrak{H}}^2] = \frac{p^2}{2}\,E[F^{p-2}\|DF\|_{\mathfrak{H}}^2]$$

$$= \frac{p^2}{2(p-1)}E[F^{p-1}\delta(DF)] = -\frac{p^2}{2(p-1)}E[F^{p-1}LF],$$

which is precisely (2.8.4).

Step 2. We shall prove that

$$\frac{d}{dt} \log\left\{E\left[(P_t F)^{1+e^{2t}(p-1)}\right]^{\frac{1}{1+e^{2t}(p-1)}}\right\} \le 0, \qquad (2.8.6)$$

which readily gives (2.8.3) by integrating over $[0, t]$ (recall that $P_0 F = F$). We have

$$\frac{d}{dt} \log\left\{E\left[(P_t F)^{1+e^{2t}(p-1)}\right]^{\frac{1}{1+e^{2t}(p-1)}}\right\}$$

$$= \frac{d}{dt}\left\{\frac{1}{1+e^{2t}(p-1)} \log E\left[(P_t F)^{1+e^{2t}(p-1)}\right]\right\}$$

$$= -\frac{2e^{2t}(p-1)}{\left(1+e^{2t}(p-1)\right)^2} \log E\left[(P_t F)^{1+e^{2t}(p-1)}\right] + \frac{\frac{d}{dt} E\left[(P_t F)^{1+e^{2t}(p-1)}\right]}{\left(1+e^{2t}(p-1)\right)E\left[(P_t F)^{1+e^{2t}(p-1)}\right]}$$

$$= -\frac{2e^{2t}(p-1)}{\left(1+e^{2t}(p-1)\right)^2} \log E\left[(P_t F)^{1+e^{2t}(p-1)}\right]$$

$$+ \frac{2e^{2t}(p-1)\,E\left[\log\left(P_t F\right)(P_t F)^{1+e^{2t}(p-1)}\right]}{\left(1+e^{2t}(p-1)\right)E\left[(P_t F)^{1+e^{2t}(p-1)}\right]} + \frac{E\left[L(P_t F) \times (P_t F)^{1+e^{2t}(p-1)}\right]}{\left(1+e^{2t}(p-1)\right)E\left[(P_t F)^{1+e^{2t}(p-1)}\right]}$$

$$= \frac{2e^{2t}(p-1)}{\left(1+e^{2t}(p-1)\right)^2 E\left[(P_t F)^{1+e^{2t}(p-1)}\right]}\left\{E\left[\log\left\{(P_t F)^{1+e^{2t}(p-1)}\right\}(P_t F)^{1+e^{2t}(p-1)}\right]\right.$$

$$\left. - \log\left\{E\left[(P_t F)^{1+e^{2t}(p-1)}\right]\right\}E\left[(P_t F)^{1+e^{2t}(p-1)}\right]\right\}$$

$$+ \frac{E\left[L(P_t F) \times (P_t F)^{1+e^{2t}(p-1)}\right]}{\left(1+e^{2t}(p-1)\right)E\left[(P_t F)^{1+e^{2t}(p-1)}\right]}.$$

From Step 1 (inequality (2.8.4)), it follows that (2.8.6) holds, so that the proof of the theorem is complete. $\qquad \square$

Corollary 2.8.14   *Let F be an element of the pth Wiener chaos of X, $p \geq 1$. Then, for all $r > q > 1$,*

$$E[|F|^q]^{1/q} \leq E[|F|^r]^{1/r} \leq \left(\frac{r-1}{q-1}\right)^{p/2} E[|F|^q]^{1/q}. \qquad (2.8.7)$$

This implies that, inside a fixed Wiener chaos, all the L^q-norms are equivalent.

Proof   The first inequality, $E[|F|^q]^{1/q} \leq E[|F|^r]^{1/r}$, is trivial by the Jensen inequality. Let us show the second inequality. Let $t = \frac{1}{2}\log\left(\frac{r-1}{q-1}\right)$. Theorem 2.8.12 yields $E[|P_t F|^r]^{1/r} \leq E[|F|^q]^{1/q}$. But $P_t F = e^{-pt} F$ by definition, so that

$$E[|F|^r]^{1/r} = e^{pt} E[|P_t F|^r]^{1/r} \leq \left(\frac{r-1}{q-1}\right)^{p/2} E[|F|^q]^{1/q}. \qquad \square$$

Remark 2.8.15   Similarly to Corollary 2.8.14, one can prove that inside a fixed sum of Wiener chaoses, all the L^q-norms are equivalent.

Exercise 2.8.16   Fix $q > 1$ and $t > 0$, and let r be a real number strictly greater than $1 + e^{2t}(q - 1)$. Show that there exists no constant $c > 0$ such that $E[|P_t F|^r]^{1/r} \leq c\, E[|F|^q]^{1/q}$ for all $F \in L^2(\Omega)$. Compare with Theorem 2.8.12. (Hint: Consider $F = e^{\lambda X(h)}$ with $\lambda \in \mathbb{R}$ and $h \in \mathfrak{H}$ such that $\|h\|_{\mathfrak{H}} = 1$, and let λ go to infinity.)

Exercise 2.8.17   The aim of this exercise is to provide a characterization of the possible limits in law inside a fixed Wiener chaos.

1. Let $Z \geq 0$ be a random variable with finite variance. Prove Paley's inequality: for every $\theta \in (0, 1)$,

$$P(Z > \theta E[Z]) \geq (1 - \theta)^2 \frac{E[Z]^2}{E[Z^2]}.$$

   (Hint: Decompose Z as $Z = Z\mathbf{1}_{\{Z > \theta E[Z]\}} + Z\mathbf{1}_{\{Z \leq \theta E[Z]\}}$ and apply Cauchy–Schwarz.)

2. Let F be an element of the pth Wiener chaos of X, $p \geq 1$. Combine Paley's inequality with the result of Corollary 2.8.14 to deduce that

$$P(F^2 > E[F^2]/2) \geq 9^{-p}/4.$$

3. Let $(F_n)_{n \geq 1}$ be a tight sequence belonging to the pth Wiener chaos of X, $p \geq 1$. Deduce from part 2 that $\sup_{n \geq 1} E[F_n^2] < \infty$, implying in turn, thanks to Corollary 2.8.14, that $\sup_{n \geq 1} E[|F_n|^r] < \infty$ for all $r > 0$.

4. Let $(F_n)_{n\geq 1}$ be a sequence belonging to the pth Wiener chaos of X, $p \geq 1$. Assume that F_n converges in law to U as $n \to \infty$. Show that U admits moments of all orders.

2.9 An integration by parts formula

The next result will play a fundamental role in this book:

Theorem 2.9.1 *Let* $F, G \in \mathbb{D}^{1,2}$, *and let* $g : \mathbb{R} \to \mathbb{R}$ *be a* C^1 *function having a bounded derivative. Then*

$$E[Fg(G)] = E[F]E[g(G)] + E[g'(G)\langle DG, -DL^{-1}F\rangle_{\mathfrak{H}}]. \qquad (2.9.1)$$

Proof We have

$$E\big[(F - E[F])g(G)\big]$$
$$= E\big[LL^{-1}F \times g(G)\big] = E\big[\delta(-DL^{-1}F)g(G)\big] \quad \text{(by Proposition 2.8.8)}$$
$$= E\big[\langle Dg(G), -DL^{-1}F\rangle_{\mathfrak{H}}\big] \quad \text{(by the duality formula (2.5.2))}$$
$$= E\big[g'(G)\langle DG, -DL^{-1}F\rangle_{\mathfrak{H}}\big] \quad \text{(by the chain rule (2.3.4))}. \qquad \square$$

Remark 2.9.2 If $\|h\|_{\mathfrak{H}} = 1$, then $X(h) \sim \mathcal{N}(0, 1)$ and $L^{-1}X(h) = -X(h)$, so that $DX(h) = -DL^{-1}X(h) = h$. Therefore, when applied to $F = G = X(h)$, formula (2.9.1) yields that

$$E[X(h)g(X(h))] = E[g'(X(h)) \times \|h\|_{\mathfrak{H}}^2] = E[g'(X(h))],$$

which is equivalent to (1.1.1).

It is interesting to observe that the quantity $-DL^{-1}F$ can be re-expressed in (at least) two ways.

Proposition 2.9.3 *Let* $F \in \mathbb{D}^{1,2}$ *with* $E[F] = 0$. *Then*

$$- DL^{-1}F = \int_0^\infty e^{-t} P_t DF \, dt \qquad (2.9.2)$$

$$= -(L - I)^{-1}DF. \qquad (2.9.3)$$

Proof Thanks to an approximation argument, we can assume that $F = I_p(f)$ for some $p \geq 1$ and $f \in \mathfrak{H}^{\odot p}$. We have that $L^{-1}F = -\frac{1}{p}F$ and therefore, according to Proposition 2.7.4, $-DL^{-1}F = I_{p-1}(f) = \frac{1}{p}DF$. Equation (2.9.2) follows immediately from the relation $P_t DF = pe^{-(p-1)t}I_{p-1}(f)$,

while (2.9.3) is a consequence of the identities $(L - I)^{-1}DF = p(L - I)^{-1}$
$I_{p-1}(f) = \frac{p}{-(p-1)-1}I_{p-1}(f) = -I_{p-1}(f)$. $\qquad\square$

Also, it is an important fact that the conditional expectation $E[\langle DF, -DL^{-1}F\rangle_{\mathfrak{H}}|F]$ is almost surely non-negative:

Proposition 2.9.4 *Let $F \in \mathbb{D}^{1,2}$ with $E[F] = 0$. Then $E[\langle DF, -DL^{-1}F\rangle_{\mathfrak{H}} |F] \geq 0$ almost surely.*

Proof Let g be a smooth non-negative real function, and set $G(x) = \int_0^x g(t)dt$, with the usual convention that $\int_0^x = -\int_x^0$ for $x < 0$. Since G is increasing and vanishing at zero, we have $xG(x) \geq 0$ for all $x \in \mathbb{R}$. In particular, $E[FG(F)] \geq 0$. Moreover, due to Theorem 2.9.1, we have $E[FG(F)] = E[E[\langle DF, -DL^{-1}F\rangle_{\mathfrak{H}}|F]g(F)]$. By approximation, we deduce that $E[E[\langle DF, -DL^{-1}F\rangle_{\mathfrak{H}}|F]1_A] \geq 0$ for any $\sigma\{F\}$-measurable set A. This implies the desired conclusion. $\qquad\square$

2.10 Absolute continuity of the laws of multiple integrals

The aim of this final section is to prove the following interesting theorem, which will be used several times throughout this book.

Theorem 2.10.1 *Let $q \geq 1$ be an integer, and $f \in \mathfrak{H}^{\odot q}$ be a symmetric kernel such that $\|f\|_{\mathfrak{H}^{\otimes q}} > 0$. Then the law of $F = I_q(f)$ is absolutely continuous with respect to the Lebesgue measure.*

Proof The proof is by induction on q. When $q = 1$, the desired property is readily checked because $I_1(f) \sim \mathcal{N}(0, \|f\|_{\mathfrak{H}^{\otimes q}})$. Now let $q \geq 2$, and let $f \in \mathfrak{H}^{\odot q}$ with $\|f\|_{\mathfrak{H}^{\otimes q}} > 0$. We assume that the desired property holds for $q - 1$, that is, the law of $I_{q-1}(g)$ is absolutely continuous for any $g \in \mathfrak{H}^{\odot(q-1)}$ such that $\|g\|_{\mathfrak{H}^{\otimes(q-1)}} > 0$. Assume for the moment that $\langle f, h\rangle_{\mathfrak{H}} = 0$ for all $h \in \mathfrak{H}$. Let $(e_j)_{j \geq 1}$ be an orthonormal basis of $\mathfrak{H}$. By writing

$$f = \sum_{j_1,\dots,j_q=1}^{\infty} a(j_1,\dots,j_q)e_{j_1} \otimes \dots \otimes e_{j_q}$$

as in (B.4.1), we have

$$\langle f, h\rangle_{\mathfrak{H}} = \sum_{j_1,\dots,j_q=1}^{\infty} a(j_1,\dots,j_q)\langle e_{j_1}, h\rangle_{\mathfrak{H}}e_{j_2} \otimes \dots \otimes e_{j_q},$$

implying in turn, because $\langle f, h \rangle_{\mathfrak{H}} = 0$ for all $h \in \mathfrak{H}$, that

$$\sum_{j_2,\ldots,j_q=1}^{\infty} \left(\sum_{j_1=1}^{\infty} a(j_1,\ldots,j_q)\langle e_{j_1}, h \rangle_{\mathfrak{H}} \right)^2 = 0 \quad \text{for all } h \in \mathfrak{H}.$$

By choosing $h = e_k$, $k = 1, 2, \ldots$, we get that $a(j_1,\ldots,j_q) = 0$ for any $j_1,\ldots,j_q \geq 1$, that is, $f = 0$. The latter fact being in contradiction to our assumption, it implies that there exists $h \in \mathfrak{H}$ so that $\langle f, h \rangle_{\mathfrak{H}} \neq 0$. Using the induction assumption, we have that the law of $I_{q-1}(\langle f, h \rangle_{\mathfrak{H}})$ has a density with respect to the Lebesgue measure. But $DF = qI_{q-1}(f)$, so that $\langle DF, h \rangle_{\mathfrak{H}} = qI_{q-1}(\langle f, h \rangle_{\mathfrak{H}})$. Thus, we have that $P(\langle DF, h \rangle_{\mathfrak{H}} = 0) = 0$. Consequently, $P(\|DF\|_{\mathfrak{H}} = 0) = 0$, because $\{\|DF\|_{\mathfrak{H}} = 0\} \subset \{\langle DF, h \rangle_{\mathfrak{H}} = 0\}$.

Now, let B be a Borel set of $\mathbb{R}$. Using (2.9.1) together with an approximation argument (because $\int_{-\infty}^{\cdot} \mathbf{1}_{B \cap [-n,n]}(y)dy$ is just Lipschitz and not $\mathcal{C}^1$), we can write, for all $n \geq 1$,

$$E\left[\mathbf{1}_{B \cap [-n,n]}(F) \frac{1}{q} \|DF\|_{\mathfrak{H}}^2 \right] = E\left[\mathbf{1}_{B \cap [-n,n]}(F)\langle DF, -DL^{-1}F \rangle_{\mathfrak{H}} \right]$$

$$= E\left[F \int_{-\infty}^{F} \mathbf{1}_{B \cap [-n,n]}(y)dy \right].$$

Assume that the Lebesgue measure of B is zero. The previous equality implies that

$$E\left[\mathbf{1}_{B \cap [-n,n]}(F) \frac{1}{q} \|DF\|_{\mathfrak{H}}^2 \right] = 0.$$

But $P(\|DF\|_{\mathfrak{H}} > 0) = 1$, so $P(F \in B \cap [-n,n]) = 0$. By letting $n \to \infty$, $P(F \in B) = 0$. That is, the law of F is absolutely continuous with respect to the Lebesgue measure. The proof of the theorem is complete. $\qquad \square$

Remark 2.10.2 More generally, as proved by Shigekawa in [132], a random variable living in a finite sum of Wiener chaoses is either constant or such that its law admits a density with respect to the Lebesgue measure.

2.11 Exercises

2.11.1 (**Poincaré inequality**) Let $F \in \mathbb{D}^{1,2}$. Prove that $\mathrm{Var}[F] \leq E[\|DF\|_{\mathfrak{H}}^2]$. (Hint: Generalize the proof of Proposition 1.3.7, or use the chaotic decomposition (2.7.8) together with the isometry formula (2.7.3).)

2.11.2 Let $F \in \mathbb{D}^{\infty,2}$. Prove that

$$\text{Var}[F] = \sum_{p=1}^{\infty} \frac{1}{p!} \|E[D^p F]\|^2_{\mathfrak{H}^{\otimes p}}.$$

If, moreover, $E[\|D^p F\|^2_{\mathfrak{H}^{\otimes p}}]/p! \to 0$ as $p \to \infty$, prove that

$$\text{Var}[F] = \sum_{p=1}^{\infty} \frac{(-1)^{p+1}}{p!} E[\|D^p F\|^2_{\mathfrak{H}^{\otimes p}}].$$

(Hint: Generalize the proof of Proposition 1.5.1.)

2.11.3 1. If $F, G \in \mathbb{D}^{1,2}$, compute the 'carré-du-champ' operator (squared field operator) defined by $\Gamma_1(F, G) := \frac{1}{2}\big(L(FG) - F\,LG - G\,LF\big)$. Show, moreover, that

$$\Gamma_2(F, G) := \frac{1}{2}\big(L(\Gamma_1(F, G)) - \Gamma_1(F, LG) - \Gamma_1(G, LF)\big)$$

satisfies $\Gamma_2(F, G) = \langle D^2 F, D^2 G\rangle_{\mathfrak{H}^{\otimes 2}} + \langle DF, DG\rangle_{\mathfrak{H}}$ for all $F, G \in \mathbb{D}^{2,2}$.

2. If $F \in \mathbb{D}^{2,2}$, show that $E[\Gamma_2(F, F)] = \text{Var}(LF)$. Deduce that $\text{Var}(F) \geq E\big[\|DL^{-1}F\|^2_{\mathfrak{H}}\big]$ for all $F \in L^2(\Omega)$.

3. Let $F \in \mathbb{D}^{1,2}$.

   (a) Use (2.9.1) to check that $\text{Var}(F) = E[\langle DF, -DL^{-1}F\rangle_{\mathfrak{H}}]$.

   (b) Deduce that $\text{Var}(F) \leq \sqrt{E[\|DF\|^2_{\mathfrak{H}}]}\sqrt{E[\|DL^{-1}F\|^2_{\mathfrak{H}}]}$.

   (c) Combine the inequalities appearing in parts 2 and 3(b) to get that $E[\|DL^{-1}F\|^2_{\mathfrak{H}}] \leq E[\|DF\|^2_{\mathfrak{H}}]$.

   (d) Deduce another proof (with respect to Exercise 2.11.1) of the Poincaré inequality.

2.11.4 Let $p \geq 1$ be an integer. Suppose that $F \in \mathbb{D}^{p,2}$ and that $u \in \text{Dom}\,\delta^p$ is symmetric. For any $0 \leq r + j \leq p$, assume that $\langle D^r F, \delta^j(u)\rangle_{\mathfrak{H}^{\otimes r}}$ belongs to $L^2(\Omega, \mathfrak{H}^{\otimes(p-r-j)})$. Prove that, for any $r = 0, \ldots, p-1$, $\langle D^r F, u\rangle_{\mathfrak{H}^{\otimes r}}$ belongs to $\text{Dom}\,\delta^{p-r}$ and

$$F\delta^p(u) = \sum_{r=0}^{p} \binom{p}{r} \delta^{p-r}\big(\langle D^r F, u\rangle_{\mathfrak{H}^{\otimes r}}\big). \tag{2.11.1}$$

(Hint: Proceed by induction, with the help of Proposition 2.5.4.)

2.11.5 Let $W = (W_t)_{t \in [0,T]}$ be a standard Brownian motion. We regard the Gaussian space generated by W as an isonormal Gaussian process over $\mathfrak{H} = L^2([0, T])$ (see Example 2.1.4). Consider the random variable $F = \sup_{[0,T]} W = W_\tau$, where τ denotes the (unique) random point where W attains its maximum on $[0, T]$. Show that $F \in \mathbb{D}^{1,2}$ and that $DF = \mathbf{1}_{[0,\tau]}$.

(Hint: Approximate F by the maximum on a finite set and apply the result of Example 2.3.9.)

2.12 Bibliographic comments

The concept of 'isonormal Gaussian process' dates back to Dudley's seminal paper [31]. Our reference to Brownian motion and stochastic calculus is the book by Revuz and Yor [119]. Random fields of the type discussed in Example 2.1.6 appear, for example, in Breuer and Major [18], Chambers and Slud [19], Giraitis and Surgailis [41] and Major [68]. For an introduction to the Gaussian free field and its many applications in modern probability, see Sheffield [131]. What we now call the 'Wiener chaos' was first studied by Wiener (in the specific case of Brownian motion) in [152]. The birth of the so-called 'Malliavin calculus' dates from the groundbreaking paper [69], written by Paul Malliavin in 1978. Our main reference for Malliavin calculus is Nualart's monograph [98]. Here, we have tried to provide the reader with a presentation as self-contained as possible of the material that is useful for our purposes (with some unavoidable exceptions, such as Meyer's inequalities – see Theorem 2.5.5). We have not followed [98] in three respects: the definition of the multiple Wiener–Itô integrals, the statement of Theorem 2.9.1 (which already appears in Üstünel's book [149]), and the proof of the multiplication formula. Other important accounts of Malliavin calculus are contained in the books by Janson [57], Malliavin [70] and Üstünel [149]. See also Nualart's lecture notes [97] and monograph [99], as well as Malliavin and Thalmaier [72]. The fact that every $F \in \mathbb{D}^{\infty,2}$ is such that $F = E[F] + \sum_{p=1}^{\infty} \frac{1}{p!} I_p(E[D^p F])$ was first proved in [137] and is known as *Stroock's formula*. Several properties of multiple stochastic integrals (with respect to not necessarily Gaussian random measures) can be found in Engel [33], Janson [57, Chapters V and VI], Kwapień and Woyczyński [63], Major [68], Peccati and Taqqu [110] and Rota and Wallstrom [123]. In particular, [110] contains a detailed account of the combinatorial structures underlying the construction of the Gaussian Wiener chaos. We refer the reader to Janson [57, Chapter V] and Nualart [98, Section 1.4.3] for a study of hypercontractivity on a Gaussian space. Theorem 2.8.12 was proved by Nelson in [81]. The fact that all L^p topologies are equivalent on a fixed Wiener chaos (and actually equivalent to the topology induced by convergence in probability) was first proved by Schreiber in [130]. The fact that the non-constant elements of a finite sum of Wiener chaoses have an absolutely continuous law was proved in [132] by Shigekawa. The infinite-dimensional Poincaré inequality appearing in Exercise 2.11.1 was proved by Houdré and Pérez-Abreu in [51]. We should also mention that, in the case where $\mathfrak{H} = L^2(A, \mathscr{A}, \mu)$, there exists a neat interpretation of the divergence δ as an anticipative *Skorohod stochastic integral*, generalizing the usual Itô integral to the case of possibly non-adapted integrands – this point is explained, for example, in [98, section 1.3.2].

3

Stein's method for one-dimensional normal approximations

The so-called 'Stein's method' is a collection of probabilistic techniques, allowing the distance between two probability distributions to be assessed by means of differential operators. The aim of this chapter is to introduce the reader to the use of Stein's method in the specific framework of one-dimensional *normal approximations*. As anticipated, we shall make use of several results from Chapter 1. Stein's method for multidimensional normal approximations is discussed in Chapter 4. All random objects are defined on an adequate probability space $(\Omega, \mathscr{F}, P)$.

3.1 Gaussian moments and Stein's lemma

As in Chapter 1, given a probability measure ν on $\mathbb{R}$, we denote by

$$m_n(\nu) = \int_{\mathbb{R}} x^n d\nu(x), \quad n \geq 0,$$

the sequence of its moments (whenever they are well defined). We now provide a quick self-contained proof of the fact that the Gaussian distribution γ is determined by its moments.

Lemma 3.1.1 *Let γ denote the standard Gaussian probability distribution, that is, $d\gamma(x) = (2\pi)^{-1/2}e^{-x^2/2}dx$. Then, γ is determined by its moments. This means that, if ν is another probability measure (having moments of all orders) satisfying $m_n(\nu) = m_n(\gamma)$ for every $n \geq 1$, then necessarily $\nu = \gamma$.*

Proof Let ν be as in the statement. We have to show that, for every $t \in \mathbb{R}$, $\int_{\mathbb{R}} e^{itx} d\nu(x) = \int_{\mathbb{R}} e^{itx} d\gamma(x)$. Since $m_k(\nu) = m_k(\gamma)$ for every k, we can write, by virtue of Taylor's expansion as well as the triangle and Cauchy–Schwarz inequalities,

$$\left| \int_{\mathbb{R}} e^{itx} d\nu(x) - \int_{\mathbb{R}} e^{itx} d\gamma(x) \right| \leq \int_{\mathbb{R}} \left| e^{itx} - \sum_{k=0}^{n} \frac{(itx)^k}{k!} \right| d\nu(x)$$

$$+ \int_{\mathbb{R}} \left| e^{itx} - \sum_{k=0}^{n} \frac{(itx)^k}{k!} \right| d\gamma(x)$$

$$\leq \left(\int_{\mathbb{R}} \frac{|tx|^{2n+2}}{(n+1)!^2} d\nu(x) \right)^{1/2}$$

$$+ \left(\int_{\mathbb{R}} \frac{|tx|^{2n+2}}{(n+1)!^2} d\gamma(x) \right)^{1/2}$$

$$= \sqrt{\frac{|t|^{2n+2} m_{2n+2}(\nu)}{(n+1)!^2}} + \sqrt{\frac{|t|^{2n+2} m_{2n+2}(\gamma)}{(n+1)!^2}}$$

$$= 2\sqrt{\frac{|t|^{2n+2} m_{2n+2}(\gamma)}{(n+1)!^2}},$$

for every $n \geq 1$. The conclusion is obtained by observing that, by Corollary 1.1.3 and Stirling's formula,

$$\lim_{n \to \infty} \frac{|t|^{2n+2} m_{2n+2}(\gamma)}{(n+1)!^2} = 0. \qquad \square$$

The next result, whose proof uses both Lemmas 1.1.1 and 3.1.1, is universally known as *Stein's lemma*. It provides a useful characterization of the one-dimensional standard Gaussian distribution.

Lemma 3.1.2 (Stein's lemma) *A real-valued random variable N has the standard Gaussian $\mathcal{N}(0, 1)$ distribution if and only if, for every differentiable function $f : \mathbb{R} \to \mathbb{R}$ such that $f' \in L^1(\gamma)$, the expectations $E[Nf(N)]$ and $E[f'(N)]$ are finite and*

$$E[Nf(N)] = E[f'(N)]. \tag{3.1.1}$$

Proof If $N \sim \mathcal{N}(0, 1)$, then relation (3.1.1) is satisfied, by virtue of Lemma 1.1.1. Now suppose that N satisfies (3.1.1). By selecting $f(x) = x^n, n \geq 0$, we deduce that $E(N) = 0$, $E(N^2) = 1$ and $E(N^{n+1}) = n \times E(N^{n-1})$ for every integer $n \geq 2$. By virtue of Corollary 1.1.3, this implies that $E(N^n) = m_n(\gamma)$ for every $n \geq 1$. Lemma 3.1.1 therefore yields that $N \sim \mathcal{N}(0, 1)$. $\qquad \square$

Remark 3.1.3 According to Lemma 1.1.1, if $N \sim \mathcal{N}(0, 1)$, then relation (3.1.1) holds for every absolutely continuous function f such that $f' \in L^1(\gamma)$.

A natural (albeit still vague) question is now the following.

(Stein's heuristic) *Suppose F is a random variable such that the expectation $E[Ff(F) - f'(F)]$ is close to zero for a large class of smooth functions f. In view of Lemma 3.1.2, is it possible to conclude that the law of F is close (in some meaningful probabilistic sense) to a $\mathcal{N}(0, 1)$ distribution? In other words, is there a quantitative version of Stein's lemma?*

As demonstrated below, the answer to this question is positive in a variety of important cases. To see this, one needs to introduce two crucial concepts: the notion of *distance* between two probability distributions, and the explicit expression for the so-called *Stein's equation* associated with a given function $h \in L^1(\gamma)$. We recall that, for any separating class of real-valued functions $\mathcal{H}$ (see Definition C.1.1 in Appendix C), the *distance* $d_{\mathcal{H}}(F, G)$, between the laws of two real-valued random variables F and G (satisfying $h(F), h(G) \in L^1(\Omega)$ for every $h \in \mathcal{H}$), is given by

$$d_{\mathcal{H}}(F, G) = \sup\{|E[h(F)] - E[h(G)]| : h \in \mathcal{H}\}.$$

Several properties of the distances $d_{\mathcal{H}}$ are discussed in Appendix C. In particular, Sections C.2 and C.3 deal with the definitions and properties of the four distances d_{Kol} (Kolmogorov distance), d_{TV} (total variation distance), d_{W} (Wasserstein distance) and d_{FM} (Fortet–Mourier distance). Stein's equations are the main object of the following section.

Exercise 3.1.4 Fix $\sigma > 0$ and $\mu \in \mathbb{R}$. Prove that the real-valued random variable N has the $\mathcal{N}(\mu, \sigma^2)$ distribution if and only if $E[(N - \mu)f(N)] = \sigma^2 E[f'(N)]$ for every differentiable function $f : \mathbb{R} \to \mathbb{R}$ such that $E|f'(N)| < \infty$.

Exercise 3.1.5 Fix $d \geq 2$ and consider a d-dimensional centered Gaussian vector $(N_1, \ldots, N_d)$ with (not necessarily positive definite) covariance matrix $C = \{C(i, j) : i, j = 1, \ldots, d\}$. Assume that the function $f : \mathbb{R}^d \to \mathbb{R}$ has bounded and continuous first partial derivatives. Prove that, for every $i = 1, \ldots, d$,

$$E\big[N_i f(N_1, \ldots, N_d)\big] = \sum_{j=1}^{d} C(i, j) E\left[\frac{\partial f}{\partial x_j}(N_1, \ldots, N_d)\right]. \tag{3.1.2}$$

Formula (3.1.2) is one of the key elements in the proof of a multidimensional version of Stein's lemma, and will be used in Chapter 4.

3.2 Stein's equations

Definition 3.2.1   Let $N \sim \mathcal{N}(0, 1)$, and let $h : \mathbb{R} \to \mathbb{R}$ be a Borel function such that $E|h(N)| < \infty$. The **Stein's equation** associated with h is the ordinary differential equation

$$f'(x) - xf(x) = h(x) - E[h(N)]. \tag{3.2.1}$$

A solution to equation (3.2.1) is a function f that is absolutely continuous and such that there exists a version of the derivative f' satisfying (3.2.1) for every $x \in \mathbb{R}$.

The next result shows that solving (3.2.1) is actually an easy task.

Proposition 3.2.2   *Every solution to (3.2.1) has the form*

$$f(x) = ce^{x^2/2} + e^{x^2/2} \int_{-\infty}^{x} \{h(y) - E[h(N)]\} e^{-y^2/2} dy, \quad x \in \mathbb{R}, \quad (3.2.2)$$

where $c \in \mathbb{R}$. In particular, by writing

$$f_h(x) := e^{x^2/2} \int_{-\infty}^{x} \{h(y) - E[h(N)]\} e^{-y^2/2} dy, \tag{3.2.3}$$

we have that f_h is the unique solution f to (3.2.1) satisfying $\lim_{x \to \pm\infty} e^{-x^2/2} f(x) = 0$.

Remark 3.2.3 (On derivatives)   For the rest of the book, whenever considering the function f_h appearing in (3.2.3), we shall write f'_h in order to indicate the version of the derivative of f_h satisfying (3.2.1) for every $x \in \mathbb{R}$, that is, given by $f'_h(x) = xf_h(x) + h(x) - E[h(N)]$ for every real x.

Remark 3.2.4   Since $\int_{-\infty}^{\infty} \{h(y) - E[h(N)]\} e^{-y^2/2} dy = 0$, for every $x \in \mathbb{R}$ we have

$$\int_{-\infty}^{x} \{h(y) - E[h(N)]\} e^{-y^2/2} dy = -\int_{x}^{\infty} \{h(y) - E[h(N)]\} e^{-y^2/2} dy. \tag{3.2.4}$$

Proof of Proposition 3.2.2   Equation (3.2.1) can be rewritten as

$$e^{x^2/2} \frac{d}{dx} [e^{-x^2/2} f(x)] = h(x) - E[h(N)],$$

so that it is immediate that any solution to (3.2.1) has the form (3.2.2). By dominated convergence, one also infers that

$$\lim_{x \to \pm\infty} \int_{-\infty}^{x} \{h(y) - E[h(N)]\} e^{-y^2/2} dy = 0,$$

yielding that the asymptotic property $\lim_{x\to\pm\infty} e^{-x^2/2} f(x) = 0$ is satisfied if and only if $c = 0$. $\qquad\qquad\square$

As already stated, for the rest of the chapter we shall use Stein's equations in order to assess the distance between the law of a random variable F and the law of a standard Gaussian $N \sim \mathcal{N}(0, 1)$. To do this, consider a function $h : \mathbb{R} \to \mathbb{R}$ such that $E|h(N)| < \infty$ and $E|h(F)| < \infty$. Since the function f_h defined in (3.2.3) is a solution to (3.2.1), by taking expectations on both sides of (3.2.1) with respect to the law of F, we deduce that

$$E[h(F)] - E[h(N)] = E[f_h'(F) - F f_h(F)]. \qquad (3.2.5)$$

In particular, if $\mathcal{H}$ is a separating class of functions such that $E|h(N)| < \infty$ and $E|h(F)| < \infty$ for every $h \in \mathcal{H}$, we infer that

$$d_{\mathcal{H}}(F, N) = \sup_{h \in \mathcal{H}} |E[f_h'(F) - F f_h(F)]|. \qquad (3.2.6)$$

Note that the right-hand side of (3.2.6) does not involve the 'target' random variable N, and also that we still do not have enough information on the functions f_h in order for (3.2.6) to be effectively put into use. Our next step is therefore the derivation of some explicit universal estimates on the functions f_h, for several special choices of the separating class $\mathcal{H}$.

3.3 Stein's bounds for the total variation distance

We shall first focus on the total variation distance.

Theorem 3.3.1 *Let $h : \mathbb{R} \to [0, 1]$ be a Borel function (observe that h is assumed to take values in $[0, 1]$). Then, the solution to Stein's equation (3.2.1) given by f_h in (3.2.3) is such that*

$$\|f_h\|_\infty \le \sqrt{\frac{\pi}{2}} \quad \text{and} \quad \|f_h'\|_\infty \le 2. \qquad (3.3.1)$$

In particular, for $N \sim \mathcal{N}(0, 1)$ and for any integrable random variable F,

$$d_{\mathrm{TV}}(F, N) \le \sup_{f \in \mathscr{F}_{\mathrm{TV}}} |E[f'(F)] - E[F f(F)]|, \qquad (3.3.2)$$

where $\mathscr{F}_{\mathrm{TV}} = \{ f : \|f\|_\infty \le \sqrt{\pi/2}, \|f'\|_\infty \le 2 \}$.

Remark 3.3.2 (Important!) The notation adopted in Theorem 3.3.1 must be understood as follows:

– The class $\mathscr{F}_{\mathrm{TV}}$ is the collection of all absolutely continuous functions $f :$ $\mathbb{R} \to \mathbb{R}$ that are: (i) bounded by $\sqrt{\pi/2}$; and (ii) such that there exists a *version* of f' that is bounded by 2.
– The supremum $\sup_{f \in \mathscr{F}_{\mathrm{TV}}} |E[f'(F)] - E[Ff(F)]|$ stands for the quantity

$$\sup |E[g(F)] - E[Ff(F)]|,$$

where the supremum is taken over all pairs (f, g) of Borel functions such that $f \in \mathscr{F}_{\mathrm{TV}}$ and g is a version of f' such that $\|g\|_\infty \leq 2$.

Plainly, the expectation $E[f'(F)]$ is unambiguously defined whenever f is almost everywhere differentiable and the law of F has a density with respect to the Lebesgue measure. $\qquad\square$

Proof of Theorem 3.3.1 For h as in the statement, $|h(x) - E[h(N)]| \leq 1$ for all $x \in \mathbb{R}$. By using the explicit expression for f_h in (3.2.3), as well as relation (3.2.4), we infer that, for every $x \in \mathbb{R}$,

$$|f_h(x)| \leq e^{x^2/2} \min \left(\int_{-\infty}^x e^{-y^2/2} dy, \int_x^\infty e^{-y^2/2} dy \right)$$
$$= e^{x^2/2} \int_{|x|}^\infty e^{-y^2/2} dy \leq \sqrt{\frac{\pi}{2}},$$

where the last inequality is obtained by observing that the function

$$x \mapsto s(x) = e^{x^2/2} \int_{|x|}^\infty e^{-y^2/2} dy$$

attains its maximum at $x = 0$, and that $s(0) = \sqrt{\pi/2}$ (indeed, for $x > 0$ we have $s'(x) = x e^{x^2/2} \int_x^\infty e^{-y^2/2} dy - 1 \leq e^{x^2/2} \int_x^\infty y e^{-y^2/2} dy - 1 = 0$ so that s is decreasing on $\mathbb{R}_+$; similarly, s is increasing on $\mathbb{R}_-$). This yields the first estimate in (3.3.1).

To prove the second estimate, observe that, since f_h solves the Stein's equation associated with h, we have (according to Remark 3.2.3)

$$f_h'(x) = h(x) - E[h(N)] + x e^{x^2/2} \int_{-\infty}^x \{h(y) - E[h(N)]\} e^{-y^2/2} dy$$
$$= h(x) - E[h(N)] - x e^{x^2/2} \int_x^\infty \{h(y) - E[h(N)]\} e^{-y^2/2} dy,$$

for every $x \in \mathbb{R}$. It follows that, by reasoning as above,

$$|f_h'(x)| \leq 1 + |x| e^{x^2/2} \int_{|x|}^\infty e^{-y^2/2} dy \leq 1 + e^{x^2/2} \int_{|x|}^\infty y e^{-y^2/2} dy = 2.$$

Finally, relation (3.3.2) follows from (3.2.6) by setting $\mathscr{H} = \{1_B : B \in \mathscr{B}(\mathbb{R})\}$: indeed, in this case (3.3.1) implies that $f_h \in \mathscr{F}_{\mathrm{TV}}$ for every $h \in \mathscr{H}$. $\qquad\square$

Remark 3.3.3 1. The requirement that F is integrable ensures that the expectation $E[Ff(F)]$ is well defined for every $f \in \mathscr{F}_{\mathrm{TV}}$.
2. Inequalities of type (3.3.2) are customarily referred to as 'Stein's bounds'.

Exercise 3.3.4 Let $h : \mathbb{R} \to \mathbb{R}$ be bounded, and let f_h be given by (3.2.3).

1. By using the same arguments as in the proof of Theorem 3.3.1, prove that:
   (a) $\|f_h\|_\infty \le \sqrt{\frac{\pi}{2}} \|h - E[h(N)]\|_\infty$;
   (b) $\|f_h'\|_\infty \le 2\|h - E[h(N)]\|_\infty$.
2. Prove that, if h is also absolutely continuous, then f_h is C^1 with f_h' absolutely continuous, and (using for f_h'' a similar convention to the one introduced in Remark 3.2.3 for f_h')

$$\|f_h\|_\infty \le 2\|h'\|_\infty, \quad \|f_h'\|_\infty \le 4\|h'\|_\infty, \quad \|f_h''\|_\infty \le 2\|h'\|_\infty.$$

3. Use part 1(a) above as well as Proposition C.3.5 of Appendix C in order to deduce an alternative proof of (3.3.2).
4. Use parts 1–3 to deduce an appropriate bound for the Fortet–Mourier distance $d_{\mathrm{FM}}(F, N)$, where F is in $L^1(\Omega)$ and $N \sim \mathscr{N}(0, 1)$. See also Section 3.5.

3.4 Stein's bounds for the Kolmogorov distance

We now turn to the Kolmogorov distance, and demonstrate that one can deduce better estimates than those following from the straightforward relation $d_{\mathrm{Kol}} \le d_{\mathrm{TV}}$. For every $z \in \mathbb{R}$, we write $f_z = f_{\mathbf{1}_{(-\infty,z]}}$, that is, f_z is obtained from f_h in (3.2.3) by setting $h = \mathbf{1}_{(-\infty,z]}$. Also, $E[h(N)] = P(N \le z) = \Phi(z)$, where $\Phi(\cdot)$ stands for the cumulative distribution function of a $\mathscr{N}(0, 1)$ random variable. Simple computations show that, for every $z \in \mathbb{R}$,

$$f_z(x) = \begin{cases} \sqrt{2\pi}\, e^{x^2/2}\Phi(x)[1 - \Phi(z)] & \text{if } x \le z \\ \sqrt{2\pi}\, e^{x^2/2}\Phi(z)[1 - \Phi(x)] & \text{if } x \ge z. \end{cases} \tag{3.4.1}$$

Remark 3.4.1 According to the previous conventions, for every $z \in \mathbb{R}$, we shall write f_z' in order to denote the version of the derivative of f_z given by $f_z'(x) = xf_z(x) + \mathbf{1}_{(-\infty,z]}(x) - \Phi(z)$ for every real x.

The next statement (whose proof is the object of Exercise 3.4.4 below) provides neat bounds on normal approximations in the Kolmogorov distance.

Theorem 3.4.2 *Let $z \in \mathbb{R}$. Then the function f_z is such that*

$$\|f_z\|_\infty \le \frac{\sqrt{2\pi}}{4} \quad \text{and} \quad \|f_z'\|_\infty \le 1. \tag{3.4.2}$$

In particular, for $N \sim \mathcal{N}(0, 1)$ and for any integrable random variable F,

$$d_{\mathrm{Kol}}(F, N) \leq \sup_{f \in \mathscr{F}_{\mathrm{Kol}}} |E[f'(F)] - E[Ff(F)]|, \qquad (3.4.3)$$

where $\mathscr{F}_{\mathrm{Kol}} = \left\{ f : \|f\|_\infty \leq \frac{\sqrt{2\pi}}{4}, \ \|f'\|_\infty \leq 1 \right\}$.

Remark 3.4.3 (Important!) The notation adopted in Theorem 3.3.1 must be understood as follows:

– The class $\mathscr{F}_{\mathrm{Kol}}$ is the collection of all absolutely continuous functions $f : \mathbb{R} \to \mathbb{R}$ that are: (i) bounded by $\frac{\sqrt{2\pi}}{4}$; (ii) differentiable except for at most a finite number of points; and (iii) such that there exists a version of f' that is bounded by 1.
– The supremum $\sup_{f \in \mathscr{F}_{\mathrm{Kol}}} |E[f'(F)] - E[Ff(F)]|$ stands for the quantity

$$\sup |E[g(F)] - E[Ff(F)]|,$$

where the supremum is taken over all pairs (f, g) of Borel functions such that $f \in \mathscr{F}_{\mathrm{Kol}}$ and g is a version of f' such that $\|g\|_\infty \leq 1$.

Observe that the expectation $E[f'(F)]$ is unambiguously defined whenever f is differentiable except at most for a finite number of points and the law of F has a density with respect to the Lebesgue measure. $\qquad \square$

Exercise 3.4.4 (Proof of Theorem 3.4.2) Fix $z \in \mathbb{R}$ and retain the notation introduced above.

1. Prove that, for every $x \in \mathbb{R}$, $f_z(x) = f_{-z}(-x)$ (this implies that, in the estimates below, one can assume that $z \geq 0$ without loss of generality).
2. Compute the derivative $\frac{d}{dx}[xf_z(x)]$, and deduce that the mapping $x \mapsto xf_z(x)$ is increasing.
3. Show that $\lim_{x \to -\infty} xf_z(x) = \Phi(z) - 1$ and also that $\lim_{x \to \infty} xf_z(x) = \Phi(z)$.
4. Use the explicit expression for f_z provided in (3.4.1) to prove that

$$f_z'(x) = \begin{cases} \left[\sqrt{2\pi}\, x e^{x^2/2} \Phi(x) + 1\right]\left[1 - \Phi(z)\right] & \text{if } x < z \\ \left[\sqrt{2\pi}\, x e^{x^2/2} (1 - \Phi(x)) - 1\right]\Phi(z) & \text{if } x > z. \end{cases} \qquad (3.4.4)$$

5. Use (3.4.4) in order to prove the two relations

$$0 < f_z'(x) \leq zf_z(x) + 1 - \Phi(z) < 1, \quad \text{if } x < z$$

and

$$-1 < zf_z(x) - \Phi(z) \leq f_z'(x) < 0, \quad \text{if } x > z.$$

Deduce the second estimate in (3.4.2).

6. Use part 5 to show that $x \mapsto f_z(x)$ attains its maximum in $x = z$. Compute $f_z(z)$ and prove that $f_z(z) \leq \frac{\sqrt{2\pi}}{4}$ for every $z \in \mathbb{R}$. This yields the first estimate in (3.4.2), and therefore the bound (3.4.3) (see [23] for more details).

Further properties of the functions f_z will be discussed in Chapter 9.

3.5 Stein's bounds for the Wasserstein distance

We start with a preliminary result. We recall that, according to Rademacher's theorem (Section E.2), a function which is Lipschitz continuous on $\mathbb{R}$ is almost everywhere differentiable.

Proposition 3.5.1 *Let $h : \mathbb{R} \to \mathbb{R}$ be a Lipschitz function with constant $K > 0$. Then the function f_h given by (3.2.3) admits the representation*

$$f_h(x) = - \int_0^\infty \frac{e^{-t}}{\sqrt{1 - e^{-2t}}} E[h(e^{-t}x + \sqrt{1 - e^{-2t}}N)N]dt. \qquad (3.5.1)$$

Moreover, f_h is of class C^1 and $\|f_h'\|_\infty \leq \sqrt{\frac{2}{\pi}} K$.

Proof The fact that $f_h \in C^1$ follows immediately from the explicit expression given in (3.2.3) and the properties of h. Set

$$\widetilde{f_h}(x) = - \int_0^\infty \frac{e^{-t}}{\sqrt{1 - e^{-2t}}} E[h(e^{-t}x + \sqrt{1 - e^{-2t}}N)N]dt.$$

By dominated convergence, we have

$$\widetilde{f_h}'(x) = - \int_0^\infty \frac{e^{-2t}}{\sqrt{1 - e^{-2t}}} E[h'(e^{-t}x + \sqrt{1 - e^{-2t}}N)N]dt, \quad x \in \mathbb{R}$$

(note that, in the previous expression, there is no ambiguity in the choice of the version of h', since N has an absolutely continuous distribution) so that

$$\left| \widetilde{f_h}'(x) \right| \leq K \, E[|N|] \int_0^\infty \frac{e^{-2t}dt}{\sqrt{1 - e^{-2t}}} = K \sqrt{\frac{2}{\pi}} \int_0^1 \frac{dv}{2\sqrt{1 - v}} = \sqrt{\frac{2}{\pi}} K.$$

Now let $\widetilde{F_h} : \mathbb{R} \to \mathbb{R}$ be defined as

$$\widetilde{F_h}(x) = \int_0^\infty \left(E[h(N) - h(e^{-t}x + \sqrt{1 - e^{-2t}}N)] \right) dt, \quad x \in \mathbb{R}.$$

Observe that $\widetilde{F_h}$ is well defined since $h(N) - h(e^{-t}x + \sqrt{1 - e^{-2t}}N)$ is integrable due to

$$\left| h(N) - h(e^{-t}x + \sqrt{1 - e^{-2t}}N) \right| \leq e^{-t}K|x| + \left(1 - \sqrt{1 - e^{-2t}}\right)K|N|$$
$$\leq e^{-t}K|x| + e^{-2t}K|N|,$$

where the last inequality follows from $1 - \sqrt{1 - u} = u/(\sqrt{1 - u} + 1) \leq u$ if $u \in [0, 1]$. By dominated convergence, we immediately see that $\widetilde{F}_h$ is differentiable with

$$\widetilde{F}_h'(x) = -\int_0^\infty e^{-t} E[h'(e^{-t}x + \sqrt{1 - e^{-2t}}N)]dt.$$

In fact, by applying Lemma 1.1.1, we see that $\widetilde{F}_h'(x) = \widetilde{f}_h(x)$. By using the notation introduced in Chapter 1 and since $\widetilde{f}_h'$ is bounded, we have, by Proposition 1.3.6 and since $\mathbb{D}^{1,2} \subset \text{Dom}\delta$, that $\widetilde{F}_h \in \text{Dom}L$. Consequently, we can write

$$\widetilde{f}_h(x) - x\widetilde{f}_h(x) = L\widetilde{F}_h(x) \quad \text{(see Proposition 1.3.6)}$$
$$= -\int_0^\infty LP_t h(x)dt \quad \text{(since } \widetilde{F}_h(x) = \int_0^\infty \left(E[h(N)] - P_t h(x)\right)dt)$$
$$= -\int_0^\infty \frac{d}{dt} P_t h(x)dt$$
$$= P_0 h(x) - P_\infty h(x) = h(x) - E[h(N)],$$

that is, $\widetilde{f}_h$ satisfies (3.2.1). All that now remains is to prove that $\widetilde{f}_h = f_h$ with f_h defined by (3.2.3). To do so, simply observe that

$$|\widetilde{f}_h(x)| \leq \int_0^\infty \frac{e^{-t}}{\sqrt{1 - e^{-2t}}} \left| E[h(e^{-t}x + \sqrt{1 - e^{-2t}}N)N] \right| dt$$
$$\leq |h(0)| \int_0^\infty \frac{e^{-t}}{\sqrt{1 - e^{-2t}}} dt$$
$$+ K \int_0^\infty \frac{e^{-t}}{\sqrt{1 - e^{-2t}}} \left(e^{-t}|x|E[|N|] + \sqrt{1 - e^{-2t}}E[N^2]\right) dt$$
$$= \alpha|x| + \beta,$$

for some finite constants $\alpha, \beta \geq 0$ depending only on $h(0)$ and K. In particular, we have that $\lim_{x \to \pm\infty} e^{-x^2/2}\widetilde{f}_h(x) = 0$, so that Proposition 3.2.2 allows us to conclude that $\widetilde{f}_h = f_h$. $\square$

The following statement allows to deal with normal approximations in the Wasserstein and Fortet–Mourier distances.

Theorem 3.5.2 *For $N \sim \mathcal{N}(0, 1)$ and for any square-integrable random variable F,*

$$d_{\text{FM}}(F, N) \leq d_{\text{W}}(F, N) \leq \sup_{f \in \mathscr{F}_{\text{W}}} |E[f'(F)] - E[Ff(F)]|, \qquad (3.5.2)$$

where $\mathscr{F}_{\text{W}} := \{ f : \mathbb{R} \to \mathbb{R} \in \mathcal{C}^1 : \| f' \|_\infty \leq \sqrt{2/\pi} \}$.

Proof of Theorem 3.5.2 In what follows, we write (as usual) $h \in \text{Lip}(c)$ to denote the fact that h is Lipschitz with constant less than or equal to c. Note that $\mathscr{F}_{\text{W}} \subset \text{Lip}(\sqrt{2/\pi})$. Using Proposition 3.5.1, we deduce that

$$
\begin{aligned}
d_{\text{FM}}(F, N) \leq d_{\text{W}}(F, N) &= \sup_{h \in \text{Lip}(1)} \left| E[h(N)] - E[h(F)] \right| \\
&= \sup_{h \in \text{Lip}(1)} \left| E[f_h'(F)] - E[Ff_h(F)] \right| \\
&\leq \sup_{f \in \mathscr{F}_{\text{W}}} |E[f'(F)] - E[Ff(F)]|. \qquad \square
\end{aligned}
$$

Remark 3.5.3 The requirement that F is square-integrable ensures that the expectation $E[Ff(F)]$ is well defined for every $f \in \mathscr{F}_{\text{W}}$. Note that the bound on $d_{\text{FM}}(F, N)$ mentioned in part 4 of Exercise 3.3.4 only requires that F is integrable.

3.6 A simple example

As a first elementary application of Stein's estimates, we deduce a bound between the laws of two centered Gaussian random variables.

Proposition 3.6.1 *Let N_1, N_2 be two centered Gaussian random variables, with standard deviations $\sigma_1, \sigma_2 > 0$. Then*

$$d_{\text{TV}}(N_1, N_2) \leq \frac{2}{\sigma_1^2 \vee \sigma_2^2} \times |\sigma_1^2 - \sigma_2^2|,$$

$$d_{\text{Kol}}(N_1, N_2) \leq \frac{1}{\sigma_1^2 \vee \sigma_2^2} \times |\sigma_1^2 - \sigma_2^2|,$$

$$d_{\text{FM}}(N_1, N_2) \leq d_{\text{W}}(N_1, N_2) \leq \frac{\sqrt{2/\pi}}{\sigma_1 \vee \sigma_2} \times |\sigma_1^2 - \sigma_2^2|.$$

Proof We assume that $\sigma_1 \leq \sigma_2$, and we denote by $N_0 \sim \mathscr{N}(0, 1)$ an auxiliary centered Gaussian random variable with unit variance. We start with the bound on d_{TV}. Elementary considerations show that $d_{\text{TV}}(N_1, N_2) = d_{\text{TV}}(N_1/\sigma_2, N_0)$. Also, according to (3.3.2) (here, Remark 3.3.2 is immaterial because N_1/σ_2 has a density) and Exercise 3.1.4,

$$d_{\mathrm{TV}}(N_1/\sigma_2, N_0) \leq \sup_{f \in \mathscr{F}_{\mathrm{TV}}} |E[f'(N_1/\sigma_2)] - E[N_1 f(N_1/\sigma_2)/\sigma_2]|$$

$$= \sup_{f \in \mathscr{F}_{\mathrm{TV}}} |E[f'(N_1/\sigma_2)(1 - \sigma_1^2/\sigma_2^2)]|$$

$$\leq 2|1 - \sigma_1^2/\sigma_2^2|.$$

The bound on d_{Kol} is proved analogously. To deal with the Wasserstein distance, observe first that

$$d_{\mathrm{W}}(N_1, N_2) = \sup_{h \in \mathrm{Lip}(\sigma_2)} |E[h(N_1/\sigma_2)] - E[h(N_0)]|.$$

Proposition 3.5.1 and Lemma 1.1.1 eventually yield

$$\sup_{h \in \mathrm{Lip}(\sigma_2)} |E[h(N_1/\sigma_2)] - E[h(N_0)]|$$

$$= \sup_{h \in \mathrm{Lip}(\sigma_2)} |E[f'_h(N_1/\sigma_2) - N_1 f_h(N_1/\sigma_2)/\sigma_2]|$$

$$\leq \sup_{f \in \mathcal{C}^1 \cap \mathrm{Lip}\left(\sigma_2 \sqrt{\frac{2}{\pi}}\right)} |E[f'(N_1/\sigma_2)] - E[N_1 f(N_1/\sigma_2)/\sigma_2]|$$

$$\leq \frac{\sigma_2 \sqrt{2}}{\sqrt{\pi}} |1 - \sigma_1^2/\sigma_2^2|,$$

where f_h is the solution (3.5.1) to the Stein's equation associated with h. The proof is concluded. $\square$

Remark 3.6.2   The quantity $\sigma_1 \vee \sigma_2$ appears in the bounds on $d_{\mathrm{TV}}(N_1, N_2)$ and $d_{\mathrm{W}}(N_1, N_2)$ with two different powers. This fact is consistent with the fact that the topologies induced by d_{TV} and d_{W} (on the probability measures on $\mathbb{R}$) cannot be compared.

3.7 The Berry–Esseen theorem

Let $Y = \{Y_k : k \geq 1\}$ be a sequence of i.i.d. random variables, such that $E[Y_1] = 0$ and $E[Y_1^2] = 1$, and define

$$V_n = \frac{1}{\sqrt{n}} \sum_{k=1}^{n} Y_k, \quad n \geq 1,$$

to be the associated sequence of normalized partial sums. The aim of this section is to use Stein's method to provide a proof of the following fundamental statement:

Theorem 3.7.1 (CLT and Berry–Esseen bounds) *As $n \to \infty$,*

$$V_n \xrightarrow{\text{Law}} N \sim \mathcal{N}(0, 1). \tag{3.7.1}$$

Moreover,

$$d_{\text{Kol}}(V_n, N) \le \frac{C\, E[|Y_1|^3]}{\sqrt{n}}, \quad n \ge 1, \tag{3.7.2}$$

where $C > 0$ is a universal constant depending neither on n nor on Y.

Relation (3.7.1) is of course the usual CLT, whereas estimate (3.7.2) corresponds to the classical *Berry–Esseen inequality*: both results represent the quintessential example of the kind of probabilistic approximations that are at the core of this book.

Remark 3.7.2 One may actually be more accurate, and show that (3.7.2) holds with the constant $C = 0.4785$. This was proved in 2009 by Tyurin [148], and is indeed the best current bound. On the other hand, according to Esseen [35], it is impossible to choose for C a constant smaller than 0.40973. See the bibliographic comments at the end of the chapter for further details and references.

Proof of (3.7.2) For each $n \ge 2$, let $C_n > 0$ be the best possible constant satisfying, for all i.i.d. random variables $Y_1, \ldots, Y_n$ with $E[|Y_1|^3] < \infty$, $\text{Var}(Y_1) = 1$ and $E[Y_1] = 0$, that

$$d_{\text{Kol}}(V_n, N) \le \frac{C_n\, E[|Y_1|^3]}{\sqrt{n}}. \tag{3.7.3}$$

As a first (rough) estimation, since Y_1 is centered with $E[Y_1^2] = 1$, we observe that $E[|Y_1|^3] \ge E[Y_1^2]^{\frac{3}{2}} = 1$, so that $C_n \le \sqrt{n}$. This is of course not sufficient to conclude the proof, since we need to show that $\sup_{n \ge 2} C_n < \infty$. (We will actually show that $\sup_{n \ge 2} C_n \le 33$.)

For any $z \in \mathbb{R}$ and $\varepsilon > 0$, introduce the function (see the figure below)

$$h_{z,\varepsilon}(x) = \begin{cases} 1 & \text{if } x \le z - \varepsilon \\ \text{linear} & \text{if } z - \varepsilon < x < z + \varepsilon \\ 0 & \text{if } x \ge z + \varepsilon. \end{cases}$$

It is immediately obvious that, for all $n \geq 2$, $\varepsilon > 0$ and $z \in \mathbb{R}$, we have

$$E[h_{z-\varepsilon,\varepsilon}(V_n)] \leq P(V_n \leq z) \leq E[h_{z+\varepsilon,\varepsilon}(V_n)].$$

Moreover, for $N \sim \mathcal{N}(0, 1)$, $\varepsilon > 0$ and $z \in \mathbb{R}$, we have, using the fact that the density of N is bounded by $\frac{1}{\sqrt{2\pi}}$,

$$E[h_{z+\varepsilon,\varepsilon}(N)] - \frac{4\varepsilon}{\sqrt{2\pi}} \leq E[h_{z-\varepsilon,\varepsilon}(N)] \leq P(N \leq z)$$

$$\leq E[h_{z+\varepsilon,\varepsilon}(N)] \leq E[h_{z-\varepsilon,\varepsilon}(N)] + \frac{4\varepsilon}{\sqrt{2\pi}}.$$

Therefore, for all $n \geq 2$ and $\varepsilon > 0$, we have

$$d_{\text{Kol}}(V_n, N) \leq \sup_{z \in \mathbb{R}} \left| E[h_{z,\varepsilon}(V_n)] - E[h_{z,\varepsilon}(N)] \right| + \frac{4\varepsilon}{\sqrt{2\pi}}.$$

Assume for the moment that, for all $\varepsilon > 0$,

$$\sup_{z \in \mathbb{R}} \left| E[h_{z,\varepsilon}(V_n)] - E[h_{z,\varepsilon}(N)] \right| \leq \frac{6\,E[|Y_1|^3]}{\sqrt{n}} + \frac{3\,C_{n-1}\,E[|Y_1|^3]^2}{\varepsilon\,n}. \quad (3.7.4)$$

We deduce that, for all $\varepsilon > 0$,

$$d_{\text{Kol}}(V_n, N) \leq \frac{6\,E[|Y_1|^3]}{\sqrt{n}} + \frac{3\,C_{n-1}\,E[|Y_1|^3]^2}{\varepsilon\,n} + \frac{4\varepsilon}{\sqrt{2\pi}}.$$

By choosing $\varepsilon = \sqrt{\frac{C_{n-1}}{n}}\,E[|Y_1|^3]$, we get that

$$d_{\text{Kol}}(V_n, N) \leq \frac{E[|Y_1|^3]}{\sqrt{n}} \left[6 + \left(3 + \frac{4}{\sqrt{2\pi}} \right) \sqrt{C_{n-1}} \right],$$

so that $C_n \leq 6 + \left(3 + \frac{4}{\sqrt{2\pi}} \right) \sqrt{C_{n-1}}$. It follows by induction (recall that $C_n \leq \sqrt{n}$) that $C_n \leq 33$, which implies the desired conclusion.

We shall now use Stein's method to prove that (3.7.4) holds. Fix $z \in \mathbb{R}$ and $\varepsilon > 0$, and let $f = f_{z,\varepsilon}$ denote Stein's solution associated with $h = h_{z,\varepsilon}$, that is, f satisfies (3.2.1). Observe that h is continuous, and therefore f is C^1. Recall from Theorem 3.3.1 that $\|f\|_\infty \leq \sqrt{\frac{\pi}{2}}$ and $\|f'\|_\infty \leq 2$. Set also $\tilde{f}(x) = xf(x)$, $x \in \mathbb{R}$. We then have

$$\left| \tilde{f}(x) - \tilde{f}(y) \right| = \left| f(x)(x - y) + (f(x) - f(y))y \right| \leq \left(\sqrt{\frac{\pi}{2}} + 2|y| \right) |x - y|. \tag{3.7.5}$$

On the other hand, set

$$V_n^i = V_n - \frac{Y_i}{\sqrt{n}}, \quad i = 1, \ldots, n.$$

Observe that V_n^i and Y_i are independent by construction. We can thus write

$$E[h(V_n)] - E[h(N)] = E[f'(V_n) - V_n f(V_n)]$$

$$= \sum_{i=1}^{n} E\left[f'(V_n)\frac{1}{n} - f(V_n)\frac{Y_i}{\sqrt{n}}\right]$$

$$= \sum_{i=1}^{n} E\left[f'(V_n)\frac{1}{n} - (f(V_n) - f(V_n^i))\frac{Y_i}{\sqrt{n}}\right]$$

(because $E[f(V_n^i)Y_i] = E[f(V_n^i)]E[Y_i] = 0$)

$$= \sum_{i=1}^{n} E\left[f'(V_n)\frac{1}{n} - f'\left(V_n^i + \theta\frac{Y_i}{\sqrt{n}}\right)\frac{Y_i^2}{n}\right]$$

(with $\theta \sim \mathcal{U}_{[0,1]}$ independent of $Y_1, \ldots, Y_n$).

We have $f'(x) = \widetilde{f}(x) + h(x) - E[h(N)]$, so that

$$E[h(V_n)] - E[h(N)] = \sum_{i=1}^{n} \left(a_i(\widetilde{f}) - b_i(\widetilde{f}) + a_i(h) - b_i(h)\right), \qquad (3.7.6)$$

where

$$a_i(g) = E[g(V_n) - g(V_n^i)]\frac{1}{n}$$

$$b_i(g) = E\left[\left(g\left(V_n^i + \theta\frac{Y_i}{\sqrt{n}}\right) - g(V_n^i)\right)Y_i^2\right]\frac{1}{n}.$$

(Here again we have used the fact that V_n^i and Y_i are independent.) Hence, to prove that (3.7.4) holds true, we must bound four terms.

First term Using (3.7.5) as well as $E[|Y_1|] \leq E[Y_1^2]^{\frac{1}{2}} = 1$ and $E[|V_n^i|] \leq E[(V_n^i)^2]^{\frac{1}{2}} \leq 1$, we have

$$|a_i(\widetilde{f})| \leq \frac{1}{n\sqrt{n}}\left(E[|Y_1|]\sqrt{\frac{\pi}{2}} + 2E[|Y_1|]E[|V_n^i|]\right) \leq \left(\sqrt{\frac{\pi}{2}} + 2\right)\frac{1}{n\sqrt{n}}.$$

Second term Similarly, and because $E[\theta] = \frac{1}{2}$, we have

$$|b_i(\widetilde{f})| \le \frac{1}{n\sqrt{n}} \left(E[\theta]E[|Y_1|^3]\sqrt{\frac{\pi}{2}} + 2E[\theta]E[|Y_1|^3]E[|V_n^i|] \right)$$

$$\le \left(\frac{1}{2}\sqrt{\frac{\pi}{2}} + 1 \right) \frac{E[|Y_1|^3]}{n\sqrt{n}}.$$

Third term By definition of h, we have

$$h(y) - h(x) = (y - x) \int_0^1 h'(x + s(y - x))ds$$

$$= -\frac{y - x}{2\varepsilon} E\left[\mathbf{1}_{[z-\varepsilon, z+\varepsilon]}(x + \widehat{\theta}(y - x)) \right],$$

with $\widehat{\theta} \sim \mathscr{U}_{[0,1]}$ independent of θ and $Y_1, \ldots, Y_n$, so that

$$|a_i(h)| \le \frac{1}{2\varepsilon\, n\sqrt{n}} E\left[|Y_i| \mathbf{1}_{[z-\varepsilon, z+\varepsilon]} \left(V_n^i + \widehat{\theta}\frac{Y_i}{\sqrt{n}} \right) \right]$$

$$= \frac{1}{2\varepsilon\, n\sqrt{n}} E\left[|Y_i|\, P\left(z - t\frac{y}{\sqrt{n}} - \varepsilon \le V_n^i \le z - t\frac{y}{\sqrt{n}} + \varepsilon \right) \bigg|_{t=\widehat{\theta},\, y=Y_i} \right]$$

$$\le \frac{1}{2\varepsilon\, n\sqrt{n}} \sup_{t\in[0,1]} \sup_{y\in\mathbb{R}} P\left(z - t\frac{y}{\sqrt{n}} - \varepsilon \le V_n^i \le z - t\frac{y}{\sqrt{n}} + \varepsilon \right).$$

We are thus left to bound $P(a \le V_n^i \le b)$ for all $a, b \in \mathbb{R}$ with $a \le b$. To do so, set $\widetilde{V}_n^i = \frac{1}{\sqrt{n-1}} \sum_{j \ne i} Y_j$, so that $V_n^i = \sqrt{1 - \frac{1}{n}}\, \widetilde{V}_n^i$. We then have, using in particular (3.7.3) (with $n - 1$ instead of n) and the fact that the standard Gaussian density is bounded by $\frac{1}{\sqrt{2\pi}}$,

$$P(a \le V_n^i \le b) = P\left(\frac{a}{\sqrt{1 - \frac{1}{n}}} \le \widetilde{V}_n^i \le \frac{b}{\sqrt{1 - \frac{1}{n}}} \right)$$

$$= P\left(\frac{a}{\sqrt{1 - \frac{1}{n}}} \le N \le \frac{b}{\sqrt{1 - \frac{1}{n}}} \right) + P\left(\frac{a}{\sqrt{1 - \frac{1}{n}}} \le \widetilde{V}_n^i \le \frac{b}{\sqrt{1 - \frac{1}{n}}} \right)$$

$$- P\left(\frac{a}{\sqrt{1 - \frac{1}{n}}} \le N \le \frac{b}{\sqrt{1 - \frac{1}{n}}} \right) \le \frac{b - a}{\sqrt{2\pi}\sqrt{1 - \frac{1}{n}}} + \frac{2\, C_{n-1}\, E[|Y_1|^3]}{\sqrt{n-1}}.$$

We deduce that

$$|a_i(h)| \le \frac{1}{\sqrt{2\pi} n\sqrt{n-1}} + \frac{C_{n-1}\, E[|Y_1|^3]}{n\sqrt{n}\sqrt{n-1}\,\varepsilon}.$$

Fourth term Similarly, we have

$$
\begin{aligned}
|b_i(h)| &= \frac{1}{2n\sqrt{n}\varepsilon} \left| E\left[Y_i^3\, \theta\, \mathbf{1}_{[z-\varepsilon, z+\varepsilon]} \left(V_n^i + \widehat{\theta}\, \theta\, \frac{Y_i}{\sqrt{n}} \right) \right] \right| \\
&\leq \frac{E[|Y_1|^3]}{4n\sqrt{n}\varepsilon} \sup_{t\in[0,1]} \sup_{y\in\mathbb{R}} P\left(z - t\frac{y}{\sqrt{n}} - \varepsilon \leq V_n^i \leq z - t\frac{y}{\sqrt{n}} + \varepsilon \right) \\
&\leq \frac{E[|Y_1|^3]}{2\sqrt{2\pi}n\sqrt{n-1}} + \frac{C_{n-1}\, E[|Y_1|^3]^2}{2n\sqrt{n}\sqrt{n-1}\,\varepsilon}.
\end{aligned}
$$

Plugging these four estimates into (3.7.6) and using the fact that $n \geq 2$ (and therefore $n - 1 \geq \frac{n}{2}$) and $E[|Y_1|^3] \geq 1$, we deduce the desired conclusion. $\square$

3.8 Exercises

3.8.1 Let $N \sim \mathcal{N}(0, 1)$. Write an alternative proof of Stein's lemma by using the fact that the mapping $\lambda \mapsto \phi(\lambda) := E[e^{i\lambda N}] = e^{-\frac{\lambda^2}{2}}$ is the unique solution to the ordinary differential equation

$$
\lambda\phi(\lambda) + \phi'(\lambda) = 0, \qquad \phi(0) = 1. \tag{3.8.1}
$$

(Hint: Choose $f(x) = \cos(\lambda x)$ and $f(x) = \sin(\lambda x)$ in (3.1.1).)

3.8.2 ('Hermite–Stein lemma') Denote by $\{H_m : m \geq 0\}$ the collection of the Hermite polynomials (see Definition 1.4.1), and fix two integers k, m such that $1 \leq k \leq m$. Prove the following variation of Stein's lemma: a random variable N has the $\mathcal{N}(0, 1)$ distribution if and only if, for every k-times differentiable function $f : \mathbb{R} \to \mathbb{R}$ whose derivatives up to order k have polynomial growth,

$$
E[f^{(k)}(N)H_{m-k}(N)] = E[f(N)H_m(N)].
$$

3.8.3 (Zero-bias transforms) Let the real-valued random variable F be such that $E[F] = 0$ and $E[F^2] = 1$.

1. Show that there exists a random variable F^* satisfying $E[Ff(F)] = E[f'(F^*)]$, for every smooth function f such that $E|Ff(F)| < \infty$. Prove that the law of F^* is uniquely determined by the previous equality. The law of F^* is called the F-**zero biased distribution**.
2. Prove that the law of F^* is absolutely continuous with respect to the Lebesgue measure, with density $g(x) = E[F\mathbf{1}_{F>x}]$, $x \in \mathbb{R}$. Deduce that the support of F^* is the closed convex hull of the support of F.
3. Prove that F has the $\mathcal{N}(0, 1)$ distribution if and only if F and F^* have the same law.

4. Assume, in addition, that F is as in Theorem 2.9.1 and that the laws of F and F^* are equivalent. Prove that the mapping $x \mapsto E[\langle DF, -DL^{-1}F \rangle_{\mathfrak{H}} | F = x]$ provides a version of the density of the law of F^* with respect to the law of F.

3.8.4 (Chen–Stein method) The aim of this exercise is to familiarize the reader with a simple version of the powerful **Chen–Stein method**, which is an analog of Stein's method in the framework of Poisson and compound Poisson approximations. A detailed introduction to this subject can be found in Erhardsson's survey [34]. We work in a probability space $(\Omega, \mathscr{F}, P)$, and consider a random variable Z with values in $\mathbb{N}$, having a Poisson distribution with parameter $\lambda > 0$, that is: for every integer $k \geq 0$, $P(Z = k) = e^{-\lambda} \lambda^k / k!$.

1. Prove that, for every bounded function $f : \mathbb{N} \to \mathbb{N}$,

$$E\left[Zf(Z) \right] = \lambda E\left[f(Z+1) \right].$$

2. Fix $A \subset \mathbb{N}$, and recursively define the function $f_A : \mathbb{N} \to \mathbb{N}$ as follows: $f_A(0) = 0$ and, for every integer $k \geq 0$,

$$\lambda f_A(k+1) - kf_A(k) = \mathbf{1}_A(k) - \mathbf{P}[Z \in A]. \tag{3.8.2}$$

Show that, for every $k \geq 0$,

$$f_A(k+1) = \frac{P[\{Z \leq k\} \cap \{Z \in A\}] - P[Z \leq k] P[Z \in A]}{\lambda P[Z = k]}. \tag{3.8.3}$$

Deduce that f_A is bounded for every $A \subset \mathbb{N}$, and also that

$$f_A(k+1) - f_A(k) = f_{A^c}(k) - f_{A^c}(k+1). \tag{3.8.4}$$

3. Prove the following **Chen–Stein lemma**. *Let W be a random variable with values in $\mathbb{N}$. Then W has a Poisson law with parameter $\lambda > 0$ if and only if, for every bounded $f : \mathbb{N} \to \mathbb{N}$,*

$$E\left[Wf(W) - \lambda f(W+1) \right] = 0.$$

4. For a fixed integer $j \geq 1$, denote by $f_{\{j\}}$ the function obtained by setting $A = \{j\}$ in (3.8.3). Show that, for every $k \geq 0$,

$$f_{\{j\}}(k+1) - f_{\{j\}}(k) > 0$$

if and only if $j = k$. Deduce that $|f_A(k+1) - f_A(k)| \leq \frac{1 - e^{-\lambda}}{\lambda}$ and

$$|f_A(j) - f_A(i)| \leq \frac{1 - e^{-\lambda}}{\lambda} \times |j - i|.$$

5. Prove the following **Chen–Stein bound**: for every random variable W with values in $\mathbb{N}$,

$$\sup_{A \subset \mathbb{N}} |P(Z \in A) - P(W \in A)| \leq \sup_{f \in \Psi(\lambda)} |E[Wf(W) - \lambda f(W + 1)]|,$$

where $\Psi(\lambda)$ stands for the collection of all bounded $f : \mathbb{N} \to \mathbb{N}$ satisfying $|f(j) - f(i)| \leq \frac{1 - e^{-\lambda}}{\lambda} \times |j - i|$.

3.8.5 (An application of the Chen–Stein bounds) Let $n \geq 1$ be an integer, and let $\{X_i : i = 1, \ldots, n\}$ be independent Bernoulli random variables such that

$$p_i = P[X_i = 1] = 1 - P[X_i = 0], \quad 0 < p_i < 1, \ i = 1, \ldots, n.$$

We write $\lambda = \sum_{i=1}^{n} p_i$. We want to explicitly evaluate the distance between the law of $W = \sum_{j=1}^{n} X_j$ and the law of a Poisson random variable with parameter λ. For $i = 1, \ldots, n$, we set

$$V_i = W - X_i = \sum_{j \neq i} X_j.$$

1. Show that, if f is bounded, then

$$E[Wf(W)] = p_i E[f(V_i + 1)].$$

With the same notation as in part 5 of Exercise 3.8.4, deduce that

$$\sup_{A \subset \mathbb{N}} |P(Z \in A) - P(W \in A)|$$

$$\leq \sup_{f \in \Phi(\lambda)} \sum_{i=1}^{n} p_i |E[f(V_i + 1) - f(W + 1)]|,$$

and infer that

$$\sup_{A \subset \mathbb{N}} |P(Z \in A) - P(W \in A)| \leq \min\left(1; \frac{1}{\lambda}\right) \sum_{i=1}^{n} p_i^2. \quad (3.8.5)$$

2. Let $\{Y_n : n \geq 1\}$ be a sequence of binomial random variables, such that each Y_n has parameters n and p_n, with $0 < p_n < 1$. Assume that, as $n \to \infty$, $n p_n \to \lambda > 0$ and $p_n \to 0$. Show that the law of Y_n converges in the sense of total variation to the law of a Poisson random variable with parameter λ.

3.8.6   We use the same notation as in the statement of Proposition 3.6.1. Moreover, without loss of generality, we assume that $\sigma_1 < \sigma_2$. Show that

$$
d_{\mathrm{Kol}}(N_1, N_2) = F\left(\sigma_2 \sqrt{\frac{\log(\sigma_2^2/\sigma_1^2)}{\sigma_2^2 - \sigma_1^2}}\right) - F\left(\sigma_1 \sqrt{\frac{\log(\sigma_2^2/\sigma_1^2)}{\sigma_2^2 - \sigma_1^2}}\right),
$$

where $F(x) = \frac{1}{\sqrt{2\pi}} \int_{-\infty}^{x} e^{-u^2/2} du$ stands for the cumulative distribution function of the $\mathcal{N}(0, 1)$ law.

3.9 Bibliographic comments

Charles Stein introduced his 'method' for normal approximation in the seminal paper [135]. His 1986 monograph [136] provides a formal unified discussion, as well as several examples. The standard modern reference concerning Stein's method for normal approximations is the monograph [22], by Chen, Goldstein and Shao. Theorem 3.3.1 and Exercise 3.3.4 are proved in [136]. The proof of Theorem 3.4.2 proposed in Exercise 3.4.4 is taken from the survey by Chen and Shao [23]. A detailed (and delightful) account of the emergence of the central limit theorem (3.7.1) can be found in Adams's monograph [1], and a modern discussion is provided by Dudley in [32, section 9.5]. Our proof of the Berry–Esseen inequality (3.7.2) by means of Stein's method is based on an idea introduced by Ho and Chen in [48] (see also Bolthausen [16]). A more standard proof of the Berry–Esseen estimates is presented in Feller's book [38, section XVI.5]. The upper bound on the smallest possible value of C in (3.7.2) has decreased from Esseen's original estimate of 7.59 (see [35]) to its current value of 0.4785 by Tyurin [148]. According to Esseen [35], it is impossible to replace the universal constant 0.4785 with a number smaller than 0.40973. As we shall see later on, Stein's method extends also to non-normal approximations. Three excellent references providing a unified view of normal and non-normal approximations are the papers by Diaconis and Zabell [29], Reinert [117] and Schoutens [129]. See Arratia, Barbour and Tavaré [5], as well as the already quoted survey by Erhardsson [34], for discussions of the Chen–Stein method for (compound) Poisson approximations. A very readable introduction to the Stein and Chen–Stein methods can also be found in the book by Ross and Peköz [122]. Exercise 3.8.2 and parts 1–3 of Exercise 3.8.3 are taken from two papers by Goldstein and Reinert, respectively [42] and [43]. A bound analogous to the estimate (3.8.5) in Exercise 3.8.5 was first proved by Le Cam (with different constants) in [64].

4

Multidimensional Stein's method

We will now show how to extend the results of the previous chapter to deal with the normal approximation of d-dimensional random vectors. These results are fully exploited in Chapter 6, where they are applied to random vectors defined on the Wiener space.

For the rest of this chapter, we fix an integer $d \geq 2$, and we denote by $\mathcal{M}_d(\mathbb{R})$ the collection of all real $d \times d$ matrices. Also, given a positive symmetric matrix $C \in \mathcal{M}_d(\mathbb{R})$, we denote by $\mathcal{N}_d(0, C)$ the law of an $\mathbb{R}^d$-valued Gaussian vector with zero mean and covariance matrix C. The identity matrix of $\mathcal{M}_d(\mathbb{R})$ is written I_d. As usual, all random objects are defined on an appropriate probability space $(\Omega, \mathscr{F}, P)$.

4.1 Multidimensional Stein's lemmas

Let us start by introducing two norms on $\mathcal{M}_d(\mathbb{R})$.

Definition 4.1.1   1. The **Hilbert–Schmidt inner product** and the **Hilbert–Schmidt norm** on $\mathcal{M}_d(\mathbb{R})$, denoted respectively by $\langle \cdot, \cdot \rangle_{\mathrm{HS}}$ and $\| \cdot \|_{\mathrm{HS}}$, are defined as follows: for every pair of matrices A and B,

$$\langle A, B \rangle_{\mathrm{HS}} = \mathrm{Tr}(AB^T), \quad \text{and} \quad \|A\|_{\mathrm{HS}} = \sqrt{\langle A, A \rangle_{\mathrm{HS}}},$$

where $\mathrm{Tr}(\cdot)$ and ^{T} denote the usual trace and transposition operators, respectively.

2. The **operator norm** of $A \in \mathcal{M}_d(\mathbb{R})$, denoted by $\| \cdot \|_{\mathrm{op}}$, is given by

$$\|A\|_{\mathrm{op}} = \sup \left\{ \|Ax\|_{\mathbb{R}^d} : x \in \mathbb{R}^d \text{ such that } \|x\|_{\mathbb{R}^d} = 1 \right\},$$

with $\| \cdot \|_{\mathbb{R}^d}$ the usual Euclidean norm on $\mathbb{R}^d$.

Remark 4.1.2 If one canonically identifies $\mathcal{M}_d(\mathbb{R})$ with $\mathbb{R}^{d^2}$, then $\langle A, B\rangle_{\mathrm{HS}} = \langle A, B\rangle_{\mathbb{R}^{d^2}}$. Other useful properties of $\langle \cdot, \cdot \rangle_{\mathrm{HS}}$, $\| \cdot \|_{\mathrm{HS}}$ and $\| \cdot \|_{\mathrm{op}}$ (which are exploited throughout this chapter) are reviewed in Exercises 4.5.1 and 4.5.2.

The next result is the exact multidimensional counterpart of Stein's Lemma 3.1.2.

Lemma 4.1.3 (Multidimensional Stein's lemma) *Let $C = \{C(i, j) : i, j = 1, \ldots, d\}$ be a non-negative definite $d \times d$ matrix (we stress that C need not be positive definite). Let $N = (N_1, \ldots, N_d)$ be a random vector with values in $\mathbb{R}^d$. Then, N has the Gaussian $\mathcal{N}_d(0, C)$ distribution if and only if*

$$E\big[\langle N, \nabla f(N)\rangle_{\mathbb{R}^d}\big] = E\big[\langle C, \mathrm{Hess}\, f(N)\rangle_{\mathrm{HS}}\big], \qquad (4.1.1)$$

*for every C^2 function $f : \mathbb{R}^d \to \mathbb{R}$ having bounded first and second derivatives. Here, as usual, $\mathrm{Hess}\, f$ denotes the **Hessian** of f, that is, the $d \times d$ matrix whose entries are given by $(\mathrm{Hess}\, f)_{i,j} = \partial^2 f / \partial x_i \partial x_j$.*

Proof If $N \sim \mathcal{N}_d(0, C)$, then (4.1.1) follows from an integration by parts. Indeed, we can write, by using (3.1.2), among others,

$$
\begin{aligned}
E\big[\langle N, \nabla f(N)\rangle_{\mathbb{R}^d}\big] &= \sum_{i=1}^d E\left[N_i \frac{\partial f}{\partial x_i}(N_1, \ldots, N_d) \right] \\
&= \sum_{i,j=1}^d C(i, j) E\left[\frac{\partial^2 f}{\partial x_i \partial x_j}(N_1, \ldots, N_d) \right] \\
&= E\big[\langle C, \mathrm{Hess}\, f(N)\rangle_{\mathrm{HS}}\big]. \qquad (4.1.2)
\end{aligned}
$$

Now suppose that N satisfies (4.1.1), and let $G \sim \mathcal{N}_d(0, C)$ be independent of N. For $f : \mathbb{R}^d \to \mathbb{R}$ as in the statement, we can write, using the elementary relation $\varphi(1) - \varphi(0) = \int_0^1 \varphi'(t) dt$ in the case $\varphi(t) = E\big[f(\sqrt{t}N + \sqrt{1-t}G)\big]$,

$$
\begin{aligned}
E[f(N)] - E[f(G)] &= \int_0^1 E\big[\langle \nabla f(\sqrt{t}N + \sqrt{1-t}\,G), N\rangle_{\mathbb{R}^d}\big] \frac{dt}{2\sqrt{t}} \\
&\quad - \int_0^1 E\big[\langle \nabla f(\sqrt{t}N + \sqrt{1-t}\,G), G\rangle_{\mathbb{R}^d}\big] \frac{dt}{2\sqrt{1-t}} \\
&= \int_0^1 E\left[E\big[\langle \nabla f(\sqrt{t}N + \sqrt{1-t}\,x), N\rangle_{\mathbb{R}^d}\big]\big|_{x=G} \right] \frac{dt}{2\sqrt{t}} \\
&\quad - \int_0^1 E\left[E\big[\langle \nabla f(\sqrt{t}x + \sqrt{1-t}\,G), G\rangle_{\mathbb{R}^d}\big]\big|_{x=N} \right] \frac{dt}{2\sqrt{1-t}}.
\end{aligned}
$$

By (4.1.1), we have that, for every $x \in \mathbb{R}^d$,

$$E\left[\langle \nabla f(\sqrt{t}N + \sqrt{1-t}\,x), N\rangle_{\mathbb{R}^d}\right] = \sqrt{t}\,E\left[\langle C, \mathrm{Hess}\,f(\sqrt{t}N + \sqrt{1-t}\,x)\rangle_{\mathrm{HS}}\right]$$
$$= h_1(x).$$

Moreover, since $G \sim \mathscr{N}_d(0, C)$, we deduce from (4.1.2) that, for every $x \in \mathbb{R}^d$,

$$E\left[\langle \nabla f(\sqrt{t}x + \sqrt{1-t}\,G), G\rangle_{\mathbb{R}^d}\right]$$
$$= \sqrt{1-t}\,E\left[\langle C, \mathrm{Hess}\,f(\sqrt{t}x + \sqrt{1-t}\,G)\rangle_{\mathrm{HS}}\right] = h_2(x).$$

As a consequence, by integrating $h_1(x)$ and $h_2(x)$, respectively, with respect to the law of G and the law of N, we infer that $E[f(N)] - E[f(G)] = 0$. Since the collection of all C^2 functions with bounded derivatives is separating (see Definition C.1.1 in Appendix C), we deduce that N and G have the same law, that is, $N \sim \mathscr{N}_d(0, C)$. $\qquad\square$

The proof of Lemma 4.1.3 is based on a quite effective interpolation technique, which we will use at several other points in the book. This technique is close in spirit to the so-called 'smart path method' in the theory of spin glasses (see, for example, [140]). In the following exercise, we point out that there exist other standard proofs of the multidimensional Stein's Lemma 4.1.3.

Exercise 4.1.4 1. Prove that the $\mathscr{N}_d(0, C)$ distribution is determined by its (joint) moments, and deduce a proof of Lemma 4.1.3 based on this fact.
2. Let $N \sim \mathscr{N}_d(0, C)$. Write a system of partial differential equations whose unique solution is given by the mapping $\lambda \mapsto E[e^{i\langle \lambda, N\rangle_{\mathbb{R}^d}}]$, $\lambda \in \mathbb{R}^d$. Deduce a proof of Lemma 4.1.3 based on this result.

4.2 Stein's equations for identity matrices

Definition 4.2.1 Let $N \sim \mathscr{N}_d(0, I_d)$ be a standard Gaussian random vector. Let $h : \mathbb{R}^d \to \mathbb{R}$ be such that $E|h(N)| < \infty$. The **Stein's equation** associated with h and N is the partial differential equation

$$\Delta f(x) - \langle x, \nabla f(x)\rangle_{\mathbb{R}^d} = h(x) - E[h(N)]. \qquad (4.2.1)$$

A solution to equation (4.2.1) is a C^2 function f satisfying (4.2.1) for every $x \in \mathbb{R}^d$.

The next result highlights a particular solution of (4.2.1).

Proposition 4.2.2 *Let* $N = (N_1, \ldots, N_d) \sim \mathscr{N}_d(0, I_d)$ *be a standard Gaussian random vector. Let* $h : \mathbb{R}^d \to \mathbb{R}$ *be a Lipschitz function with constant* $K := \sup_{x \neq y} \frac{|h(x)-h(y)|}{\|x-y\|_{\mathbb{R}^d}} > 0$. *Then, the function* $f_h : \mathbb{R}^d \to \mathbb{R}$ *given by*

$$f_h(x) = \int_0^\infty E[h(N) - h(e^{-t}x + \sqrt{1 - e^{-2t}}N)]dt$$

is well defined, belongs to $\mathcal{C}^2$, satisfies (4.2.1) for every $x \in \mathbb{R}^d$ and is such that

$$\sup_{x \in \mathbb{R}^d} \|\operatorname{Hess} f_h(x)\|_{HS} \le \sqrt{d}\, K.$$

Proof We adapt the proof of Proposition 3.5.1 to this multidimensional setting. First, observe that f_h is well defined because $h(N) - h(e^{-t}x + \sqrt{1 - e^{-2t}}N)$ is integrable, due to

$$\frac{\left| h(N) - h(e^{-t}x + \sqrt{1 - e^{-2t}}N) \right|}{K} \le e^{-t}\|x\|_{\mathbb{R}^d} + \left(1 - \sqrt{1 - e^{-2t}}\right)\|N\|_{\mathbb{R}^d}$$

$$\le e^{-t}\|x\|_{\mathbb{R}^d} + e^{-2t}\|N\|_{\mathbb{R}^d},$$

where the last inequality follows from $1 - \sqrt{1 - u} = u/(\sqrt{1 - u} + 1) \le u$ if $u \in [0, 1]$. Moreover, by differentiating under the integral sign (recall that, according to Rademacher's theorem (Section E.2), a Lipschitz function is almost everywhere differentiable) and by integrating by parts, we infer that f_h is $\mathcal{C}^1$ and, moreover,

$$\frac{\partial f_h}{\partial x_i}(x) = -\int_0^\infty e^{-t} E\left[\frac{\partial h}{\partial x_i}(e^{-t}x + \sqrt{1 - e^{-2t}}N)\right]dt$$

$$= -\int_0^\infty \frac{e^{-t}}{\sqrt{1 - e^{-2t}}} E[h(e^{-t}x + \sqrt{1 - e^{-2t}}N)N_i]dt,$$

for any $i = 1, \ldots, d$. By differentiating once more under the integral sign, we get that f_h is twice differentiable with, for any $i, j = 1, \ldots, d$,

$$\frac{\partial^2 f_h}{\partial x_i \partial x_j}(x) = -\int_0^\infty \frac{e^{-2t}}{\sqrt{1 - e^{-2t}}} E\left[\frac{\partial h}{\partial x_j}(e^{-t}x + \sqrt{1 - e^{-2t}}N)N_i\right]dt.$$

Since

$$E\left[\frac{\partial h}{\partial x_j}(e^{-t}x + \sqrt{1 - e^{-2t}}N)N_i\right]$$

$$= \begin{cases} E[h(e^{-t}x + \sqrt{1 - e^{-2t}}N)(N_i^2 - 1)] & \text{if } i = j \\\\ E[h(e^{-t}x + \sqrt{1 - e^{-2t}}N)N_i N_j] & \text{if } i \ne j, \end{cases}$$

we deduce that f_h is actually C^2. Also, for any fixed $x \in \mathbb{R}^d$ and $B \in \mathcal{M}_d(\mathbb{R})$, we have

$$|\langle \operatorname{Hess} f_h(x), B \rangle_{\mathrm{HS}}|$$

$$= \left| \int_0^\infty \frac{e^{-2t}}{\sqrt{1 - e^{-2t}}} E[\langle BN, \nabla h(e^{-t}x + \sqrt{1 - e^{-2t}}N) \rangle_{\mathbb{R}^d}] dt \right|$$

$$\leq \|\nabla h\|_\infty E[\|BN\|_{\mathbb{R}^d}] \int_0^\infty \frac{e^{-2t}}{\sqrt{1 - e^{-2t}}} dt \leq \sqrt{d} \, K \sqrt{E[\|BN\|_{\mathbb{R}^d}^2]},$$

the last inequality being a consequence of the relations $\|\nabla h\|_\infty \leq \sqrt{d} \, K$ and $\int_0^\infty \frac{e^{-2t}}{\sqrt{1-e^{-2t}}} dt = 1$. On the other hand,

$$E[\|BN\|_{\mathbb{R}^d}^2] = \sum_{i=1}^d E\left[\left(\sum_{j=1}^d b_{ij} N_j \right)^2 \right] = \sum_{i,j=1}^d b_{ij}^2 = \|B\|_{\mathrm{HS}}^2,$$

so that $|\langle \operatorname{Hess} f_h(x), B \rangle_{\mathrm{HS}}| \leq \sqrt{d} \, K \|B\|_{\mathrm{HS}}$ for every $B \in \mathcal{M}_d(\mathbb{R})$ and $x \in \mathbb{R}^d$, from which we deduce that $\|\operatorname{Hess} f_h(x)\|_{\mathrm{HS}} \leq \sqrt{d} \, K$ for every $x \in \mathbb{R}^d$.

Finally, in order to prove that f_h satisfies (4.2.1), we may and will assume (for the rest of the proof and without loss of generality) that $N = (N_1, \dots, N_d) = (X(h_1), \dots, X(h_d))$, where X is an isonormal Gaussian process over some Hilbert space $\mathfrak{H}$, the functions h_i belong to $\mathfrak{H}$ and $\langle h_i, h_j \rangle_{\mathfrak{H}} = E[N_i N_j] = \delta_{ij}$ (Kronecker's symbol). By using the notation introduced in Section 2.8.2 and since we have proved that $\operatorname{Hess} f_h$ is bounded, we deduce that $f_h(N) \in \operatorname{Dom} L$. Hence,

$$\Delta f_h(N) - \langle N, \nabla f_h(N) \rangle_{\mathbb{R}^d} \tag{4.2.2}$$

$$= L f_h(N) \quad \text{(by Theorem 2.8.9 and an approximation argument)}$$

$$= -\int_0^\infty L P_t h(N) dt \quad \text{(since } f_h(N) = \int_0^\infty \left(E[h(N)] - P_t h(N) \right) dt \text{)}$$

$$= -\int_0^\infty \frac{d}{dt} P_t h(N) dt$$

$$= P_0 h(N) - P_\infty h(x) = h(N) - E[h(N)].$$

Since the support of the law of N coincides with $\mathbb{R}^d$, we obtain immediately (e.g. by a continuity argument) that f_h must satisfy (4.2.1) for every $x \in \mathbb{R}^d$. $\qquad \square$

Remark 4.2.3 In view of Theorem 2.8.9, the chain of equalities (4.2.2) implies that

$$f_h(N) = f_h(X(h_1), \dots, X(h_d)) = L^{-1} h(X(h_1), \dots, X(h_d)).$$

In the next section, we extend the above results to the case of a Gaussian vector with a general positive definite covariance matrix.

4.3 Stein's equations for general positive definite matrices

For this section, we fix a *positive definite* $d \times d$ matrix C. We also denote by A a non-singular symmetric matrix such that $A^2 = C$ (note that the matrix A always exists).

Definition 4.3.1   Let $N \sim \mathcal{N}_d(0, C)$. Let $h : \mathbb{R}^d \to \mathbb{R}$ be such that $E|h(N)| < \infty$. The **Stein's equation** associated with h and N is the partial differential equation

$$\langle C, \operatorname{Hess} f(x) \rangle_{\mathrm{HS}} - \langle x, \nabla f(x) \rangle_{\mathbb{R}^d} = h(x) - E[h(N)]. \tag{4.3.1}$$

A solution to the equation (4.3.1) is a C^2 function f satisfying (4.3.1) for every $x \in \mathbb{R}^d$.

The next statement is a generalization of Proposition 4.2.2.

Proposition 4.3.2   *Let $N \sim \mathcal{N}_d(0, C)$. Let $h : \mathbb{R}^d \to \mathbb{R}$ be a Lipschitz function with constant $K > 0$. Then the function $f_h : \mathbb{R}^d \to \mathbb{R}$ given by*

$$f_h(x) = \int_0^\infty E[h(N) - h(e^{-t}x + \sqrt{1 - e^{-2t}}N)]dt \tag{4.3.2}$$

is well defined, belongs to C^2, satisfies (4.3.1) for every $x \in \mathbb{R}^d$ and is such that

$$\sup_{x \in \mathbb{R}^d} \|\operatorname{Hess} f_h(x)\|_{\mathrm{HS}} \le \sqrt{d}\, K \times \|C^{-1}\|_{\mathrm{op}} \times \|C\|_{\mathrm{op}}^{1/2}. \tag{4.3.3}$$

Proof   We start by representing f_h in terms of the solution of an equation of type (4.2.1). Indeed, we have that $f_h(x) = V(A^{-1}x)$, where

$$V(x) = \int_0^\infty E[h_A(A^{-1}N) - h_A(e^{-t}x + \sqrt{1 - e^{-2t}}A^{-1}N)]dt$$

and $h_A(x) = h(Ax)$. By virtue of Proposition 4.2.2, this implies that f_h is C^2. Moreover, since $A^{-1}N \sim \mathcal{N}_d(0, I_d)$, the function V satisfies the Stein's equation

$$\Delta V(x) - \langle x, \nabla V(x) \rangle_{\mathbb{R}^d} = h_A(x) - E[h_A(Y)], \tag{4.3.4}$$

where $Y \sim \mathcal{N}_d(0, I_d)$. Now rewrite the four objects appearing in (4.3.4) as follows:

$$\Delta V(x) = \text{Tr}(\text{Hess}\, V(x)) = \text{Tr}(A\, \text{Hess}\, f_h(Ax)\, A) = \langle C, \text{Hess}\, f_h(Ax) \rangle_{\text{HS}},$$

$$\langle x, \nabla V(x) \rangle_{\mathbb{R}^d} = \langle Ax, \nabla f_h(Ax) \rangle_{\mathbb{R}^d},$$

$$h_A(x) = h(Ax),$$

$$E[h_A(Y)] = E[h(N)].$$

By plugging these equalities into (4.3.4), and since A is non-singular, we deduce immediately that f_h satisfies (4.3.1) for every $x \in \mathbb{R}^d$.

All that remains the proof of the estimate (4.3.3). Start by observing that, by Proposition 4.2.2,

$$\sup_{x \in \mathbb{R}^d} \|\text{Hess}\, V(x)\|_{\text{HS}} \leq \|h_A\|_{\text{Lip}} \leq \|A\|_{\text{op}} \sqrt{d}\, K. \tag{4.3.5}$$

On the other hand,

$$\sup_{x \in \mathbb{R}^d} \|\text{Hess}\, f_h(x)\|_{\text{HS}}$$

$$= \sup_{x \in \mathbb{R}^d} \|A^{-1}\text{Hess}\, V(A^{-1}x)A^{-1}\|_{\text{HS}}$$

$$= \sup_{x \in \mathbb{R}^d} \|A^{-1}\text{Hess}\, V(x)A^{-1}\|_{\text{HS}} \leq \|A^{-2}\|_{\text{op}} \sup_{x \in \mathbb{R}^d} \|\text{Hess}\, V(x)\|_{\text{HS}}$$

$$\leq \|A^{-2}\|_{\text{op}} \|A\|_{\text{op}} \sqrt{d}\, K$$

$$= \|C^{-1}\|_{\text{op}} \|C\|_{\text{op}}^{1/2} \sqrt{d}\, K \quad \text{(see Exercise 4.5.1, part 2)},$$

thus concluding the proof. □

Exercise 4.3.3 Let C be a positive definite matrix, let $h : \mathbb{R}^d \to \mathbb{R}$ be Lipschitz, and let f_h be given by (4.3.2). Consider an isonormal Gaussian process $X = \{X(h) : h \in \mathfrak{H}\}$, and assume that the vector $(X(h_1), \ldots, X(h_d))$ is such that $E[X(h_i)X(h_j)] = C(i, j)$. Prove that $f_h(X(h_1), \ldots, X(h_d)) = L^{-1}h(X(h_1), \ldots, X(h_d))$.

4.4 Bounds on the Wasserstein distance

The following statement allows us to deal with normal approximations in the Wasserstein (and Fortet–Mourier) distance – see Definition C.2.1 in Appendix C. It represents a quantitative version of Lemma 4.1.3.

Theorem 4.4.1 *Let C be a positive definite $d \times d$ matrix, and let $N \sim \mathcal{N}_d(0, C)$. For any square-integrable $\mathbb{R}^d$-valued random vector F, we have*

$$d_{\text{FM}}(F, N) \leq d_{\text{W}}(F, N)$$
$$\leq \sup_{f \in \mathscr{F}_{\text{W}}^d(C)} |E[\langle C, \text{Hess } f(F)\rangle_{\text{HS}}] - E[\langle F, \nabla f(F)\rangle_{\mathbb{R}^d}]|,$$

$$(4.4.1)$$

where $\mathscr{F}_{\text{W}}^d(C) := \left\{ f : \mathbb{R}^d \to \mathbb{R} \in \mathcal{C}^2 : \sup_{x \in \mathbb{R}^d} \|\text{Hess} f(x)\|_{\text{HS}} \leq \sqrt{d} \, \|C^{-1}\|_{\text{op}} \right.$
$\left. \|C\|_{\text{op}}^{1/2} \right\}.$

Proof   We use Proposition 4.3.2, and deduce that

$$d_{\text{FM}}(F, N) \leq d_{\text{W}}(F, N) = \sup_{h \in \text{Lip}(1)} \left| E[h(N)] - E[h(F)] \right|$$
$$= \sup_{h \in \text{Lip}(1)} \left| E[\langle C, \text{Hess } f_h(F)\rangle_{\text{HS}}] - E[\langle F, \nabla f_h(F)\rangle_{\mathbb{R}^d}] \right|$$
$$\leq \sup_{f \in \mathscr{F}_{\text{W}}^d(C)} \left| E[\langle C, \text{Hess } f(F)\rangle_{\text{HS}}] - E[\langle F, \nabla f(F)\rangle_{\mathbb{R}^d}] \right|. \quad \square$$

Remark 4.4.2   The requirement that F is square-integrable ensures that the expectation $E[\langle F, \nabla f(F)\rangle_{\mathbb{R}^d}]$ is well defined for every $f \in \mathscr{F}_{\text{W}}^d(C)$.

4.5 Exercises

4.5.1   1. Check that the two norms $\| \cdot \|_{\text{HS}}$ and $\| \cdot \|_{\text{op}}$ satisfy the five axioms of a **matrix norm** on $\mathcal{M}_d(\mathbb{R})$, that is (with $\| \cdot \|$ standing for either $\| \cdot \|_{\text{HS}}$ or $\| \cdot \|_{\text{op}}$:

   (i) $\|A\| \geq 0$;

   (ii) $\|A\| = 0$ if and only if $A = 0$;

   (iii) $\|cA\| = |c| \times \|A\|$ for every real c;

   (iv) $\|A + B\| \leq \|A\| + \|B\|$;

   (v) $\|AB\| \leq \|A\| \times \|B\|$.

2. Let $A \in \mathcal{M}_d(\mathbb{R})$, and let A^T denote the transpose of A. Check that $\|A\|_{\text{op}}$ coincides with the largest **singular value** of A, that is, with the square root of the largest eigenvalue of $A^T A$.

3. Deduce from part 2 that, if $A \in \mathcal{M}_d(\mathbb{R})$ is normal (that is, $A^T A = AA^T$), then $\|A\|_{\text{op}}$ coincides with the **spectral radius** of A, i.e.,

$$\|A\|_{\text{op}} = \max\{|\lambda| : \lambda \text{ is an eigenvalue of } A\}.$$

Deduce that, if $A \in \mathcal{M}_d(\mathbb{R})$ is symmetric, then $\|A^k\|_{op} = \|A\|_{op}^k$ for any integer $k \geq 1$.

4. Prove that $\|AB\|_{HS} \leq \min\{\|A\|_{HS}\|B\|_{op}; \|A\|_{op}\|B\|_{HS}\}$.

4.5.2 1. Let $f : \mathbb{R}^d \to \mathbb{R}$ be differentiable, let $A \in \mathcal{M}_d(\mathbb{R})$ and set $f_A(x) = f(Ax)$. Prove that $\langle Ax, \nabla f(Ax) \rangle_{\mathbb{R}^d} = \langle x, \nabla f_A(x) \rangle_{\mathbb{R}^d}$, for every $x \in \mathbb{R}^d$.

2. Assume that f is twice differentiable, and denote by Hess f the Hessian matrix of f. For every $A \in \mathcal{M}_d(\mathbb{R})$ and every $x \in \mathbb{R}^d$, prove that Hess $f_A(x) = A^T$ Hess $f(Ax)A$.

4.5.3 Let $Y \sim \mathcal{N}_d(0, K)$ and $Z \sim \mathcal{N}_d(0, C)$, where K and C are two positive definite covariance matrices. Prove that $d_W(Y, Z) \leq Q(C, K) \times \|C - K\|_{HS}$, where

$$Q(C, K) := \sqrt{d} \min\{\|C^{-1}\|_{op} \|C\|_{op}^{1/2}; \|K^{-1}\|_{op} \|K\|_{op}^{1/2}\}. \quad (4.5.1)$$

The estimate (4.5.1) is the d-dimensional analog of Proposition 3.6.1.

4.5.4 (**The generator approach – see [7, 44]**) The 'generator approach' to Stein's method uses the properties of Markov processes to deduce and (formally) solve Stein's equations associated with general (not necessarily normal) probability distributions. Let $Y = \{Y_t : t \geq 0\}$ be a Markov process with values in a Polish space $\mathbb{U}$, and define the semigroup $\{T_t : t \geq 0\}$ associated with Y by the relation $T_t f(x) = E[f(Y_t)|X_0 = x]$. Write $\mathcal{L}f(x) = \lim_{t \to 0} \frac{T_t f(x) - f(x)}{t}$ for the generator of T. Denote by μ the stationary distribution of Y.

1. Show that a random element Z with values in $\mathbb{U}$ has the distribution μ if and only if $E[\mathcal{L}f(Z)] = 0$ for every function $f : \mathbb{U} \to \mathbb{R}$ in the domain of $\mathcal{L}$.

2. Prove that, for every bounded function $h : \mathbb{U} \to \mathbb{R}$, $T_t h - h = \mathcal{L}\left(\int_0^t T_t h dt\right)$.

3. The **Stein's equation** associated with Y and with a test function $h : \mathbb{U} \to \mathbb{R}$ is

$$h(x) - \int_{\mathbb{U}} h d\mu = \mathcal{L}f(x). \quad (4.5.2)$$

A solution to (4.5.2) is a function f in the domain of $\mathcal{L}$ satisfying (4.5.2) for every $x \in \mathbb{U}$. Show that, if $f = \mathcal{L}^{-1}\left[h - \int_{\mathbb{U}} h d\mu\right]$ is well defined, then it is a solution to (4.5.2). Here, $\mathcal{L}^{-1}$ stands for the pseudo-inverse of $\mathcal{L}$.

4. Use Remark 2.8.5 and Theorem 2.8.9 to prove that, for every $d \geq 2$ and every non-negative definite $d \times d$ matrix C, there exists a Markov process with values in $\mathbb{R}^d$ and with generator L satisfying

$$Lf(x) = \langle C, \operatorname{Hess} f(x) \rangle_{\mathrm{HS}} - \langle x, \nabla f(x) \rangle_{\mathbb{R}^d},$$

for every f in C^2 with bounded derivatives. Prove the multidimensional Stein's lemma 4.1.3 by means of this result and of the generator approach.

5. Use Exercise 4.3.3 to solve Stein's equation (4.3.1) by means of the generator approach.

4.6 Bibliographic comments

An outstanding resource for results in matrix analysis is the monograph by Horn and Johnson [49] (in particular, [49, chapter 5] is devoted to matrix norms and their properties). There are relatively few references concerning multivariate normal approximations using Stein's method. The paper [95], by Nourdin, Peccati and Réveillac, basically contains all the results discussed in the present chapter (mostly with different proofs). See Chen, Goldstein and Shao [22, chapter 12] for a general discussion of available techniques. Other important references on the topic are the works by Chatterjee and Meckes [21], Raič [116], Reinert and Röllin [118] and Rinott and Rotar [120]. As already pointed out, the 'generator approach', as briefly described in Exercise 4.5.4, was initiated in the two papers by Barbour [7] and Götze [44]: see Reinert [117] and Schoutens [129] for further discussions and further references on the subject. A classic repository of results about Markov processes, semigroups and generators is the monograph by Ethier and Kurtz [36].

5

Stein meets Malliavin: univariate normal approximations

In this chapter, we show how Malliavin calculus and Stein's method may be combined into a powerful and flexible tool for studying probabilistic approximations. In particular, our aim is to use these two techniques to assess the distance between the laws of regular functionals of an isonormal Gaussian process and a one-dimensional normal distribution.

The highlight of the chapter is arguably Section 5.2, where we deduce a complete characterization of Gaussian approximations inside a fixed Wiener chaos. As discussed below, the approach developed in this chapter yields results that are *systematically stronger* than the so-called 'method of moments and cumulants', which is the most popular tool used in the proof of central limit theorems for functional of Gaussian fields.

Note that, in view of the chaos representation (2.7.8), any general result involving random variables in a fixed chaos is a key for studying probabilistic approximations of more general functionals of Gaussian fields. This last point is indeed one of the staples of the entire book, and will be abundantly illustrated in Section 5.3 as well as in Chapter 7.

Throughout the following, we fix an isonormal Gaussian process $X = \{X(h) : h \in \mathfrak{H}\}$, defined on a suitable probability space $(\Omega, \mathscr{F}, P)$ such that $\mathscr{F} = \sigma\{X\}$. We will also adopt the language and notation of Malliavin calculus introduced in Chapter 2.

5.1 Bounds for general functionals

We start with a useful result.

Proposition 5.1.1 *Let $F \in \mathbb{D}^{1,2}$ be such that $E[F] = 0$ and $E[F^2] = 1$, and let $f : \mathbb{R} \to \mathbb{R}$ be Lipschitz with constant $K > 0$. Assume that either (a) F*

has a density, or (b) F does not necessarily have a density but f is of class C^1.
Then $E[f'(F)]$ is defined without ambiguity and

$$\left| E[f'(F)] - E[Ff(F)] \right| \leq K \times E\left[\left| 1 - \langle DF, -DL^{-1}F \rangle_{\mathfrak{H}} \right| \right]. \quad (5.1.1)$$

If, in addition, $F \in \mathbb{D}^{1,4}$ then $\langle DF, -DL^{-1}F \rangle_{\mathfrak{H}}$ is square-integrable and

$$E\left[\left| 1 - \langle DF, -DL^{-1}F \rangle_{\mathfrak{H}} \right| \right] \leq \sqrt{\text{Var}[\langle DF, -DL^{-1}F \rangle_{\mathfrak{H}}]}. \quad (5.1.2)$$

Proof   Assume first that f is C^1. In this case, (2.9.1) with $G = F$ and $g = f$
yields

$$\begin{aligned}
\left| E[f'(F)] - E[Ff(F)] \right| &= \left| E[f'(F)(1 - \langle DF, -DL^{-1}F \rangle_{\mathfrak{H}})] \right| \\
&\leq K \times E\left[\left| 1 - \langle DF, -DL^{-1}F \rangle_{\mathfrak{H}} \right| \right],
\end{aligned}$$

which is (5.1.1). When f is not necessarily C^1 but F has a density, then (5.1.1)
continues to hold, as we infer by using an approximation argument (e.g. by
convoluting f with an approximation of the identity).

To prove the second part of the statement, observe on the one hand that,
thanks to Proposition 2.9.3 among others, we can write

$$\begin{aligned}
E\left[\|DL^{-1}F\|_{\mathfrak{H}}^4 \right] &= E\left[\left\| \int_0^\infty e^{-t} P_t DF dt \right\|_{\mathfrak{H}}^4 \right] \\
&\leq E\left[\left(\int_0^\infty e^{-t} \|P_t DF\|_{\mathfrak{H}} dt \right)^4 \right] \\
&\leq \int_0^\infty e^{-t} E\left[\|P_t DF\|_{\mathfrak{H}}^4 \right] dt \leq E\left[\|DF\|_{\mathfrak{H}}^4 \right] \int_0^\infty e^{-t} dt \\
&= E\left[\|DF\|_{\mathfrak{H}}^4 \right].
\end{aligned}$$

Therefore

$$\begin{aligned}
E\left[\langle DF, -DL^{-1}F \rangle_{\mathfrak{H}}^2 \right] &\leq E\left[\|DF\|_{\mathfrak{H}}^2 \|DL^{-1}F\|_{\mathfrak{H}}^2 \right] \\
&\leq \sqrt{E\left[\|DF\|_{\mathfrak{H}}^4 \right]} \sqrt{E\left[\|DL^{-1}F\|_{\mathfrak{H}}^4 \right]} \leq E\left[\|DF\|_{\mathfrak{H}}^4 \right],
\end{aligned}$$

which implies that $\langle DF, -DL^{-1}F \rangle_{\mathfrak{H}}^2$ is square-integrable if $F \in \mathbb{D}^{1,4}$. On
the other hand, by choosing $G = F$ and $g(x) = x$ in (2.9.1), we get
$E[\langle DF, -DL^{-1}F \rangle_{\mathfrak{H}}] = 1$. Therefore, the desired inequality is obtained by
applying the Cauchy–Schwarz inequality:

$$\begin{aligned}
E[|1 - \langle DF, -DL^{-1}F \rangle_{\mathfrak{H}}|] &\leq \sqrt{E[(1 - \langle DF, -DL^{-1}F \rangle_{\mathfrak{H}})^2])} \\
&= \sqrt{\text{Var}[\langle DF, -DL^{-1}F \rangle_{\mathfrak{H}}]}. \quad \square
\end{aligned}$$

Remark 5.1.2 If F is in $\mathbb{D}^{1,2}$, but not necessarily in $\mathbb{D}^{1,4}$, then one can only deduce that $\langle DF, -DL^{-1}F\rangle_{\mathfrak{H}} \in L^1(\Omega)$.

Proposition 5.1.1 is indeed the key to proving the next statement, providing upper bounds on some distances between the law of a regular centered F and a Gaussian random variable with the same variance.

Theorem 5.1.3 *Let $F \in \mathbb{D}^{1,2}$ with $E[F] = 0$ and $E[F^2] = \sigma^2 > 0$, and let $N \sim \mathcal{N}(0, \sigma^2)$. Then*

$$d_{\mathrm{FM}}(F, N) \le d_{\mathrm{W}}(F, N) \le \frac{\sqrt{2}}{\sigma\sqrt{\pi}} \, E\big[\big|\sigma^2 - \langle DF, -DL^{-1}F\rangle_{\mathfrak{H}}\big|\big]. \quad (5.1.3)$$

If, in addition, F has a density, then

$$d_{\mathrm{TV}}(F, N) \le \frac{2}{\sigma^2} \, E\big[\big|\sigma^2 - \langle DF, -DL^{-1}F\rangle_{\mathfrak{H}}\big|\big], \quad (5.1.4)$$

$$d_{\mathrm{Kol}}(F, N) \le \frac{1}{\sigma^2} \, E\big[\big|\sigma^2 - \langle DF, -DL^{-1}F\rangle_{\mathfrak{H}}\big|\big]. \quad (5.1.5)$$

Proof The proof consists of three steps.
Step 1: Case $\sigma = 1$. To get inequality(5.1.3) ((5.1.4), (5.1.5)) in this case, it suffices to combine Proposition 5.1.1 with (3.5.2) ((3.3.2), (3.4.3)).
Step 2: General σ for (5.1.3). Elementary computations yield

$$d_{\mathrm{W}}(F, N) = \sup_{h\in\mathrm{Lip}(\sigma)} \left| E\left[h\left(\frac{F}{\sigma}\right)\right] - E\left[h\left(\frac{N}{\sigma}\right)\right]\right|.$$

Since $N/\sigma \sim \mathcal{N}(0, 1)$, Proposition 3.5.1 yields

$$\sup_{h\in\mathrm{Lip}(\sigma)} \left| E\left[h\left(\frac{F}{\sigma}\right)\right] - E\left[h\left(\frac{N}{\sigma}\right)\right]\right|$$

$$= \sup_{h\in\mathrm{Lip}(\sigma)} \left| E\left[f'_h\left(\frac{F}{\sigma}\right) - \frac{F}{\sigma}f_h\left(\frac{F}{\sigma}\right)\right]\right|$$

$$\le \sup_{f\in\mathcal{C}^1\cap\mathrm{Lip}\left(\sigma\sqrt{\frac{2}{\pi}}\right)} \left| E\left[f'\left(\frac{F}{\sigma}\right) - \frac{F}{\sigma}f\left(\frac{F}{\sigma}\right)\right]\right|$$

$$\le \frac{\sigma\sqrt{2}}{\sqrt{\pi}} E\left[\left|1 - \frac{1}{\sigma^2}\langle DF, -DL^{-1}F\rangle_{\mathfrak{H}}\right|\right],$$

where f_h is the solution of the Stein's equation associated with h – as given in (3.5.1). The proof is deduced immediately.

Step 3: General σ for (5.1.4) and (5.1.5). We have

$$
\begin{aligned}
d_{\mathrm{TV}}(F, N) &= \sup_{B \in \mathscr{B}(\mathbb{R})} \left| P(F \in B) - P(N \in B) \right| \\
&= \sup_{B \in \mathscr{B}(\mathbb{R})} \left| P(F \in \sigma B) - P(N \in \sigma B) \right| \\
&= d_{\mathrm{TV}}\left(\frac{F}{\sigma}, \frac{N}{\sigma} \right) \leq 2E\left[\left| 1 - \frac{1}{\sigma^2} \langle DF, -DL^{-1}F \rangle_{\mathfrak{H}} \right| \right] \quad \text{(by Step 1)} \\
&= \frac{2}{\sigma^2} E[|\sigma^2 - \langle DF, -DL^{-1}F \rangle_{\mathfrak{H}}|],
\end{aligned}
$$

which is (5.1.4). The proof of (5.1.5) is similar. $\qquad\square$

Remark 5.1.4 1. By combining Theorem 5.1.3 with Proposition 3.6.1 and the triangle inequality, one can easily deduce bounds involving centered random variables with different variances. For instance, if $F \in \mathbb{D}^{1,2}$ is centered with variance σ^2, and if $N \sim \mathcal{N}(0, \gamma^2)$, then

$$
d_{\mathrm{TV}}(F, N) \leq \frac{2}{\sigma^2 \vee \gamma^2} |\sigma^2 - \gamma^2| + \frac{2}{\sigma^2} E\left[|\sigma^2 - \langle DF, -DL^{-1}F \rangle_{\mathfrak{H}}| \right].
$$

$$
\tag{5.1.6}
$$

Analogous estimates can be deduced for the three distances d_{Kol}, d_{FM} and d_{W}.

2. Under the assumptions of Theorem 5.1.3, we have that $E[\langle DF, -DL^{-1}F \rangle_{\mathfrak{H}}] = \sigma^2$. Moreover, by following the same line of reasoning as in the proof of the second part of Lemma 5.1.1, we immediately deduce that, if $F \in \mathbb{D}^{1,4}$, then $\langle DF, -DL^{-1}F \rangle_{\mathfrak{H}} \in L^2(\Omega)$, and therefore, by Cauchy–Schwarz,

$$
E\left[|\sigma^2 - \langle DF, -DL^{-1}F \rangle_{\mathfrak{H}}| \right] \leq \sqrt{\mathrm{Var}(\langle DF, -DL^{-1}F \rangle_{\mathfrak{H}})} < \infty.
$$

When variances are difficult to estimate, it is often useful to deal with bounds where one does not divide by σ^2, as happens in (5.1.3), (5.1.4) and (5.1.5). In Theorem 5.1.5, we show that bounds of this kind can be deduced by means of a 'smart-path method', similar to that already encountered in the proof of the multidimensional Stein's Lemma 4.1.3. Observe that the next statement involves test functions h that are more regular than that entering in the definition of the distances d_{TV}, d_{Kol}, d_{FM} and d_{W}. Theorem 5.1.5 will be generalized in Chapter 6 to the multidimensional case (see Theorem 6.1.2).

Theorem 5.1.5 *Let $F \in \mathbb{D}^{1,2}$ with $E[F] = 0$ and $E[F^2] = \sigma^2 > 0$, and let $N \sim \mathcal{N}(0, \sigma^2)$. Let $h : \mathbb{R} \to \mathbb{R}$ be C^2 with $\|h''\|_\infty < \infty$. Then,*

$$E[h(N)] - E[h(F)]$$
$$= \frac{1}{\sqrt{2\pi}} \int_0^\infty dt\, e^{-2t} \int_{-\infty}^\infty dx\, e^{-x^2/2}\, E[h''(e^{-t}\sigma x + \sqrt{1-e^{-2t}}\,F)$$
$$\times (\langle DF, -DL^{-1}F\rangle_\mathfrak{H} - \sigma^2)]. \qquad (5.1.7)$$

In particular,

$$\left|E[h(N)] - E[h(F)]\right| \le \frac{1}{2}\|h''\|_\infty E[\left|\langle DF, -DL^{-1}F\rangle_\mathfrak{H} - \sigma^2\right|]. \quad (5.1.8)$$

Proof For $t \ge 0$, set

$$\varphi(t) = \frac{1}{\sqrt{2\pi}} \int_{-\infty}^\infty E[h(e^{-t}\sigma x + \sqrt{1-e^{-2t}}\,F)]e^{-x^2/2}dx,$$

and observe that $E[h(F)] - E[h(N)] = \int_0^\infty \varphi'(t)dt$. We have

$$\varphi'(t) = -\frac{e^{-t}\sigma}{\sqrt{2\pi}} E\left[\int_{-\infty}^\infty h'(e^{-t}\sigma x + \sqrt{1-e^{-2t}}\,F)\, x e^{-x^2/2}dx\right]$$
$$+\frac{e^{-2t}}{\sqrt{1-e^{-2t}}} \times \frac{1}{\sqrt{2\pi}} \int_{-\infty}^\infty E[h'(e^{-t}\sigma x + \sqrt{1-e^{-2t}}\,F)F]e^{-x^2/2}dx.$$

Performing an integration by parts on the first integral, and plugging identity (2.9.1) into the second expectation, yields

$$\varphi'(t) = -\frac{e^{-2t}\sigma^2}{\sqrt{2\pi}} E\left[\int_{-\infty}^\infty h''(e^{-t}\sigma x + \sqrt{1-e^{-2t}}\,F)e^{-x^2/2}dx\right]$$
$$+ \frac{e^{-2t}}{\sqrt{2\pi}} \int_{-\infty}^\infty E[h''(e^{-t}\sigma x + \sqrt{1-e^{-2t}}\,F)\langle DF, -DL^{-1}F\rangle_\mathfrak{H}]e^{-x^2/2}dx$$
$$= \frac{e^{-2t}}{\sqrt{2\pi}} \int_{-\infty}^\infty E[h''(e^{-t}\sigma x + \sqrt{1-e^{-2t}}\,F)(\langle DF, -DL^{-1}F\rangle_\mathfrak{H} - \sigma^2)]e^{-x^2/2}dx,$$

from which (5.1.7) follows. By applying the inequality $|h''(e^{-t}\sigma x + \sqrt{1-e^{-2t}}\,F)| \le \|h''\|_\infty$ to (5.1.7), we deduce estimate (5.1.8). $\qquad\square$

5.2 Normal approximations on Wiener chaos

5.2.1 Some preliminary considerations

Fix $q \ge 2$, and suppose that $F_n = I_q(f_n)$, $n \ge 1$, is a sequence of random variables in the qth Wiener chaos of X, with variances converging to one. When dealing with limit theorems for functionals of Gaussian fields, one is often asked to determine conditions on the sequence $\{F_n : n \ge 1\}$ so that, as

$n \to \infty$, the law of F_n converges weakly to a standard $\mathcal{N}(0, 1)$ distribution. This task is by no means trivial. Indeed, although the case of double integrals could in principle be dealt with by using the results of Section 2.7.4, there is in general no explicit characterization of the laws of random variables belonging to a chaos of order $q \geq 2$. Due to this lack of information, one of the most popular ways of deducing CLTs on a Wiener chaos is the so-called 'method of moments and cumulants', basically involving proving that the moments (or cumulants) of F_n converge to those of a standard Gaussian random variable. The theoretical background to this technique is the object of Proposition 5.2.2 (see also Theorem A.3.1).

Remark 5.2.1 In what follows, given a random variable F with finite moments of all orders, we denote by $\{\kappa_j(F) : j \geq 1\}$ the collection of its cumulants, as defined in Section A.2.2.

Proposition 5.2.2 (Method of moments and cumulants) *Fix $q \geq 2$ and let $F_n = I_q(f_n)$, $n \geq 1$, be a sequence of random variables in the qth chaos of X, such that $E(F_n^2) = q! \|f_n\|_{\mathfrak{H}^{\otimes q}}^2 \to \sigma^2 > 0$, as $n \to \infty$. Then, as $n \to \infty$, the following conditions are equivalent:*

1. *The sequence F_n converges in distribution to $N \sim \mathcal{N}(0, \sigma^2)$.*
2. *For every integer $j \geq 3$, $E(F_n^j) \to E(N^j)$.*
3. *For every integer $j \geq 3$, $\kappa_j(F_n) \to 0$.*

Proof Note that $\kappa_1(F_n) = E(F_n) = E(N) = 0$ and $\kappa_2(F_n) = E(F_n^2) \to \sigma^2 = \kappa_2(N) = E(N^2)$. Since $\kappa_j(N) = 0$ for every $j \geq 3$, and since every $\kappa_j(F_n)$ can be expressed as a finite linear combination of products of moments of F_n of order less or equal to j and vice versa (see Corollary A.2.4), conditions 2 and 3 are easily seen to be equivalent.

We now deal with the equivalence between conditions 1 and 2. First observe that, since $E(F_n^2) \to \sigma^2$, the sequence of the laws of the random variables F_n is relatively compact, and moreover, by virtue of the hypercontractivity property stated in Theorem 2.7.2, for every real $\eta \geq 2$,

$$\sup_{n \geq 1} E|F_n|^\eta < \infty. \qquad (5.2.1)$$

Now suppose that condition 2 is satisfied. Relative compactness implies that every subsequence $\{n(r)\}$ contains a further subsequence $\{n(r')\}$ such that the law of $F_{n(r')}$ converges weakly to some probability measure μ. Since (5.2.1) holds, we also have that, for every $j \geq 1$,

$$E\left(F_{n(r')}^j\right) \to \int_{\mathbb{R}} x^j d\mu(x),$$

and therefore $\int_{\mathbb{R}} x^j \mu(dx) = E(N^j)$. Since the law of N is determined by its moments (thanks to Lemma 3.1.1), we deduce that $d\mu(x) = (2\pi)^{-1/2}e^{-x^2/2}dx$. This shows that condition 2 implies condition 1. On the other hand, if condition 1 is satisfied, then relation (5.2.1) and the continuous mapping theorem together imply that $E(F_n^j) \to E(N^j)$ for every $j \geq 3$, thus concluding the proof. □

There are several ways, mostly combinatorial, of computing the moments and cumulants of chaotic random variables: see, for example, Peccati and Taqqu [110], Major [68] or Surgailis [139] (as well as the bibliographic comments at the end of the present chapter) for an overview of combinatorial techniques based on *diagrams*; see Chapter 8 for several explicit formulae for cumulants based on a recursive use of Malliavin operators. All these results yield expressions for moments and cumulants whose complexity significantly increases with j. It follows that the verification of conditions 2 and 3 in Proposition 5.2.2 is in general a hard and computationally demanding task.

We shall show that all these difficulties can be successfully overcome by using Stein's method and Malliavin calculus. Indeed, in Section 5.2.2 it is proved that, by specializing the estimates of Section 5.1 to chaotic random variables, one can deduce the following drastic (and surprising) simplification of Proposition 5.2.2 (see Theorem 5.2.7 for a precise statement):

In order that condition 1 in the statement of Proposition 5.2.2 is satisfied (that is, in order that $\{F_n\}$ obeys a CLT) it is necessary and sufficient that $E(F_n^4) \to E(N^4) = 3$.

In other words, the method of moments and cumulants for normal approximations boils down to a 'fourth-moment method' inside a fixed Wiener chaos. Observe that, since they are strongly based on Stein's method, the results proved below also yield explicit estimates on the four distances d_{TV}, d_{Kol}, d_{FM} and d_W.

Remark 5.2.3 As a by-product of our analysis, we will prove in Corollary 5.2.11 that a non-zero random variable belonging to a Wiener chaos of order $q \geq 2$ cannot be Gaussian.

5.2.2 Estimates, fourth moments and CLTs

Our first result is a technical lemma.

Lemma 5.2.4 *Fix an integer $q \geq 2$, and let $F = I_q(f)$ take the form of a multiple integral of order q. We have the identities:*

$$\frac{1}{q}\|DF\|_{\mathfrak{H}}^2 = E[F^2] + q\sum_{r=1}^{q-1}(r-1)!\binom{q-1}{r-1}^2 I_{2q-2r}(f\widetilde{\otimes}_r f), \qquad (5.2.2)$$

$$\mathrm{Var}\left(\frac{1}{q}\|DF\|_{\mathfrak{H}}^2\right) = \frac{1}{q^2}\sum_{r=1}^{q-1} r^2 r!^2 \binom{q}{r}^4 (2q-2r)!\,\|f\widetilde{\otimes}_r f\|_{\mathfrak{H}^{\otimes 2q-2r}}^2, \quad (5.2.3)$$

$$E[\|D^2 F \otimes_1 D^2 F\|_{\mathfrak{H}^{\otimes 2}}^2]$$

$$\leq q^4(q-1)^4 \sum_{r=1}^{q-1} (r-1)!^2 \binom{q-2}{r-1}^4 (2q-2-2r)!\,\|f\otimes_r f\|_{\mathfrak{H}^{\otimes 2q-2r}}^2,$$
$$(5.2.4)$$

and

$$E[F^4] - 3E[F^2]^2 = \frac{3}{q}\sum_{r=1}^{q-1} r r!^2 \binom{q}{r}^4 (2q-2r)!\,\|f\widetilde{\otimes}_r f\|_{\mathfrak{H}^{\otimes 2q-2r}}^2 \quad (5.2.5)$$

$$= \sum_{r=1}^{q-1} q!^2 \binom{q}{r}^2 \left\{\|f\otimes_r f\|_{\mathfrak{H}^{\otimes 2q-2r}}^2 + \binom{2q-2r}{q-r}\|f\widetilde{\otimes}_r f\|_{\mathfrak{H}^{\otimes 2q-2r}}^2\right\}.$$
$$(5.2.6)$$

In particular,

$$\mathrm{Var}\left(\frac{1}{q}\|DF\|_{\mathfrak{H}}^2\right) \leq \frac{q-1}{3q}\left(E[F^4] - 3E[F^2]^2\right) \leq (q-1)\mathrm{Var}\left(\frac{1}{q}\|DF\|_{\mathfrak{H}}^2\right).$$
$$(5.2.7)$$

Remark 5.2.5 Since $E(F) = 0$, we have that $E[F^4] - 3E[F^2]^2 = \kappa_4(F)$. Relations (5.2.5) and (5.2.6) therefore imply the following non-trivial fact: for every random variable F belonging to some Wiener chaos,

$$E(F^4) - 3E(F^2)^2 = \kappa_4(F) \geq 0. \quad (5.2.8)$$

Proof of Lemma 5.2.4 We have $DF = qI_{q-1}(f)$, so that

$$\frac{1}{q}\|DF\|_{\mathfrak{H}}^2 = q\|I_{q-1}(f)\|_{\mathfrak{H}}^2$$

$$= q\sum_{r=0}^{q-1} r!\binom{q-1}{r}^2 I_{2q-2-2r}(f\widetilde{\otimes}_{r+1}f) \quad \text{(by (2.7.9))}$$

$$= q\sum_{r=1}^{q} (r-1)!\binom{q-1}{r-1}^2 I_{2q-2r}(f\widetilde{\otimes}_r f).$$

$$= q!\|f\|_{\mathfrak{H}^{\otimes q}}^2 + q\sum_{r=1}^{q-1} (r-1)!\binom{q-1}{r-1}^2 I_{2q-2r}(f\widetilde{\otimes}_r f).$$

Since $E[F^2] = q!\|f\|^2_{\mathfrak{H}^{\otimes q}}$, the proof of (5.2.2) is concluded. The identity (5.2.3) follows from (5.2.2) and the orthogonality properties of multiple integrals. We have $D^2 F = q(q-1)I_{q-2}(f)$. Hence, using the multiplication formula (2.7.9),

$$D^2 F \otimes_1 D^2 F = q^2(q-1)^2 \sum_{r=0}^{q-2} r! \binom{q-2}{r}^2 I_{2q-4-2r}\left(f \widetilde{\otimes}_{r+1} f\right)$$

$$= q^2(q-1)^2 \sum_{r=1}^{q-1} (r-1)! \binom{q-2}{r-1}^2 I_{2q-2-2r}\left(f \widetilde{\otimes}_r f\right).$$

Using the orthogonality and isometry properties of the integrals I_q, we get (5.2.4). Using (2.8.8), together with $D(F^3) = 3F^2 DF$, yields

$$E[F^4] = \frac{1}{q}E[\delta DF \times F^3] = \frac{1}{q}E[\langle DF, D(F^3)\rangle_{\mathfrak{H}}] = \frac{3}{q}E[F^2\|DF\|^2_{\mathfrak{H}}].$$
(5.2.9)

Moreover, we infer from the product formula (2.7.9) that

$$F^2 = I_q(f)^2 = \sum_{r=0}^{q} r! \binom{q}{r}^2 I_{2q-2r}(f \widetilde{\otimes}_r f).$$ (5.2.10)

Relation (5.2.5) is deduced by combining (5.2.10) with (5.2.2) and (5.2.9), and by using once again the orthogonality properties of multiple integrals to compute the expectation $E[F^2\|DF\|^2_{\mathfrak{H}}]$. Relation (5.2.7) follows by comparing (5.2.3) and (5.2.5).

It remains to prove (5.2.6). Without loss of generality, we can assume that $\mathfrak{H}$ is equal to $L^2(A, \mathcal{A}, \mu)$, where $(A, \mathcal{A})$ is a measurable space and μ is a σ-finite measure without atoms. Let σ be a permutation of $\{1, \ldots, 2q\}$ (this fact is written in symbols as $\sigma \in \mathfrak{S}_{2q}$). If $r \in \{0, \ldots, q\}$ denotes the cardinality of $\{\sigma(1), \ldots, \sigma(q)\} \cap \{1, \ldots, q\}$ then it is readily checked that r is also the cardinality of $\{\sigma(q+1), \ldots, \sigma(2q)\} \cap \{q+1, \ldots, 2q\}$ and that

$$\int_{A^{2q}} f(t_1, \ldots, t_q) f(t_{\sigma(1)}, \ldots, t_{\sigma(q)}) f(t_{q+1}, \ldots, t_{2q}) f(t_{\sigma(q+1)}, \ldots, t_{\sigma(2q)})$$

$$\times d\mu(t_1) \ldots d\mu(t_{2q}) = \int_{A^{2q-2r}} f \otimes_r f(x_1, \ldots, x_{2q-2r})^2 d\mu(x_1) \ldots d\mu(x_{2q-2r})$$

$$= \|f \otimes_r f\|^2_{\mathfrak{H}^{\otimes(2q-2r)}}.$$ (5.2.11)

Moreover, for any fixed $r \in \{0, \ldots, q\}$, there are $\binom{q}{r}^2 (q!)^2$ permutations $\sigma \in \mathfrak{S}_{2q}$ such that $\#\{\sigma(1), \ldots, \sigma(q)\} \cap \{1, \ldots, q\} = r$. (Indeed, such a permutation is completely determined by the choice of: (a) r distinct elements $y_1, \ldots, y_r$ of $\{1, \ldots, q\}$; (b) $q-r$ distinct elements $y_{r+1}, \ldots, y_q$ of

$\{q + 1, \ldots, 2q\}$; (c) a bijection between $\{1, \ldots, q\}$ and $\{y_1, \ldots, y_q\}$; (d) a bijection between $\{q + 1, \ldots, 2q\}$ and $\{1, \ldots, 2q\} \setminus \{y_1, \ldots, y_q\}$.) Now, observe that the symmetrization of $f \otimes f$ is given by

$$f \widetilde{\otimes} f(t_1, \ldots, t_{2q}) = \frac{1}{(2q)!} \sum_{\sigma \in \mathfrak{S}_{2q}} f(t_{\sigma(1)}, \ldots, t_{\sigma(q)}) f(t_{\sigma(q+1)}, \ldots, t_{\sigma(2q)}).$$

Therefore,

$$\|f \widetilde{\otimes} f\|^2_{\mathfrak{H}^{\otimes 2q}} = \frac{1}{(2q)!^2} \sum_{\sigma, \sigma' \in \mathfrak{S}_{2q}} \int_{A^{2q}} f(t_{\sigma(1)}, \ldots, t_{\sigma(q)}) f(t_{\sigma(q+1)}, \ldots, t_{\sigma(2q)})$$

$$\times f(t_{\sigma'(1)}, \ldots, t_{\sigma'(q)}) f(t_{\sigma'(q+1)}, \ldots, t_{\sigma'(2q)}) d\mu(t_1) \ldots d\mu(t_{2q})$$

$$= \frac{1}{(2q)!} \sum_{\sigma \in \mathfrak{S}_{2q}} \int_{A^{2q}} f(t_1, \ldots, t_q) f(t_{q+1}, \ldots, t_{2q})$$

$$\times f(t_{\sigma(1)}, \ldots, t_{\sigma(q)}) f(t_{\sigma(q+1)}, \ldots, t_{\sigma(2q)}) d\mu(t_1) \ldots d\mu(t_{2q})$$

$$= \frac{1}{(2q)!} \sum_{r=0}^{q} \sum_{\substack{\sigma \in \mathfrak{S}_{2q} \\ \{\sigma(1), \ldots, \sigma(q)\} \cap \{1, \ldots, q\} = r}} \int_{A^{2q}} f(t_1, \ldots, t_q) f(t_{q+1}, \ldots, t_{2q})$$

$$\times f(t_{\sigma(1)}, \ldots, t_{\sigma(q)}) f(t_{\sigma(q+1)}, \ldots, t_{\sigma(2q)}) d\mu(t_1) \ldots d\mu(t_{2q}).$$

Using (5.2.11), we deduce that

$$(2q)! \|f \widetilde{\otimes} f\|^2_{\mathfrak{H}^{\otimes 2q}} = 2(q!)^2 \|f\|^4_{\mathfrak{H}^{\otimes q}} + (q!)^2 \sum_{r=1}^{q-1} \binom{q}{r}^2 \|f \otimes_r f\|^2_{\mathfrak{H}^{\otimes(2q-2r)}}.$$

$$(5.2.12)$$

Using the orthogonality and isometry properties of the integrals I_q, the identity (5.2.10) yields

$$E[F^4] = \sum_{r=0}^{q} (r!)^2 \binom{q}{r}^4 (2q - 2r)! \|f \widetilde{\otimes}_r f\|^2_{\mathfrak{H}^{\otimes(2q-2r)}}$$

$$= (2q)! \|f \widetilde{\otimes} f\|^2_{\mathfrak{H}^{\otimes(2q)}} + (q!)^2 \|f\|^4_{\mathfrak{H}^{\otimes q}}$$

$$+ \sum_{r=1}^{q-1} (r!)^2 \binom{q}{r}^4 (2q - 2r)! \|f \widetilde{\otimes}_r f\|^2_{\mathfrak{H}^{\otimes(2q-2r)}}.$$

By inserting (5.2.12) in the previous identity (and because $(q!)^2 \|f\|^4_{\mathfrak{H}^{\otimes q}} = E[F^2]^2$), we get (5.2.6). $\square$

By combining Lemma 5.2.4 with Theorem 5.1.3, we are finally able to explicitly relate norms of Malliavin operators with moments and cumulants, in order to deduce bounds on normal approximations inside a Wiener chaos.

Theorem 5.2.6 *Let $q \geq 2$ be an integer, and let $F = I_q(f)$ have the form of a multiple integral of order q such that $E(F^2) = \sigma^2 > 0$. Then, for $N \sim \mathcal{N}(0, \sigma^2)$,*

$$d_{\mathrm{TV}}(F, N) \leq \frac{2}{\sigma^2}\sqrt{\mathrm{Var}\left(\frac{1}{q}\|DF\|_{\mathfrak{H}}^2\right)} \leq \frac{2}{\sigma^2}\sqrt{\frac{q-1}{3q}[E[F^4] - 3\sigma^4]};$$

$$(5.2.13)$$

$$d_{\mathrm{Kol}}(F, N) \leq \frac{1}{\sigma^2}\sqrt{\mathrm{Var}\left(\frac{1}{q}\|DF\|_{\mathfrak{H}}^2\right)} \leq \frac{1}{\sigma^2}\sqrt{\frac{q-1}{3q}[E[F^4] - 3\sigma^4]};$$

$$(5.2.14)$$

$$d_{\mathrm{W}}(F, N) \leq \frac{1}{\sigma}\sqrt{\mathrm{Var}\left(\frac{2}{q\pi}\|DF\|_{\mathfrak{H}}^2\right)} \leq \frac{1}{\sigma}\sqrt{\frac{2q-2}{3\pi q}[E[F^4] - 3\sigma^4]}.$$

$$(5.2.15)$$

Proof Since $L^{-1}F = -\frac{1}{q}F$, we have $\langle DF, -DL^{-1}F \rangle_{\mathfrak{H}} = \frac{1}{q}\|DF\|_{\mathfrak{H}}^2$. Moreover, it follows from (5.2.2) that $\frac{1}{q}E[\|DF\|_{\mathfrak{H}}^2] = E[F^2]$. Hence, since F has a density by Theorem 2.10.1, Theorem 5.1.3 yields

$$d_{\mathrm{TV}}(F, N) \leq \frac{2}{\sigma^2}E\left[\left|\frac{1}{q}\|DF\|_{\mathfrak{H}}^2 - \frac{1}{q}E[\|DF\|_{\mathfrak{H}}^2]\right|\right].$$

By Cauchy–Schwarz, we have

$$E\left[\left|\frac{1}{q}\|DF\|_{\mathfrak{H}}^2 - \frac{1}{q}E[\|DF\|_{\mathfrak{H}}^2]\right|\right] \leq \sqrt{\mathrm{Var}\left(\frac{1}{q}\|DF\|_{\mathfrak{H}}^2\right)}.$$

Inequality (5.2.7) allows us to conclude that (5.2.13) holds. The proofs of (5.2.14) and (5.2.15) are similar. $\qquad\square$

A fundamental consequence of the bounds (5.2.13)–(5.2.15) is the following simplification of the method of moments and cumulants, as stated in Proposition 5.2.2.

Theorem 5.2.7 (Fourth-moment theorem) *Let $F_n = I_q(f_n)$, $n \geq 1$, be a sequence of random variables belonging to the qth chaos of X, for some fixed integer $q \geq 2$ (so that $f_n \in \mathfrak{H}^{\odot q}$). Assume, moreover, that $E[F_n^2] \to \sigma^2 > 0$ as $n \to \infty$. Then, as $n \to \infty$, the following five assertions are equivalent:*

(i) F_n *converges in distribution to* $N \sim \mathscr{N}(0, \sigma^2)$.

(ii) $E[F_n^4] \to 3\sigma^4 = E(N^4)$ *(or, equivalently,* $\kappa_4(F_n) \to 0$).

(iii) $\mathrm{Var}\left(\|DF_n\|_{\mathfrak{H}}^2\right) \to 0$.

(iv) $\|f_n \widetilde{\otimes}_r f_n\|_{\mathfrak{H}^{\otimes(2q-2r)}} \to 0$, *for all* $r = 1, \ldots, q-1$.

(v) $\|f_n \otimes_r f_n\|_{\mathfrak{H}^{\otimes(2q-2r)}} \to 0$, *for all* $r = 1, \ldots, q-1$.

Proof Since $E[F_n^2] \to \sigma^2$, we have by Theorem 2.7.2 that, for every real $r \geq 2$, $\sup_n E|F_n|^r < \infty$. This yields immediately that, if (i) holds, then $E(F_n^j) \to E(N^j)$, for every $j \geq 3$, and therefore that (i) implies (ii). That (ii) implies (iii) follows from (5.2.7), whereas the implication (iii)→(iv) is a consequence of (5.2.3). The implication (iv)→(v) comes from (5.2.5) and (5.2.6). The fact that (v) implies (i) follows from the inequality $\|f_n \widetilde{\otimes} f_n\|_{\mathfrak{H}^{\otimes(2n-2r)}} \leq \|f_n \otimes f_n\|_{\mathfrak{H}^{\otimes(2n-2r)}}$, combined with (5.2.3) and Theorem 5.2.6. $\qquad\square$

In subsequent parts of the book, we shall study many consequences and generalizations of Theorems 5.2.6 and 5.2.7. To conclude this section, we present three immediate interesting corollaries. The first shows that, inside a Wiener chaos, the total variation and Wasserstein distances metrize the convergence towards a Gaussian distribution.

Corollary 5.2.8 *Let the assumptions of Theorem 5.2.7 prevail. As* $n \to \infty$, *the following assertions are equivalent:*

1. $F_n \xrightarrow{\text{Law}} N \sim \mathscr{N}(0, \sigma^2)$.
2. $d_{\text{TV}}(F_n, N) \to 0$.
3. $d_{\text{W}}(F_n, N) \to 0$.

Proof It suffices to prove that condition 1 in the statement implies both conditions 2 and 3. Assume that condition 1 holds. By Theorem 5.2.7, we have that $E[F_n^4] - 3\sigma^4 \to 0$. The estimates in Theorem 5.2.6 show that conditions 2 and 3 are therefore necessarily satisfied. $\qquad\square$

Remark 5.2.9 Since the law of N is absolutely continuous, we always have that, if F_n convergence to N in distribution, then $d_{\text{Kol}}(F_n, N) \to 0$ (see Proposition C.3.2).

The next result is a 'simplified method of moments' for sequences of chaotic random variables with possibly different orders.

Corollary 5.2.10 *Let* $q(n)$, $n \geq 1$, *be a sequence of natural numbers such that* $q(n) \geq 1$. *Let* $F_n = I_{q(n)}(f_n)$, $n \geq 1$, *be a sequence of multiple integrals, with* $f_n \in \mathfrak{H}^{\odot q(n)}$. *Assume that, as* $n \to \infty$, $E(F_n^2) \to \sigma^2 > 0$ *and*

$E(F_n^4) \to 3\sigma^4$. Then, as $n \to \infty$, F_n converges to $N \sim \mathcal{N}(0, \sigma^2)$ in the sense of both the total variation and Wasserstein distances.

Proof This is an immediate consequence of the following estimates, which are easily deduced from Proposition 3.6.1 and Theorem 5.2.6 (the case $q(n) = 1$ must be treated separately):

$$d_{\mathrm{TV}}(F_n, N) \leq 2\sqrt{\frac{E[F_n^4] - 3E[F_n^2]^2}{3E[F_n^2]^2}} + \frac{2\left|E[F_n^2] - \sigma^2\right|}{E[F_n^2] \vee \sigma^2};$$

$$d_{\mathrm{W}}(F_n, N) \leq \sqrt{\frac{2\{E[F_n^4] - 3E[F_n^2]^2\}}{3\pi E[F_n^2]}} + \frac{\sqrt{\frac{2}{\pi}}\left|E[F_n^2] - \sigma^2\right|}{\sqrt{E[F_n^2]} \vee \sigma}. \qquad \square$$

The last consequence of Theorem 5.2.7 discussed in this section is a simple proof of the fact that Wiener chaoses of order greater than or equal to 2 do not contain Gaussian random variables.

Corollary 5.2.11 *Fix $q \geq 2$ and $f \in \mathfrak{H}^{\odot q}$ such that $E[I_q(f)^2] = \sigma^2 > 0$. Then $E[I_q(f)^4] > 3\sigma^4$; in particular, the distribution of $I_q(f)$ cannot be normal.*

Proof We already know that $E[I_q(f)^4] \geq 3\sigma^4$ (see Remark 5.2.5). Assume that $E[I_q(f)^4] = 3\sigma^4$. From (5.2.5), we deduce that $\|f \widetilde{\otimes}_{q-1} f\|_{\mathfrak{H}^{\otimes 2}} = 0$. On the other hand, by (B.4.5) (we follow here the same notation; in particular, $\{e_j\}$ stands for any orthonormal basis of $\mathfrak{H}$), we have

$$f \otimes_{q-1} f = \sum_{l_1, \dots, l_{q-1}, i, j=1}^{\infty} \langle f, e_{l_1} \otimes \dots \otimes e_{l_{q-1}} \otimes e_i \rangle_{\mathfrak{H}^{\otimes q}}$$
$$\times \langle f, e_{l_1} \otimes \dots \otimes e_{l_{q-1}} \otimes e_j \rangle_{\mathfrak{H}^{\otimes q}} e_i \otimes e_j.$$

In particular, observe that $f \otimes_{q-1} f$ is symmetric. Therefore,

$$0 = \|f \widetilde{\otimes}_{q-1} f\|_{\mathfrak{H}^{\otimes 2}}^2 = \|f \otimes_{q-1} f\|_{\mathfrak{H}^{\otimes 2}}^2$$

$$= \sum_{i,j=1}^{\infty} \left(\sum_{l_1, \dots, l_{q-1}=1}^{\infty} \langle f, e_{l_1} \otimes \dots \otimes e_{l_{q-1}} \otimes e_i \rangle_{\mathfrak{H}^{\otimes q}} \langle f, e_{l_1} \otimes \dots \otimes e_{l_{q-1}} \otimes e_j \rangle_{\mathfrak{H}^{\otimes q}} \right)^2$$

$$\geq \sum_{i=1}^{\infty} \left(\sum_{l_1, \dots, l_{q-1}=1}^{\infty} \langle f, e_{l_1} \otimes \dots \otimes e_{l_{q-1}} \otimes e_i \rangle_{\mathfrak{H}^{\otimes q}}^2 \right)^2.$$

Hence, $\langle f, e_{l_1} \otimes \dots \otimes e_{l_{q-1}} \otimes e_i \rangle_{\mathfrak{H}^{\otimes q}} = 0$ for all $i, l_1, \dots, l_{q-1} \geq 1$. This leads to $f = 0$, which is in contradiction to $E[I_q(f)^2] > 0$. Consequently, $E[I_q(f)^4] > 3\sigma^4$ and the distribution of $I_q(f)$ cannot be normal. $\qquad \square$

Remark 5.2.12   Fix an integer $q \geq 2$. As a by-product of Corollary 5.2.11 and of the fact that the qth Wiener chaos $\mathscr{H}_q$ is closed for the $L^2(\Omega)$-norm by definition, we see that it is not possible for a sequence $\{I_q(f_n) : n \geq 1\}$ of $\mathscr{H}_q$ to converge in $L^2(\Omega)$ towards a non-zero Gaussian random variable. We also recall that, according to a result by Schreiber [130], inside a finite sum of Wiener chaoses and for every $p \geq 1$, the topology induced by the convergence in $L^p(\Omega)$ (on the class of the probability measures on the real line) is equivalent to the topology induced by the convergence in probability. In particular, this proves that a sequence of random variables inside a fixed Wiener chaos cannot converge in probability to a non-zero Gaussian random variable. See also Corollary 2.8.14.

5.3 Normal approximations in the general case

5.3.1 Main results

We now focus on some criteria for the asymptotic normality of random variables having a possibly infinite chaos decomposition. Further findings in this direction (based on multidimensional normal approximations) will be discussed in Section 6.3. Our first result is a direct consequence of Theorem 5.1.3.

Theorem 5.3.1   *Let $\{F_n : n \geq 1\}$ be a sequence in $\mathbb{D}^{1,2}$ such that $E[F_n] = 0$ for all n. Assume that $E[F_n^2] \to \sigma^2 > 0$ as $n \to \infty$. If*

$$\langle DF_n, -DL^{-1}F_n \rangle_{\mathfrak{H}} \overset{L^1(\Omega)}{\longrightarrow} \sigma^2 \quad \text{as } n \to \infty, \qquad (5.3.1)$$

then $F_n \overset{\text{Law}}{\to} \mathscr{N}(0, \sigma^2)$ as $n \to \infty$.

Remark 5.3.2   If each F_n has the specific form of a multiple integral of order $q \geq 2$ then the condition (5.3.1) turns out to be also necessary for asymptotic normality, since in this case $\langle DF_n, -DL^{-1}F_n \rangle_{\mathfrak{H}} = \frac{1}{q} \|DF_n\|_{\mathfrak{H}}^2$ so that, by virtue of Theorem 5.2.7, convergence (5.3.1) is equivalent to the convergence of the fourth cumulant of F_n towards zero. Note that, if each F_n belongs to the qth Wiener chaos, then the sequence $\langle DF_n, -DL^{-1}F_n \rangle_{\mathfrak{H}}$, $n \geq 1$, lives inside a finite sum of Wiener chaoses. By Remark 2.8.15, therefore, $\langle DF_n, -DL^{-1}F_n \rangle_{\mathfrak{H}} \to \sigma^2$ in $L^1(\Omega)$ if and only if $\langle DF_n, -DL^{-1}F_n \rangle_{\mathfrak{H}} \to \sigma^2$ in $L^2(\Omega)$.

By assuming that the sequence $\{F_n : n \geq 1\}$ is contained in $\mathbb{D}^{2,4}$, we can obtain further (quite useful) sufficient conditions for asymptotic normality. A pivotal role is played by the following result, containing an infinite-dimensional analog of the second-order Poincaré inequality appearing in Proposition 1.6.1. The proof is deferred to Section 5.3.2.

Theorem 5.3.3 (Second-order Poincaré inequalities) *Let F be a centered element of* $\mathbb{D}^{2,4}$ *such that* $E[F^2] = \sigma^2 > 0$, *and let* $N \sim \mathcal{N}(0, \sigma^2)$. *Assume F has a density. Then the following estimate holds:*

$$d_{\mathrm{TV}}(F, N) \leq \frac{3}{\sigma^2} E[\|D^2 F\|_{\mathrm{op}}^4]^{1/4} \times E[\|D^2 F\|_{\mathfrak{H}}^4]^{1/4}, \tag{5.3.2}$$

where $\|D^2 F\|_{\mathrm{op}}$ *indicates the operator norm of the (random) Hilbert–Schmidt operator*

$$\mathfrak{H} \to \mathfrak{H} : f \mapsto \left\langle f, D^2 F \right\rangle_{\mathfrak{H}}.$$

Moreover,

$$E[\|D^2 F\|_{\mathrm{op}}^4] \leq E[\|D^2 F \otimes_1 D^2 F\|_{\mathfrak{H}^{\otimes 2}}^2]. \tag{5.3.3}$$

Exercise 5.3.4 Prove bounds analogous to (5.3.2) and (5.3.3) for the Wasserstein distance $d_{\mathrm{W}}(F, N)$, and recover Proposition 1.6.1.

Theorem 5.3.3 yields the following set of sufficient conditions for asymptotic normality. (See also Theorem 6.3.1 for another interesting criterion.)

Proposition 5.3.5 *Let* $(F_n)_{n \geq 1}$ *be a sequence in* $\mathbb{D}^{2,4}$ *such that* $E[F_n] = 0$ *and* F_n *has a density for all n. Suppose that:*

(i) $E[F_n^2] \to \sigma^2 > 0$ *as* $n \to \infty$;
(ii) $\sup_{n \geq 1} E[\|D F_n\|_{\mathfrak{H}}^4] < \infty$;
(iii) $E[\|D^2 F_n\|_{\mathrm{op}}^4] \to 0$ *as* $n \to \infty$.

Then $F_n \overset{\mathrm{Law}}{\to} N \sim \mathcal{N}(0, \sigma^2)$, *as* $n \to \infty$, *in the sense of the total variation distance. Moreover, a sufficient condition for* (iii) *is that* $E[\|D^2 F \otimes_1 D^2 F\|_{\mathfrak{H}^{\otimes 2}}^2] \to 0$, *as* $n \to \infty$.

Proof Let $N_n \sim \mathcal{N}(0, \sigma_n^2)$, where $\sigma_n^2 = E[F_n^2]$. If conditions (i)–(iii) are satisfied, then (5.3.2) implies immediately that $d_{\mathrm{TV}}(F_n, N_n) \to 0$. Since $d_{\mathrm{TV}}(N, N_n) \to 0$ (use, for example, Proposition 3.6.1), the conclusion is obtained by writing

$$d_{\mathrm{TV}}(F_n, N) \leq d_{\mathrm{TV}}(F_n, N_n) + d_{\mathrm{TV}}(N, N_n).$$

The last part of the statement follows from inequality (5.3.3). □

As an application of Proposition 5.3.5, we deduce the following refinement of Theorem 5.2.7.

Proposition 5.3.6 *Let the notation and assumptions of Theorem 5.2.7 prevail. Then, each of conditions* (i)–(v) *is equivalent to either of the following:*

(vi) $E[\|D^2 F_n \otimes_1 D^2 F_n\|^2_{\mathfrak{H}^{\otimes 2}}] \to 0$;

(vii) $E[\|D^2 F_n\|^4_{op}] \to 0$.

Proof We shall prove that (v) $\to$ (vi) $\to$ (vii) $\to$ (i). The fact that (v) implies (vi) follows from (5.2.4), while (5.3.3) yields that (vi) $\to$ (vii). Suppose that (vii) is satisfied. Because $E[\|DF_n\|^2_{\mathfrak{H}}] = qE[F_n^2] \to q\sigma^2$, the sequence $E[\|DF_n\|^2_{\mathfrak{H}}]$, $n \geq 1$, is bounded. Moreover, the random variables $\|DF_n\|^2_{\mathfrak{H}}$ live inside a *finite* sum of Wiener chaoses. Hence, Theorem 2.7.2 implies that the sequence $E[\|DF_n\|^4_{\mathfrak{H}}]$, $n \geq 1$, is bounded as well. Since convergence in total variation implies convergence in distribution, Proposition 5.3.5 implies therefore that (i) holds.   □

Another application of Theorem 5.3.3 is the object of Exercise 5.4.5.

5.3.2 Proof of Theorem 5.3.3

We first state a lemma containing, among other things, a general version (see (5.3.6)) of the infinite-dimensional Poincaré inequality shown in Exercise 2.11.1.

Lemma 5.3.7 *Fix $p \geq 2$ and let $F \in \mathbb{D}^{1,p}$.*

1. *The following estimate holds:*

$$E\left\|DL^{-1}F\right\|^p_{\mathfrak{H}} \leq E\|DF\|^p_{\mathfrak{H}}. \qquad (5.3.4)$$

2. *If in addition $F \in \mathbb{D}^{2,p}$, then*

$$E\left\|D^2 L^{-1}F\right\|^p_{op} \leq \frac{1}{2^p} E\left\|D^2 F\right\|^p_{op}, \qquad (5.3.5)$$

where $\left\|D^2 F\right\|_{op}$ denotes the operator norm of the random Hilbert–Schmidt operator $\mathfrak{H} \to \mathfrak{H}: f \mapsto \langle f, D^2 F \rangle_{\mathfrak{H}}$ (and similarly for $\|D^2 L^{-1}F\|_{op}$).

3. *If p is an even integer, then*

$$E\left[(F - E[F])^p\right] \leq (p-1)^{p/2} E\left[\|DF\|^p_{\mathfrak{H}}\right]. \qquad (5.3.6)$$

Proof 1. In what follows, we denote by X' an independent copy of X, and we will sometimes write $DF = \Phi_{DF}(X)$, for a measurable function $\Phi_{DF}: \mathbb{R}^{\mathfrak{H}} \to \mathfrak{H}$, whenever the underlying isonormal Gaussian process X plays a role. Also, for a generic random element G, we write E_G to indicate that we are taking the expectation with respect to G. We can write

$$-DL^{-1}F = \int_0^\infty e^{-t} P_t DF \, dt \quad \text{(by (2.9.2))}$$

$$= \int_0^\infty e^{-t} E_{X'} \left[\Phi_{DF} \left(e^{-t} X + \sqrt{1 - e^{-2t}} X' \right) \right] dt$$

(via Mehler's formula (2.8.1))

$$= E_Y \left[E_{X'} \left[\Phi_{DF} \left(e^{-Y} X + \sqrt{1 - e^{-2Y}} X' \right) \right] \right],$$

where $Y \sim \mathcal{E}(1)$ is an independent exponential random variable with mean 1. It follows that

$$E\left[\left\| DL^{-1}F \right\|_{\mathfrak{H}}^p \right] = E_X \left[\left\| E_Y \left[E_{X'} \left[\Phi_{DF} \left(e^{-Y} X + \sqrt{1 - e^{-2Y}} X' \right) \right] \right] \right\|_{\mathfrak{H}}^p \right]$$

$$\leq E_X \left[E_Y \left[E_{X'} \left[\left\| \Phi_{DF} \left(e^{-Y} X + \sqrt{1 - e^{-2Y}} X' \right) \right\|_{\mathfrak{H}}^p \right] \right] \right]$$

$$= E_Y \left[E_X \left[E_{X'} \left[\left\| \Phi_{DF} \left(e^{-Y} X + \sqrt{1 - e^{-2Y}} X' \right) \right\|_{\mathfrak{H}}^p \right] \right] \right]$$

$$= E_Y \left[E_X \left[\|DF\|_{\mathfrak{H}}^p \right] \right] = E\left[\|DF\|_{\mathfrak{H}}^p \right],$$

using the fact that $e^{-t} X' + \sqrt{1 - e^{-2t}} X \overset{\text{Law}}{=} X$ for any $t \geq 0$.

2. By proceeding as in the proof of (2.9.2), we can show that $-D^2 L^{-1} F = \int_0^\infty e^{-2t} P_t D^2 F \, dt$. Hence, this time writing $D^2 F = \Psi_{D^2 F}(X)$ for a measurable function $\Psi_{D^2 F} : \mathbb{R}^{\mathfrak{H}} \to \mathfrak{H}^{\otimes 2}$, we have

$$-D^2 L^{-1} F = \int_0^\infty e^{-2t} P_t D^2 F \, dt$$

$$= \int_0^\infty e^{-2t} E_{X'} \left[\Psi_{D^2 F} \left(e^{-t} X + \sqrt{1 - e^{-2t}} X' \right) \right] dt \quad \text{(by (2.8.1))}$$

$$= \frac{1}{2} E_Z \left[E_{X'} \left[\Psi_{D^2 F} \left(e^{-Z} X + \sqrt{1 - e^{-2Z}} X' \right) \right] \right],$$

where $Z \sim \mathcal{E}(2)$ is an independent exponential random variable with mean $\frac{1}{2}$. Thus

$$E\left[\left\| D^2 L^{-1} F \right\|_{\text{op}}^p \right]$$

$$= \frac{1}{2^p} E_X \left[\left\| E_Z \left[E_{X'} \left[\Psi_{D^2 F} \left(e^{-Z} X + \sqrt{1 - e^{-2Z}} X' \right) \right] \right] \right\|_{\text{op}}^p \right]$$

$$\leq \frac{1}{2^p} E_X \left[E_Z \left[E_{X'} \left[\left\| \Psi_{D^2 F} \left(e^{-Z} X + \sqrt{1 - e^{-2Z}} X' \right) \right\|_{\text{op}}^p \right] \right] \right]$$

$$= \frac{1}{2^p} E_Z \left[E_X \left[E_{X'} \left[\left\| \Psi_{D^2 F} \left(e^{-Z} X + \sqrt{1 - e^{-2Z}} X' \right) \right\|_{\mathrm{op}}^p \right] \right] \right]$$

$$= \frac{1}{2^p} E_Z \left[E_X \left[\left\| \Psi_{D^2 F} (X) \right\|_{\mathrm{op}}^p \right] \right] = \frac{1}{2^p} E \left[\left\| D^2 F \right\|_{\mathrm{op}}^p \right].$$

3. Without loss of generality, we can assume that $E[F] = 0$. Writing $p = 2k$, we have

$$
\begin{aligned}
E[F^{2k}] &= E[F \times F^{2k-1}] \\
&= (2k-1) E[\langle DF, -DL^{-1}F \rangle_{\mathfrak{H}} F^{2k-2}] \quad \text{(by (2.9.1))} \\
&\le (2k-1) \left(E[|\langle DF, -DL^{-1}F \rangle_{\mathfrak{H}}|^k] \right)^{\frac{1}{k}} \left(E[F^{2k}] \right)^{1-\frac{1}{k}} \quad \text{(by Hölder)},
\end{aligned}
$$

from which we infer that

$$
\begin{aligned}
&E[F^{2k}] \\
&\le (2k-1)^k E[|\langle DF, -DL^{-1}F \rangle_{\mathfrak{H}}|^k] \le (2k-1)^k E[\|DF\|_{\mathfrak{H}}^k \|DL^{-1}F\|_{\mathfrak{H}}^k] \\
&\le (2k-1)^k \sqrt{E[\|DF\|_{\mathfrak{H}}^{2k}]} \sqrt{E[\|DL^{-1}F\|_{\mathfrak{H}}^{2k}]} \le (2k-1)^k E[\|DF\|_{\mathfrak{H}}^{2k}]. \quad \square
\end{aligned}
$$

The next two technical results will also be useful in the proof of Theorem 5.3.3.

Lemma 5.3.8 *Let* $F, G \in \mathbb{D}^{2,4}$. *Then, the two random elements* $\langle D^2 F, DG \rangle_{\mathfrak{H}}$ *and* $\langle DF, D^2 G \rangle_{\mathfrak{H}}$ *belong to* $L^2(\Omega, \mathfrak{H})$. *Moreover,* $\langle DF, DG \rangle_{\mathfrak{H}}$ *belongs to* $\mathbb{D}^{1,2}$ *and*

$$D\langle DF, DG \rangle_{\mathfrak{H}} = \langle D^2 F, DG \rangle_{\mathfrak{H}} + \langle DF, D^2 G \rangle_{\mathfrak{H}}. \tag{5.3.7}$$

Proof As usual, we associate the random Hilbert–Schmidt operator $f \mapsto \langle f, D^2 F \rangle_{\mathfrak{H}^{\otimes 2}}$ to the (symmetric) random kernel $D^2 F$ of $\mathfrak{H}^{\odot 2}$. Denote by $(\gamma_j)_{j \ge 1}$ the sequence of its (random) eigenvalues. We have that

$$\left\| D^2 F \right\|_{\mathrm{op}}^2 = \max_{j \ge 1} |\gamma_j|^2 \le \sum_{j=1}^{\infty} |\gamma_j|^2 = \left\| D^2 F \right\|_{\mathfrak{H}^{\otimes 2}}^2.$$

Hence,

$$
\begin{aligned}
E[\langle D^2 F, DG \rangle_{\mathfrak{H}}^2] &\le E[\|D^2 F\|_{\mathrm{op}}^2 \|DG\|_{\mathfrak{H}}^2] \le E[\|D^2 F\|_{\mathfrak{H}^{\otimes 2}}^2 \|DG\|_{\mathfrak{H}}^2] \\
&\le \sqrt{E[\|D^2 F\|_{\mathfrak{H}^{\otimes 2}}^4]} \sqrt{E[\|DG\|_{\mathfrak{H}}^4]} < \infty,
\end{aligned}
$$

and similarly $E\big[\langle D^2G, DF\rangle_{\mathfrak{H}}^2\big] < \infty$. By a standard approximation argument (approximating F, G by smooth functionals), we finally deduce that $\langle DF, DG\rangle_{\mathfrak{H}}^2$ belongs to $\mathbb{D}^{1,2}$ with Malliavin derivative given by (5.3.7). □

Lemma 5.3.9 **(Random contraction inequality)** *Let* $F \in \mathbb{D}^{2,4}$. *Then*

$$\left\|D^2F\right\|_{\text{op}}^4 \leq \left\|D^2F \otimes_1 D^2F\right\|_{\mathfrak{H}^{\otimes 2}}^2. \tag{5.3.8}$$

Proof As in the previous proof, we denote by $(\gamma_j)_{j\geq 1}$ the sequence of the random eigenvalues of the Hilbert–Schmidt operator associated with $D^2F \in \mathfrak{H}^{\odot 2}$. The desired conclusion follows from

$$\left\|D^2F\right\|_{\text{op}}^4 = \max_{j\geq 1} |\gamma_j|^4 \leq \sum_{j=1}^{\infty} |\gamma_j|^4 = \left\|D^2F \otimes_1 D^2F\right\|_{\mathfrak{H}^{\otimes 2}}^2. \qquad □$$

By combining the previous results, we obtain the following statement:

Proposition 5.3.10 *Let* $F, G \in \mathbb{D}^{2,4}$ *be such that* $E[F] = E[G] = 0$. *Then*

$$E\big[\big|E[FG] - \langle DF, -DL^{-1}G\rangle_{\mathfrak{H}}\big|\big] \leq E[\|D^2F\|_{\text{op}}^4]^{1/4} E[\|DG\|_{\mathfrak{H}}^4]^{1/4}$$
$$+ \frac{1}{2} E[\|D^2G\|_{\text{op}}^4]^{1/4} E[\|DF\|_{\mathfrak{H}}^4]^{1/4}. \tag{5.3.9}$$

Proof Set $W = \langle DF, -DL^{-1}G\rangle_{\mathfrak{H}}$ and notice first that, by (2.9.1),

$$E[W] = E[\langle DF, -DL^{-1}G\rangle_{\mathfrak{H}}] = E[FG].$$

By (5.3.6) with $p = 2$, we have $E[|W - E[W]|] \leq \sqrt{\text{Var}(W)} \leq \sqrt{E[\|DW\|_{\mathfrak{H}}^2]}$. So, our problem is now to evaluate $\sqrt{E[\|DW\|_{\mathfrak{H}}^2]}$. By using Lemma 5.3.8 and the triangle inequality, we deduce that

$$\sqrt{E[\|DW\|_{\mathfrak{H}}^2]}$$
$$= \sqrt{E\big[\big\|\langle D^2F, -DL^{-1}G\rangle_{\mathfrak{H}} + \langle DF, -D^2L^{-1}G\rangle_{\mathfrak{H}}\big\|_{\mathfrak{H}}^2\big]}$$
$$\leq \sqrt{E\big[\big\|\langle D^2F, -DL^{-1}G\rangle_{\mathfrak{H}}\big\|_{\mathfrak{H}}^2\big]} + \sqrt{E\big[\big\|\langle DF, -D^2L^{-1}G\rangle_{\mathfrak{H}}\big\|_{\mathfrak{H}}^2\big]}.$$

We have

$$\left\|\langle D^2F, -DL^{-1}G\rangle_{\mathfrak{H}}\right\|_{\mathfrak{H}}^2 \leq \left\|D^2F\right\|_{\text{op}}^2 \left\|DL^{-1}G\right\|_{\mathfrak{H}}^2,$$
$$\left\|\langle DF, -D^2L^{-1}G\rangle_{\mathfrak{H}}\right\|_{\mathfrak{H}}^2 \leq \|DF\|_{\mathfrak{H}}^2 \left\|D^2L^{-1}G\right\|_{\text{op}}^2.$$

It follows from Cauchy–Schwarz that

$$\sqrt{E\big[\,\|DW\|_{\mathfrak{H}}^2\,\big]} \le \left(E \left\| DL^{-1}G \right\|_{\mathfrak{H}}^4 \times E \left\| D^2 F \right\|_{\text{op}}^4 \right)^{1/4}$$
$$+ \left(E \, \|DF\|_{\mathfrak{H}}^4 \times E \left\| D^2 L^{-1} G \right\|_{\text{op}}^4 \right)^{1/4}.$$

The desired conclusion follows by using (5.3.4) and (5.3.5), respectively, with $p = 4$.   □

Conclusion of the proof of Theorem 5.3.3 Recall the estimate (5.1.4):

$$d_{\text{TV}}(F, N) \le \frac{2}{\sigma^2} \, E\big[\,|\sigma^2 - \langle DF, -DL^{-1}F\rangle_{\mathfrak{H}}|\,\big].$$

Apply (5.3.9) with $F = G$ to deduce (5.3.2), and observe that (5.3.8) is nothing more than (5.3.3).   □

5.4 Exercises

5.4.1 Prove the equivalence between conditions (i), (ii), (iv) and (v), in the statement of Theorem 5.2.7 in the case $q = 2$, by using formula (2.7.17).

5.4.2 The following argument is taken from Nualart and Ortiz-Latorre [100]. Let $q \ge 2$, and let $F_n = I_q(f_n)$ be a sequence of chaotic random variables such that $E[F_n^2] = 1$ for all n. Denote by $\phi_n(\lambda) = E[e^{i\lambda F_n}]$ the characteristic function of F_n.

1. Show that, for every real λ, $\phi_n'(\lambda) = \frac{i\lambda}{q} E[e^{i\lambda F_n} \|DF_n\|_{\mathfrak{H}}^2]$.
2. Prove the implication (iii) → (i) in Theorem 5.2.7 by using part 1, as well as the fact that the mapping $t \mapsto E[e^{itN}]$, $N \sim \mathscr{N}(0, 1)$, is the unique solution to the differential equation $\phi'(t) + t\phi(t) = 0$, $\phi(0) = 1$.

5.4.3 **(Cramér type result)** The following argument is taken from Tudor [147]. Let $F_n = I_{p_n}(f_n)$ and $G_n = I_{q_n}(g_n)$, $n \ge 1$, be two sequences of independent multiple integrals, with $p_n, q_n \ge 1$, $f_n \in \mathfrak{H}^{\odot p_n}$ and $g_n \in \mathfrak{H}^{\odot q_n}$. Assume that $\lim_{n\to\infty} E[F_n^2] = \alpha^2 > 0$, $\lim_{n\to\infty} E[G_n^2] = \beta^2 > 0$ and $F_n + G_n$ converges in law to $\mathscr{N}(0, \alpha^2 + \beta^2)$ as $n \to \infty$. By combining (5.2.13) with $\kappa_4(F_n + G_n) = \kappa_4(F_n) + \kappa_4(G_n)$ (why?), $\kappa_4(F_n) \ge 0$ and $\kappa_4(G_n) \ge 0$ (see (5.2.8)), prove that both $d_{\text{TV}}\big(F_n, \mathscr{N}(0, \alpha^2)\big)$ and $d_{\text{TV}}\big(G_n, \mathscr{N}(0, \beta^2)\big)$ tend to zero as $n \to \infty$.

5.4.4 **(Parameter estimation for Ornstein–Uhlenbeck processes)** The following arguments are taken from Hu and Nualart [52]. Let B be a standard Brownian motion, and consider the Ornstein–Uhlenbeck process X associated with B, defined as the unique solution to the Langevin equation:

$$X_t = -\theta \int_0^t X_s ds + \sigma B_t, \quad t \in [0, T].$$

If the parameter $\theta \in \mathbb{R}$ is unknown and if the process X can be observed continuously, then an important problem is to estimate the parameter θ based on these observations. Let $\widehat{\theta}_T$ be the estimator given by $\widehat{\theta}_T = -\int_0^T X_t dX_t / \int_0^T X_t^2 dt$.

Our first goal is to show that $\widehat{\theta}_T$ is strongly consistent, that is, that $\widehat{\theta}_T \xrightarrow{\text{a.s.}} \theta$ as $T \to \infty$.

1. Prove that

$$\widehat{\theta}_T = \theta - \frac{\frac{\sigma}{T} \int_0^T X_t dB_t}{\frac{1}{T} \int_0^T X_t^2 dt}.$$

2. Show that $X_t = Y_t - Y_0 e^{-\theta t}$ with $Y_t = \sigma \int_{-\infty}^t e^{-\theta(t-s)} dB_s$.

3. Verify that Y is a centered stationary Gaussian process on $\mathbb{R}_+$, and deduce that Y^2 is a strictly stationary process.

4. By using the ergodic theorem, show that $\frac{1}{T} \int_0^T Y_t^2 dt \xrightarrow{\text{a.s.}} \frac{\sigma^2}{2\theta}$ as $T \to \infty$. Deduce that $\frac{1}{T} \int_0^T X_t^2 dt \xrightarrow{\text{a.s.}} \frac{\sigma^2}{2\theta}$ as $T \to \infty$ as well.

5. Prove that $\widehat{\theta}$ is strongly consistent.

We now wish to show that $\widehat{\theta}_T$ is asymptotically normal in the sense that $\sqrt{T}(\widehat{\theta}_T - \theta) \xrightarrow{\text{Law}} \mathcal{N}(0, 2\theta)$ as $T \to \infty$.

6. By using the result of part 4, show that we are left to prove that $\frac{1}{\sqrt{T}} \int_0^T X_t dB_t \xrightarrow{\text{Law}} \mathcal{N}(0, \sigma^2/2\theta)$ as $T \to \infty$.

7. Show that $\text{Var}\left[\frac{1}{\sqrt{T}} \int_0^T X_t dB_t\right] \to \frac{\sigma^2}{2\theta}$ as $T \to \infty$.

8. Verify that

$$\frac{1}{\sqrt{T}} \int_0^T X_t dB_t = I_2(f_T)$$

with

$$f_T(t, s) = \frac{\sigma}{2\sqrt{T}} e^{-\theta|t-s|} \mathbf{1}_{[0,T]}(s) \mathbf{1}_{[0,T]}(t).$$

9. Prove that $\|f_T \otimes_1 f_T\|_{L^2}^2 \to 0$ as $T \to \infty$. From Theorem 5.2.7, deduce that $\widehat{\theta}_T$ is asymptotically normal.

10. Using similar arguments, show that $\widehat{\theta}_T$ continues to be a strongly consistent and asymptotically normal estimator of θ when B is replaced by a fractional Brownian motion of Hurst index $H > 1/2$.

5.4.5 The following arguments are taken from Nourdin, Peccati and Reinert [93]. Let $B = (B_t)_{t \in \mathbb{R}}$ denote a centered Gaussian process with stationary increments and such that $\int_{\mathbb{R}} |\rho(x)| dx < \infty$, where $\rho(u-v) :=$ $E[(B_{u+1} - B_u)(B_{v+1} - B_v)]$. (Also, in order to avoid trivialities, we assume that ρ is not identically zero.) Let $f : \mathbb{R} \to \mathbb{R}$ be a real function of class $\mathcal{C}^2$, let $N \sim \mathcal{N}(0, 1)$, and assume that f is not constant, that $E|f(N)| < \infty$ and that $E|f''(N)|^4 < \infty$. Finally, fix $a < b$ in $\mathbb{R}$ and, for any $T > 0$, consider $F_T = \frac{1}{\sqrt{T}} \int_{aT}^{bT} \left(f(B_{u+1} - B_u) - E[f(N)] \right) du$. The goal of this exercise is to prove that $F_T / \sqrt{\mathrm{Var} F_T}$ converges in distribution to $\mathcal{N}(0, 1)$ as $T \to \infty$.

1. Show that the Gaussian space generated by B can be identified with an isonormal Gaussian process of the type $X = \{X(h),\ h \in \mathfrak{H}\}$, and that we have $B_t - B_s = X(\mathbf{1}_{[s,t]})$.
2. Using (5.3.6), show that $E|f'(N)|^4 < \infty$ and $E|f(N)|^4 < \infty$.
3. Compute

$$\|DF_T\|_{\mathfrak{H}}^4 = \frac{1}{T^2} \int_{[aT,bT]^4} f'(B_{u+1} - B_u) f'(B_{v+1} - B_v) f'(B_{w+1} - B_w)$$
$$\times f'(B_{z+1} - B_z) \rho(w - z) \rho(u - v) du\,dv\,dw\,dz.$$

4. Deduce that

$$E\big[\|DF_T\|_{\mathfrak{H}}^4\big] \leq E|f'(N)|^4 \left(\frac{1}{T} \int_{aT}^{bT} du \int_{\mathbb{R}} |\rho(x)| dx \right)^2 = O(1).$$

5. Compute

$$D^2 F_T \otimes_1 D^2 F_T = \frac{1}{T} \int_{[aT,bT]^2} f''(B_{u+1} - B_u) f''(B_{v+1} - B_v)$$
$$\times \rho(u - v) \mathbf{1}_{[u,u+1]} \otimes \mathbf{1}_{[v,v+1]} du\,dv.$$

6. Deduce that $E\big[\|D^2 F_T \otimes_1 D^2 F_T\|_{\mathfrak{H}^{\otimes 2}}^2\big]$ is less than

$$E|f''(N)|^4 \frac{b-a}{T} \int_{\mathbb{R}^3} |\rho(x)||\rho(y)||\rho(t)||\rho(x-y-t)| dx\,dy\,dt = O(T^{-1}).$$

7. Conclude, by means of Proposition 5.3.5, that $F_T / \sqrt{\mathrm{Var} F_T}$ converges in distribution to $\mathcal{N}(0, 1)$ as $T \to \infty$.
8. Prove that $\lim_{T \to \infty} \mathrm{Var} F_T$ exists when f is symmetric. (Hint: Expand f in terms of Hermite polynomials.)

5.4.6 **(Convergence to centered chi-square law)** Fix an integer $v \geq 1$, and let $N_1, \ldots, N_v$ be independent $\mathcal{N}(0, 1)$ random variables. Consider $F_\infty = -v + \sum_{i=1}^{v} N_i^2$, that is, F_∞ has the centered chi-square law with v degrees of freedom. Let $q \geq 2$ be an even integer, and define

$$c_q := \frac{1}{(q/2)!\binom{q-1}{q/2-1}^2} = \frac{4}{(q/2)!\binom{q}{q/2}^2}. \qquad (5.4.1)$$

Let $F_n = I_q(f_n)$ be a sequence of chaotic random variables such that $E[F_n^2] = 2v$ for all n. Consider the following four assumptions, as $n \to \infty$:

(i) $E[F_n^4] - 12E[F_n^3] \to 12v^2 - 48v$.

(ii) $\|f_n \widetilde{\otimes}_{q/2} f_n - c_q \times f_n\|_{\mathfrak{H}^{\otimes q}} \to 0$ and $\|f_n \widetilde{\otimes}_p f_n\|_{\mathfrak{H}^{\otimes 2(q-p)}} \to 0$ for every $p = 1, \ldots, q-1$ such that $p \neq q/2$.

(iii) $\|DF_n\|_{\mathfrak{H}}^2 - 2q F_n \to 2qv$ in $L^2(\Omega)$.

(iv) $F_n \to F_\infty$ in distribution.

The aim of this exercise is to show that these four assumptions are equivalent. It is taken from Nourdin and Peccati [87].

1. Use Theorem 2.7.2 to show that $\sup_{n \geq 1} E\left[|F_n|^p\right] < \infty$ for every $p > 0$. Deduce that (iv) implies (i).

2. Use relation (5.2.10) (with F_n instead of F) to prove that

$$E[F_n^3] = q! (q/2)! \binom{q}{q/2}^2 \langle f_n, f_n \widetilde{\otimes}_{q/2} f_n \rangle_{\mathfrak{H}^{\otimes q}}. \qquad (5.4.2)$$

3. Combine (5.4.2) with (5.2.5) to deduce that

$$E[F_n^4] - 12 E[F_n^3]$$

$$= 12v^2 + \frac{3}{q} \sum_{\substack{p=1,\ldots,q-1 \\ p \neq q/2}} q^2 (p-1)! \binom{q-1}{p-1}^2 q! \binom{q}{p}^2$$

$$\times (2q - 2p)! \|f_n \widetilde{\otimes}_p f_n\|_{\mathfrak{H}^{\otimes 2(q-p)}}^2$$

$$+ 3q (q/2 - 1)! \binom{q-1}{q/2-1}^2 (q/2)! \binom{q}{q/2}^2 q! \|f_n \widetilde{\otimes}_{q/2} f_n\|_{\mathfrak{H}^{\otimes q}}^2$$

$$- 12q! (q/2)! \binom{q}{q/2}^2 \langle f_n, f_n \widetilde{\otimes}_{q/2} f_n \rangle_{\mathfrak{H}^{\otimes q}}. \qquad (5.4.3)$$

4. Using elementary simplifications, check that

$$\frac{3}{2} \frac{(q!)^5}{[(q/2)!]^6} \|f_n \widetilde{\otimes}_{q/2} f_n - f_n \times c_q\|_{\mathfrak{H}^{\otimes q}}^2$$

$$= -12q!\,(q/2)! \binom{q}{q/2}^2 \langle f_n, f_n \widetilde{\otimes}_{q/2} f_n \rangle_{\mathfrak{H}^{\otimes q}} + 24q!\,\|f_n\|_{\mathfrak{H}^{\otimes q}}^2$$

$$+3q\,(q/2-1)! \binom{q-1}{q/2-1}^2 (q/2)! \binom{q}{q/2}^2 q!\,\|f_n \widetilde{\otimes}_{q/2} f_n\|_{\mathfrak{H}^{\otimes q}}^2,$$

where c_q is defined in (5.4.1). Deduce that (i) and (ii) are equivalent.

5. Show that

$$E[F_n\|DF_n\|_{\mathfrak{H}}^2] = q^2(q/2-1)! \binom{q-1}{q/2-1}^2 q!\langle f_n \widetilde{\otimes}_{q/2} f_n, f_n \rangle_{\mathfrak{H}^{\otimes q}}.$$
$$(5.4.4)$$

6. Use (5.4.4) and (5.2.3) to show that, if (ii) holds, then $E\big[\big(\|DF_n\|_{\mathfrak{H}}^2 - 2q F_n - 2q\nu\big)^2\big] \to 0$ as $n \to \infty$, that is, (ii) implies (iii).

7. Show that (iv) holds if and only if any subsequence $(F_{n'})$ converging in distribution to some random variable G is necessarily such that $G \overset{\text{Law}}{=} F_\infty$.(Hint: Use Prokhorov's theorem.)

8. Assume that F_n converges in distribution to some G, and let $\phi_n(\lambda) = E\big[e^{i\lambda F_n}\big]$ denote the characteristic function of F_n. Prove that $\phi_n'(\lambda) \to E\big[G\,e^{i\lambda G}\big]$ as $n \to \infty$.

9. Show that $\phi_n'(\lambda) = -\frac{\lambda}{q} E\big[e^{i\lambda F_n}\|DF_n\|_{\mathfrak{H}}^2\big]$.

10. Assume that (iii) holds and that F_n converges in distribution to some G. Let $\phi_\infty(\lambda) = E\big[e^{i\lambda G}\big]$ denote the characteristic function of G. Prove that

$$(1 - 2i\lambda)\phi_\infty'(\lambda) + 2\lambda\,\nu\,\phi_\infty(\lambda) = 0.$$

Deduce that $\phi_\infty(\lambda) = \Big(\frac{e^{-i\lambda}}{\sqrt{1-i2\lambda}}\Big)^\nu$, and then $G \overset{\text{Law}}{=} F_\infty$.

11. Show that (iii) implies (iv).

5.4.7 **(Almost sure central limit theorem)** Let $\{Y_n\}_{n\geq 1}$ be a sequence of real-valued i.i.d. random variables with $E[Y_n] = 0$ and $E[Y_n^2] = 1$, and denote

$$V_n = \frac{1}{\sqrt{n}} \sum_{k=1}^n Y_k.$$

The celebrated almost sure central limit theorem (ASCLT) states that the sequence of random empirical measures associated with V_n converges almost surely to the standard Gaussian distribution as $n \to \infty$;

that is, almost surely, for any bounded and continuous function φ : $\mathbb{R} \to \mathbb{R}$,

$$\frac{1}{\log n} \sum_{k=1}^{n} \frac{1}{k} \varphi(V_k) \longrightarrow \frac{1}{2\pi} \int_{\mathbb{R}} \varphi(x) e^{-\frac{x^2}{2}} dx, \quad \text{as } n \to \infty.$$

In this exercise, we investigate the ASCLT for a sequence of functionals of general Gaussian fields, by adopting the framework and notation of the present chapter and by following the approach developed by Bercu, Nourdin and Taqqu in [11]. We will also make use of the following result, due to Ibragimov and Lifshits [54], providing a sufficient condition allowing an ASCLT to be deduced from convergence in law. Let $\{G_n\}$ be a sequence of random variables converging in distribution towards a random variable G_∞, and set

$$\Delta_n(t) = \frac{1}{\log n} \sum_{k=1}^{n} \frac{1}{k} \left(e^{itG_k} - E[e^{itG_\infty}] \right);$$

if $\sup_{|t| \le r} \sum_n E|\Delta_n(t)|^2/(n \log n) < \infty$ for all $r > 0$, then, almost surely, for all continuous and bounded function $\varphi : \mathbb{R} \to \mathbb{R}$, we have

$$\frac{1}{\log n} \sum_{k=1}^{n} \frac{1}{k} \varphi(G_k) \longrightarrow E[\varphi(G_\infty)], \quad \text{as } n \to \infty.$$

Now let $\{G_n\}$ be a sequence in $\mathbb{D}^{2,4}$ satisfying, for all $n \ge 1$, $E[G_n] = 0$ and $E[G_n^2] = 1$. Assume, moreover, that the following four assumptions are satisfied:

$(A_1) \quad \sup_{n \ge 1} E\big[\|DG_n\|_{\mathfrak{H}}^4\big] < \infty;$

$(A_2) \quad \lim_{n \to \infty} E[\|D^2 G_n \otimes_1 D^2 G_n\|_{\mathfrak{H}^{\otimes 2}}^2] = 0;$

$(A_3) \quad \sum_{n \ge 2} \frac{1}{n \log^2 n} \sum_{k=1}^{n} \frac{1}{k} E[\|D^2 G_k \otimes_1 D^2 G_k\|_{\mathfrak{H}^{\otimes 2}}^2]^{\frac{1}{4}} < \infty;$

$(A_4) \quad \sum_{n \ge 2} \frac{1}{n \log^3 n} \sum_{k,l=1}^{n} \frac{|E(G_k G_l)|}{kl} < \infty.$

Prove that $G_n \to \mathcal{N}(0,1)$ as $n \to \infty$, and that $\{G_n\}$ satisfies an ASCLT as well.

5.4.8 **(Independence of multiple integrals)** Let $F = I_p(f)$ and $G = I_q(g)$ be two multiple integrals (with $f \in \mathfrak{H}^{\odot p}$, $g \in \mathfrak{H}^{\odot q}$ and $p, q \ge 1$). The goal of this exercise is to show that F and G are independent if and only if $f \otimes_1 g = 0$. The following arguments are taken from Üstünel and Zakai [150] for the necessity and Kallenberg [59] for the sufficiency.

1. By proceeding as in the proof of (5.2.6), show that

$$(p+q)! \| f \widetilde{\otimes} g \|^2_{\mathfrak{H}^{\otimes (p+q)}} = p!q! \sum_{r=0}^{p \wedge q} \binom{p}{r} \binom{q}{r} \| f \otimes_r g \|^2_{\mathfrak{H}^{\otimes (p+q-2r)}}.$$

2. Using the multiplication formula (2.7.9), compute the chaotic decomposition of FG, and deduce that

$$E[F^2 G^2] = \sum_{r=0}^{p \wedge q} r!^2 \binom{p}{r}^2 \binom{q}{r}^2 (p+q-2r)! \| f \widetilde{\otimes}_r g \|^2_{\mathfrak{H}^{\otimes (p+q-2r)}}.$$

3. Using $E[F^2] E[G^2] = p!q! \| f \|^2_{\mathfrak{H}^{\otimes p}} \| g \|^2_{\mathfrak{H}^{\otimes q}}$ as well as the identities obtained in parts 1 and 2, deduce that

$$\mathrm{Cov}(F^2, G^2) \geq p!q! \, pq \, \| f \otimes_1 g \|^2_{\mathfrak{H}^{\otimes (p+q-2)}}.$$

4. Prove the necessity of the condition stated above, that is, $f \otimes_1 g$ vanishes whenever F and G are independent.

5. Let us now prove the sufficiency. Without loss of generality, we assume in the rest of this exercise that $\mathfrak{H}$ is equal to $L^2(A, \mathcal{A}, \mu)$, where $(A, \mathcal{A})$ is a measurable space and μ is a σ-finite measure without atoms. Let $\{e_n\}$ be any orthonormal basis in $\mathfrak{H}$, and let H_p be the Hermite polynomial of degree p. Prove that

$$I_p \big(e_1^{\otimes p_1} \otimes \ldots \otimes e_m^{\otimes p_m} \big) = H_{p_1} \big(I_1(e_1) \big) \ldots H_{p_m} \big(I_1(e_m) \big),$$

whenever $p_1 + \ldots + p_m = p$.

6. Let $\mathcal{H}_f$ denote the Hilbert space in $\mathfrak{H}$ spanned by all functions

$$s \mapsto \int_{A^{p-1}} f(x_1, \ldots, x_{p-1}, s) h(x_1, \ldots, x_{p-1}) \mu(dx_1) \ldots \mu(dx_{p-1}),$$

$$s \in A,$$

where $h \in \mathfrak{H}^{\otimes (p-1)}$, and similarly define $\mathcal{H}_g$. Show that condition $f \otimes_1 g = 0$ implies that $\mathcal{H}_f$ and $\mathcal{H}_g$ are orthogonal.

7. Let $\{\phi_n\}$ be an orthonormal basis for $\mathcal{H}_f$, and let $\{\psi_n\}$ be an orthonormal basis for $\mathcal{H}_g$. Prove that f and g can be decomposed as

$$f = \sum a_{i_1, \ldots, i_p} \phi_{i_1} \otimes \ldots \otimes \phi_{i_p}, \quad g = \sum b_{j_1, \ldots, j_q} \psi_{j_1} \otimes \ldots \otimes \psi_{j_q}.$$

8. Using the results of parts 5–7, deduce that if $f \otimes_1 g = 0$ then F and G are independent.

5.5 Bibliographic comments

The results of Section 5.1 are taken from Nourdin and Peccati [88], except for Theorem 5.1.5, which is a one-dimensional version of more general estimates proved in Nourdin, Peccati and Reinert [94]. Note that [88] was the first paper to establish an explicit connection between Malliavin calculus and Stein's method. Some of the results presented in this chapter are also discussed in the monographs by Nualart [99, chapter 9] and Chen, Goldstein and Shao [22, chapter 14]. For an introduction to the method of moments and cumulants in a Gaussian setting, see the already quoted works by Major [68], Peccati and Taqqu [110] and Surgailis [139]. For some distinguished examples of applications of this method, see Chambers and Slud [19], Ginovyan and Sahakyan [40], Giraitis and Surgailis [41], Marinucci [73], Maruyama [75, 76] and Sodin and Tsirelson [134]. The left-hand inequality in each of (5.2.13)–(5.2.15) is taken from [88] (with some slight variations), while the right-hand inequality comes from [94]. The equivalence between (i), (ii), (iv) and (v) in Theorem 5.2.7 was first proved by Nualart and Peccati in [101], by means of stochastic calculus techniques (see also Peccati [105]). The equivalence between (i) and (iii) in Theorem 5.2.7 was discovered by Nualart and Ortiz-Latorre in their fundamental paper [100]: the argument at the core of their proof is sketched in Exercise 5.4.2. Theorem 5.3.3 and Proposition 5.3.6 are taken from Nourdin, Peccati and Reinert [93]. Note that Nualart and Ortiz-Latorre [100] gave the first ever results linking Malliavin calculus and limit theorems on Wiener space. Ledoux [65] gives a general discussion of 'fourth-moment conditions' related to the normal approximation of random variables belonging to the chaos of a Markov operator. Other approaches to Theorem 5.2.7 (yielding several generalizations) are developed in Peccati and Taqqu [107–109], using concepts from decoupling. In [85], by Nourdin and Nualart, one can find extensions of the theory developed in this chapter to the case of stable convergence. The paper by Peccati, Solé, Taqqu and Utzet [106] contains versions of the results of this chapter involving random variables defined on the Poisson space, whereas [62], by Kemp, Nourdin, Peccati and Speicher, provides several extensions to the framework of free probability. See Nourdin [83] for alternative proofs, and see Viens [151] for several generalizations and applications.

6

Multivariate normal approximations

The goal of this chapter is to prove several multivariate analogs of the results discussed in Chapter 5. In what follows, we fix an isonormal Gaussian process $X = \{X(h) : h \in \mathfrak{H}\}$, defined on a suitable probability space $(\Omega, \mathscr{F}, P)$ such that $\mathscr{F} = \sigma\{X\}$. We adopt the language and notation of Malliavin calculus introduced in Chapter 2. Also, we freely use the matrix notation introduced in Chapter 4.

6.1 Bounds for general vectors

We start by proving a first multivariate version of Theorem 5.1.3. We only deal with the Wasserstein distance.

Theorem 6.1.1 *Fix $d \geq 2$, and let $F = (F_1, \ldots, F_d)$ be a random vector such that $F_i \in \mathbb{D}^{1,4}$ with $E[F_i] = 0$ for any i. Let $C \in \mathcal{M}_d(\mathbb{R})$ be a symmetric and positive definite matrix, and let $N \sim \mathcal{N}_d(0, C)$. Then*

$$
d_{\mathrm{W}}(F, N) \leq \sqrt{d} \, \|C^{-1}\|_{\mathrm{op}} \|C\|_{\mathrm{op}}^{1/2} \sqrt{\sum_{i,j=1}^{d} E\left[\left(C(i, j) - \langle DF_j, -DL^{-1}F_i \rangle_{\mathfrak{H}}\right)^2\right]}.
$$

$$(6.1.1)$$

Proof First, by reasoning as in Proposition 5.1.1, we show that $\langle DF_j, -DL^{-1}F_i \rangle_{\mathfrak{H}}$ is square-integrable for all i, j. On the other hand, for $f \in \mathcal{C}^2(\mathbb{R}^d)$ such that $\sup_{x \in \mathbb{R}^d} \|\mathrm{Hess} f(x)\|_{\mathrm{HS}} \leq \sqrt{d} \, \|C^{-1}\|_{\mathrm{op}} \|C\|_{\mathrm{op}}^{1/2}$, we can write

116

$$\left| E[\langle C, \operatorname{Hess} f(F)\rangle_{\mathrm{HS}} - \langle F, \nabla f(F)\rangle_{\mathbb{R}^d}] \right|$$

$$= \left| \sum_{i,j=1}^{d} C(i,j) E\left[\frac{\partial^2 f}{\partial x_i x_j}(F)\right] - \sum_{i=1}^{d} E\left[F_i \frac{\partial f}{\partial x_i}(F)\right] \right|$$

$$= \left| \sum_{i,j=1}^{d} C(i,j) E\left[\frac{\partial^2 f}{\partial x_i x_j}(F)\right] - \sum_{i,j=1}^{d} E\left[\frac{\partial^2 f}{\partial x_i x_j}(F)\langle DF_j, -DL^{-1}F_i\rangle_{\mathfrak{H}}\right] \right|$$

(by (2.9.1))

$$= \left| \sum_{i,j=1}^{d} E\left[\frac{\partial^2 f}{\partial x_i x_j}(F)\big(C(i,j) - \langle DF_j, -DL^{-1}F_i\rangle_{\mathfrak{H}}\big)\right] \right|$$

$$= \left| E\langle \operatorname{Hess} f(F), C - M\rangle_{\mathrm{HS}} \right| \quad \text{(with } M := \big(\langle DF_j, -DL^{-1}F_i\rangle_{\mathfrak{H}}\big)_{1 \le i, j \le d}\text{)}$$

$$\le \sqrt{E\|\operatorname{Hess} f(F)\|_{\mathrm{HS}}^2} \sqrt{E\|C - M\|_{\mathrm{HS}}^2}$$

(by the Cauchy–Schwarz inequality)

$$\le \sqrt{d}\, \|C^{-1}\|_{\mathrm{op}}\, \|C\|_{\mathrm{op}}^{1/2} \sqrt{E\|C - M\|_{\mathrm{HS}}^2},$$

so that we deduce (6.1.1) by applying Theorem 4.4.1. $\qquad\square$

In the next statement, we shall use interpolation techniques (the 'smart path method') in order to partially generalize Theorem 6.1.1 to the case where the approximating covariance matrix C is not necessarily positive definite. This additional difficulty forces us to work with functions that are smoother than those involved in the definition of the Wasserstein distance. To this end, we will adopt the following shorthand notation: for every $h : \mathbb{R}^d \to \mathbb{R}$ of class $\mathcal{C}^2$, we set

$$\|h''\|_\infty = \max_{i,j=1,\dots,d} \sup_{x \in \mathbb{R}^d} \left| \frac{\partial^2 h}{\partial x_i \partial x_j}(x) \right|.$$

Theorem 6.1.2 below is the multivariate counterpart of Theorem 5.1.5.

Theorem 6.1.2 *Fix $d \ge 2$, and let $F = (F_1, \dots, F_d)$ be a random vector such that $F_i \in \mathbb{D}^{1,4}$ with $E[F_i] = 0$ for any i. Let $C \in \mathcal{M}_d(\mathbb{R})$ be a symmetric non-negative definite matrix, and let $N \sim \mathcal{N}_d(0, C)$. Then, for any $h : \mathbb{R}^d \to \mathbb{R}$ belonging to $\mathcal{C}^2$ such that $\|h''\|_\infty < \infty$,*

$$E[h(N)] - E[h(F)] = \frac{1}{2} \sum_{i,j=1}^{d} \int_0^1 E\left[\frac{\partial^2 h}{\partial x_i \partial x_j}(\sqrt{1-t}\,F + \sqrt{t}\,N)\right.$$

$$\left. \times \Big(C(i,j) - \langle DF_i, -DL^{-1}F_j\rangle_{\mathfrak{H}}\Big)\right] dt. \quad (6.1.2)$$

As a consequence,

$$\left| E[h(F)] - E[h(N)] \right| \le \frac{1}{2} \|h''\|_\infty \sqrt{ \sum_{i,j=1}^{d} E\left[\left(C(i,j) - \langle DF_j, -DL^{-1}F_i \rangle_{\mathfrak{H}} \right)^2 \right] }.$$

$$(6.1.3)$$

Proof Without loss of generality, we assume that N is independent of the underlying isonormal Gaussian process X. Let h be as in the statement of the theorem. For any $t \in [0, 1]$, set $\Psi(t) = E\left[h\left(\sqrt{1-t}\,F + \sqrt{t}\,N\right)\right]$, so that

$$E[h(N)] - E[h(F)] = \Psi(1) - \Psi(0) = \int_0^1 \Psi'(t)\,dt.$$

We can easily see that Ψ is differentiable on $(0, 1)$ with

$$\Psi'(t) = \sum_{i=1}^{d} E\left[\frac{\partial h}{\partial x_i}\left(\sqrt{1-t}\,F + \sqrt{t}\,N\right) \left(\frac{1}{2\sqrt{t}} N_i - \frac{1}{2\sqrt{1-t}} F_i \right) \right].$$

By integrating by parts, we can write

$$E\left[\frac{\partial h}{\partial x_i}\left(\sqrt{1-t}\,F + \sqrt{t}\,N\right) N_i \right] = E\left\{ E\left[\frac{\partial h}{\partial x_i}\left(\sqrt{1-t}\,x + \sqrt{t}\,N\right) N_i \right]_{|x=F} \right\}$$

$$= \sqrt{t} \sum_{j=1}^{d} C(i,j)\, E\left\{ E\left[\frac{\partial^2 h}{\partial x_i \partial x_j}\left(\sqrt{1-t}\,x + \sqrt{t}\,N\right) \right]_{|x=F} \right\}$$

$$= \sqrt{t} \sum_{j=1}^{d} C(i,j)\, E\left[\frac{\partial^2 h}{\partial x_i \partial x_j}\left(\sqrt{1-t}\,F + \sqrt{t}\,N\right) \right].$$

By using (2.9.1) in order to perform the integration by parts, we can also write

$$E\left[\frac{\partial h}{\partial x_i}\left(\sqrt{1-t}\,F + \sqrt{t}\,N\right) F_i \right]$$

$$= E\left\{ E\left[\frac{\partial h}{\partial x_i}\left(\sqrt{1-t}\,F + \sqrt{t}\,x\right) F_i \right]_{|x=N} \right\}$$

$$= \sqrt{1-t} \sum_{j=1}^{d} E\left\{ E\left[\frac{\partial^2 h}{\partial x_i \partial x_j}\left(\sqrt{1-t}\,F + \sqrt{t}\,x\right) \langle DF_j, -DL^{-1}F_i \rangle_{\mathfrak{H}} \right]_{|x=N} \right\}$$

$$= \sqrt{1-t} \sum_{j=1}^{d} E\left[\frac{\partial^2 h}{\partial x_i \partial x_j}\left(\sqrt{1-t}\,F + \sqrt{t}\,N\right) \langle DF_j, -DL^{-1}F_i \rangle_{\mathfrak{H}} \right].$$

Hence

$$\Psi'(t) = \frac{1}{2} \sum_{i,j=1}^{d} E\left[\frac{\partial^2 h}{\partial x_i \partial x_j}\left(\sqrt{1-t}F + \sqrt{t}N\right)\left(C(i,j) - \langle DF_j, -DL^{-1}F_j\rangle_{\mathfrak{H}}\right)\right],$$

and the desired conclusion follows. $\qquad\square$

The following statement uses the previous result to provide a criterion of multivariate normal approximation, which is in the spirit of Theorem 5.3.5.

Theorem 6.1.3 *Fix $d \geq 2$. Consider vectors $F_n = (F_{1,n}, \ldots, F_{d,n})$, $n \geq 1$, with $E[F_{i,n}] = 0$ and $F_{i,n} \in \mathbb{D}^{2,4}$ for every $i = 1, \ldots, d$ and $n \geq 1$. Let $C \in \mathcal{M}_d(\mathbb{R})$ be a symmetric non-negative definite matrix, and let $N \sim \mathcal{N}_d(0, C)$. Suppose that:*

(i) *for any $i, j = 1, \ldots, d$, $E[F_{i,n}F_{j,n}] \to C(i,j)$ as $n \to \infty$;*
(ii) *for any $i = 1, \ldots, d$, $\sup_{n\geq 1} E[\|DF_{i,n}\|_{\mathfrak{H}}^4] < \infty$;*
(iii) *for any $i = 1, \ldots, d$, $E[\|D^2 F_{i,n} \otimes_1 D^2 F_{i,n}\|_{\mathfrak{H}^{\otimes 2}}^2] \to 0$ as $n \to \infty$.*

Then $F_n \overset{\text{Law}}{\to} \mathcal{N}_d(0, C)$ as $n \to \infty$.

Proof For any $n \geq 1$, let $C_n \in \mathcal{M}_d(\mathbb{R})$ denote the symmetric positive matrix defined by $C_n(i,j) = E[F_{i,n}F_{j,n}]$. Also let $N_n \sim \mathcal{N}_d(0, C_n)$. In Theorem 6.1.2, it is shown that, for any $h : \mathbb{R}^d \to$ belonging to $\mathcal{C}^2$ such that $\|h''\|_\infty < \infty$, we have

$$\left|E[h(F_n)] - E[h(N_n)]\right| \leq \frac{1}{2}\|h''\|_\infty$$

$$\times \sqrt{\sum_{i,j=1}^{d} E\left[(E[F_{i,n}F_{j,n}] - \langle DF_{i,n}, -DL^{-1}F_{j,n}\rangle_{\mathfrak{H}})^2\right]}.$$

Hence, using Proposition 5.3.10, we get

$$\left|E[h(F_n)] - E[h(N_n)]\right|$$

$$\leq \sqrt{\frac{5}{8}}\|h''\|_\infty \sqrt{\sum_{i,j=1}^{d} E\left[\|D^2 F_{i,n} \otimes_1 D^2 F_{i,n}\|_{\mathfrak{H}^{\otimes 2}}^2\right]^{1/2} E\left[\|DF_{j,n}\|_{\mathfrak{H}}^4\right]^{1/2}}$$

$$\leq \sqrt{\frac{5}{8}}\|h''\|_\infty \sum_{i,j=1}^{d} E\left[\|D^2 F_{i,n} \otimes_1 D^2 F_{i,n}\|_{\mathfrak{H}^{\otimes 2}}^2\right]^{1/4} E\left[\|DF_{j,n}\|_{\mathfrak{H}}^4\right]^{1/4}$$

$$= \sqrt{\frac{5}{8}}\|h''\|_\infty \sum_{i=1}^{d} E\left[\|D^2 F_{i,n} \otimes_1 D^2 F_{i,n}\|_{\mathfrak{H}^{\otimes 2}}^2\right]^{1/4} \times \sum_{j=1}^{d} E\left[\|DF_{j,n}\|_{\mathfrak{H}}^4\right]^{1/4}.$$

Due to assumptions (ii) and (iii), we deduce that $\left| E[h(F_n)] - E[h(N_n)] \right| \to 0$ as $n \to \infty$. On the other hand, due to assumption (i), it is clear that $\left| E[h(N_n)] - E[h(N)] \right| \to 0$ as $n \to \infty$. Hence $\left| E[h(F_n)] - E[h(N)] \right| \to 0$ as $n \to \infty$ for all $h : \mathbb{R}^d \to \mathbb{R}$ belonging to $\mathcal{C}^2$ and such that $\|h''\|_\infty < \infty$, which is enough to prove that $F_n \overset{\text{Law}}{\to} \mathcal{N}_d(0, C)$ as $n \to \infty$. $\qquad\square$

6.2 The case of Wiener chaos

Now, let us investigate what happens in the specific case of vectors of *multiple stochastic integrals*. We start with the following technical lemma. In what follows, we use the convention that $\sum_{r=1}^{0} \equiv 0$.

Lemma 6.2.1 *Let $F = I_p(f)$ and $G = I_q(g)$, with $f \in \mathfrak{H}^{\odot p}$ and $g \in \mathfrak{H}^{\odot q}$ $(p, q \geq 1)$. Let α be a real constant. If $p = q$, we have the estimate:*

$$E\left[\left(\alpha - \frac{1}{p} \langle DF, DG \rangle_\mathfrak{H} \right)^2 \right] \leq (\alpha - E[FG])^2 \tag{6.2.1}$$

$$+ \frac{p^2}{2} \sum_{r=1}^{p-1} (r-1)!^2 \binom{p-1}{r-1}^4 (2p-2r)!$$

$$\times \left(\| f \otimes_{p-r} f \|^2_{\mathfrak{H}^{\otimes 2r}} + \| g \otimes_{p-r} g \|^2_{\mathfrak{H}^{\otimes 2r}} \right).$$

If $p < q$, we have that

$$E\left[\left(\alpha - \frac{1}{q} \langle DF, DG \rangle_\mathfrak{H} \right)^2 \right]$$

$$\leq \alpha^2 + p!^2 \binom{q-1}{p-1}^2 (q-p)! \| f \|^2_{\mathfrak{H}^{\otimes p}} \| g \otimes_{q-p} g \|^2_{\mathfrak{H}^{\otimes 2p}}$$

$$+ \frac{p^2}{2} \sum_{r=1}^{p-1} (r-1)!^2 \binom{p-1}{r-1}^2 \binom{q-1}{r-1}^2 (p+q-2r)!$$

$$\times \left(\| f \otimes_{p-r} f \|^2_{\mathfrak{H}^{\otimes 2r}} + \| g \otimes_{q-r} g \|^2_{\mathfrak{H}^{\otimes 2r}} \right). \tag{6.2.2}$$

Proof Thanks to the multiplication formula (2.7.9), we can write

$$\langle DF, DG \rangle_\mathfrak{H} = p q \langle I_{p-1}(f), I_{q-1}(g) \rangle_\mathfrak{H}$$

$$= p q \sum_{r=0}^{p \wedge q - 1} r! \binom{p-1}{r} \binom{q-1}{r} I_{p+q-2-2r}(f \widetilde{\otimes}_{r+1} g)$$

$$= p q \sum_{r=1}^{p \wedge q} (r-1)! \binom{p-1}{r-1} \binom{q-1}{r-1} I_{p+q-2r}(f \widetilde{\otimes}_r g).$$

It follows that

$$
E\left[\left(\alpha - \frac{1}{q} \langle DF, DG \rangle_{\mathfrak{H}}\right)^2\right]
$$

$$
= \begin{cases}
\alpha^2 + p^2 \sum_{r=1}^{p} (r-1)!^2 \binom{p-1}{r-1}^2 \binom{q-1}{r-1}^2 (p+q-2r)! \\
\quad \times \| f \widetilde{\otimes}_r g \|^2_{\mathfrak{H}^{\otimes(p+q-2r)}} \qquad \text{if } p < q \\[12pt]
(\alpha - p! \langle f, g \rangle_{\mathfrak{H}^{\otimes p}})^2 + p^2 \sum_{r=1}^{p-1} (r-1)!^2 \binom{p-1}{r-1}^4 (2p-2r)! \\
\quad \times \| f \widetilde{\otimes}_r g \|^2_{\mathfrak{H}^{\otimes(2p-2r)}} \qquad \text{if } p = q.
\end{cases}
\tag{6.2.3}
$$

If $r < p \leq q$, then

$$
\begin{aligned}
\| f \widetilde{\otimes}_r g \|^2_{\mathfrak{H}^{\otimes(p+q-2r)}} &\leq \| f \otimes_r g \|^2_{\mathfrak{H}^{\otimes(p+q-2r)}} = \langle f \otimes_{p-r} f, g \otimes_{q-r} g \rangle_{\mathfrak{H}^{\otimes 2r}} \\
&\leq \| f \otimes_{p-r} f \|_{\mathfrak{H}^{\otimes 2r}} \| g \otimes_{q-r} g \|_{\mathfrak{H}^{\otimes 2r}} \\
&\leq \frac{1}{2} \left(\| f \otimes_{p-r} f \|^2_{\mathfrak{H}^{\otimes 2r}} + \| g \otimes_{q-r} g \|^2_{\mathfrak{H}^{\otimes 2r}} \right).
\end{aligned}
\tag{6.2.4}
$$

If $r = p < q$, then

$$
\| f \widetilde{\otimes}_p g \|^2_{\mathfrak{H}^{\otimes(q-p)}} \leq \| f \otimes_p g \|^2_{\mathfrak{H}^{\otimes(q-p)}} \leq \| f \|^2_{\mathfrak{H}^{\otimes p}} \| g \otimes_{q-p} g \|_{\mathfrak{H}^{\otimes 2p}}.
\tag{6.2.5}
$$

By plugging these two inequalities into (6.2.3), we immediately deduce the desired conclusion. $\qquad\square$

We are now in a position to state and prove the following result, which can be seen as a multivariate counterpart of Theorem 5.2.6. We recall the convention $\sum_{r=1}^{0} \equiv 0$.

Theorem 6.2.2 *Let $d \geq 2$ and $q_d, \ldots, q_1 \geq 1$ be some fixed integers. Consider the vector $F = (F_1, \ldots, F_d) = (I_{q_1}(f_1), \ldots, I_{q_d}(f_d))$ with $f_i \in \mathfrak{H}^{\odot q_i}$ for each i. Let $C \in \mathcal{M}_d(\mathbb{R})$ be the symmetric non-negative definite matrix given by $C(i,j) = E[F_i F_j]$, and let $N \sim \mathcal{N}_d(0, C)$. Set*

$$
m(F) = \psi\left(E[F_1^4] - 3E[F_1^2]^2, E[F_1^2], \ldots, E[F_d^4] - 3E[F_d^2]^2, E[F_d^2] \right)
$$

with $\psi : (\mathbb{R} \times \mathbb{R}_+)^d \to \mathbb{R}$ given by

$$
\psi(x_1, y_1, \ldots, x_d, y_d) = \sum_{i,j=1}^{d} \mathbf{1}_{\{q_i = q_j\}} \sqrt{2 \sum_{r=1}^{q_i - 1} \binom{2r}{r} |x_i|^{1/2}}
$$

$$
+ \sum_{i,j=1}^{d} \mathbf{1}_{\{q_i \neq q_j\}} \left\{ \sqrt{2} \sqrt{y_j} |x_i|^{1/4} + \sum_{r=1}^{q_i \wedge q_j - 1} \sqrt{2(q_i + q_j - 2r)!} \binom{q_j}{r} |x_i|^{1/2} \right\}.
$$

Then:

1. *for any* $h : \mathbb{R}^d \to \mathbb{R}$ *of class* C^2 *such that* $\|h''\|_\infty < \infty$,

$$\left| E[h(F)] - E[h(N)] \right| \leq \frac{1}{2} \|h''\|_\infty m(F);$$

2. *if, in addition,* C *is positive definite, then*

$$d_W(F, N) \leq \sqrt{d} \, \|C^{-1}\|_{op} \|C\|_{op}^{1/2} m(F).$$

Proof   We deduce from (5.2.6) that $E[I_q(f)^4] - 3E[I_q(f)^2]^2 \geq 0$ and

$$\|f \otimes_r f\|_{\mathfrak{H}^{\otimes 2q - 2r}}^2 \leq \frac{r!^2 (q - r)!^2}{q!^4} \left(E[I_q(f)^4] - 3E[I_q(f)^2]^2 \right)$$

for all $q \geq 2$, $f \in \mathfrak{H}^{\odot q}$ and $r \in \{1, \ldots, q - 1\}$. Therefore, if $f, g \in \mathfrak{H}^{\odot q}$, then

$$E\left[\left(E[I_q(f)I_q(g)] - \frac{1}{q} \langle DI_q(f), DI_q(g) \rangle_{\mathfrak{H}} \right)^2 \right] = 0$$

if $q = 1$, whereas inequality (6.2.1) yields that, for $q \geq 2$,

$$E\left[\left(E[I_q(f)I_q(g)] - \frac{1}{q} \langle DI_q(f), DI_q(g) \rangle_{\mathfrak{H}} \right)^2 \right]$$

$$\leq \left(E[I_q(f)^4] - 3E[I_q(f)^2]^2 + E[I_q(g)^4] - 3E[I_q(g)^2]^2 \right) \sum_{r=1}^{q-1} \frac{r^2 (2q - 2r)!}{2q^2 (q - r)!^2}$$

$$\leq \frac{1}{2} \left[E[I_q(f)^4] - 3E[I_q(f)^2]^2 + E[I_q(g)^4] - 3E[I_q(g)^2]^2 \right] \sum_{r=1}^{q-1} \binom{2r}{r}.$$

$$(6.2.6)$$

On the other hand, if $p < q$, $f \in \mathfrak{H}^{\odot p}$ and $g \in \mathfrak{H}^{\odot q}$, inequality (6.2.2) leads to

$$E\left[\left(\frac{1}{p} \langle DI_p(f), DI_q(g) \rangle_{\mathfrak{H}} \right)^2 \right] = \frac{q^2}{p^2} E\left[\left(\frac{1}{q} \langle DI_p(f), DI_q(g) \rangle_{\mathfrak{H}} \right)^2 \right]$$

$$\leq E[I_p(f)^2] \sqrt{E[I_q(g)^4] - 3E[I_q(g)^2]^2} + \frac{1}{2p^2} \sum_{r=1}^{p-1} r^2 (p + q - 2r)!$$

$$\times \left[\frac{q!^2}{(q - r)!^2 p!^2} \left(E[I_p(f)^4] - 3E[I_p(f)^2]^2 \right) \right.$$

$$\left. + \frac{p!^2}{(p - r)!^2 q!^2} \left(E[I_q(g)^4] - 3E[I_q(g)^2]^2 \right) \right]$$

$$\leq E[I_p(f)^2]\sqrt{E[I_q(g)^4] - 3E[I_q(g)^2]^2} + \frac{1}{2}\sum_{r=1}^{p-1}(p+q-2r)!$$

$$\times \left[\binom{q}{r}^2 \left(E[I_p(f)^4] - 3E[I_p(f)^2]^2\right) + \binom{p}{r}^2 \left(E[I_q(g)^4] - 3E[I_q(g)^2]^2\right)\right],$$

so that, if $p \neq q$, $f \in \mathfrak{H}^{\odot p}$ and $g \in \mathfrak{H}^{\odot q}$, we have that both $E\left[\left(\frac{1}{p}\langle DI_p(f),\right.\right.$ $\left.\left. DI_q(g)\rangle_{\mathfrak{H}}\right)^2\right]$ and $E\left[\left(\frac{1}{q}\langle DI_p(f), DI_q(g)\rangle_{\mathfrak{H}}\right)^2\right]$ are less than or equal to

$$E[I_p(f)^2]\sqrt{E[I_q(g)^4] - 3E[I_q(g)^2]^2}$$

$$+ E[I_q(g)^2]\sqrt{E[I_p(f)^4] - 3E[I_p(f)^2]^2} + \frac{1}{2}\sum_{r=1}^{p\wedge q-1}(p+q-2r)!$$

$$\times \left[\binom{q}{r}^2 \left(E[I_p(f)^4] - 3E[I_p(f)^2]^2\right) + \binom{p}{r}^2 \left(E[I_q(g)^4] - 3E[I_q(g)^2]^2\right)\right].$$

$$\tag{6.2.7}$$

Since two multiple integrals of different orders are orthogonal, we have that

$$C(i, j) = E[F_i F_j] = E[I_{q_i}(f_i)I_{q_j}(f_j)] = 0 \quad \text{whenever } q_i \neq q_j.$$

Thus, by using (6.2.6) and (6.2.7) together with $\sqrt{x_1 + \ldots + x_n} \leq \sqrt{x_1} + \ldots + \sqrt{x_n}$, we eventually get the desired conclusion, by means of (6.1.1) and (6.1.3) respectively. $\qquad \square$

A direct consequence of Theorem 6.2.2 is the following result, stating the fundamental fact that

for a sequence of vectors of multiple Wiener–Itô integrals componentwise convergence to Gaussian always implies joint convergence.

As demonstrated in the next section, this result (which can be seen as a multivariate counterpart to Theorem 5.2.7) allows us to effectively study the normal approximation of general functionals, by using their Wiener–Itô chaotic decomposition.

Theorem 6.2.3 *Let $d \geq 2$ and $q_d, \ldots, q_1 \geq 1$ be some fixed integers. Consider vectors*

$$F_n = (F_{1,n}, \ldots, F_{d,n}) = (I_{q_1}(f_{1,n}), \ldots, I_{q_d}(f_{d,n})), \quad n \geq 1,$$

with $f_{i,n} \in \mathfrak{H}^{\odot q_i}$. Let $C \in \mathcal{M}_d(\mathbb{R})$ be a symmetric non-negative definite matrix, and let $N \sim \mathcal{N}_d(0, C)$. Assume that

$$\lim_{n \to \infty} E[F_{i,n} F_{j,n}] = C(i, j), \quad 1 \le i, j \le d. \tag{6.2.8}$$

Then, as $n \to \infty$, the following two conditions are equivalent:

(a) F_n *converges in law to* N.
(b) *For every* $1 \le i \le d$, $F_{i,n}$ *converges in law to* $\mathcal{N}(0, C(i, i))$.

Proof The implication (a)→(b) is trivial, whereas the implication (b)→(a) follows directly from Theorem 6.2.2. Indeed, since $E[F_{i,n}^2] \to C(i, i)$, Theorem 2.7.2 implies that $\sup_n E|F_{i,n}|^4 < \infty$. This immediately yields that, if (b) holds, then $E[F_{i,n}^4] \to 3C(i, i)^2$, and therefore $m(F_n) \to 0$. $\quad\square$

Remark 6.2.4 If the integers $q_d, \ldots, q_1$ are pairwise disjoint in Theorem 6.2.3, then (6.2.8) is automatically satisfied with $C(i, j) = 0$ for all $i \ne j$.

6.3 CLTs via chaos decompositions

We now consider the problem of assessing the asymptotic normality of a sequence $(F_n)_{n \ge 1}$ of square-integrable random variables, starting from their *possibly infinite* chaos decomposition, that is,

$$F_n = \sum_{q=1}^{\infty} I_q(f_{n,q}) \quad \text{with } f_{n,q} \in \mathfrak{H}^{\odot q}, q \ge 1, n \ge 1. \tag{6.3.1}$$

When $F_n \in \mathbb{D}^{1,4}$, a simple computation based on the multiplication formula (2.7.9) leads to

$$\langle DF_n, -DL^{-1}F_n \rangle_{\mathfrak{H}} = \sum_{q,p=1}^{\infty} q \langle I_{q-1}(f_{n,q}), I_{p-1}(f_{n,p}) \rangle_{\mathfrak{H}}$$

$$= \sum_{q,p=1}^{\infty} q \sum_{r=0}^{p \wedge q - 1} r! \binom{p-1}{r} \binom{q-1}{r} I_{p+q-2-2r}(f_{n,q} \widetilde{\otimes}_{r+1} f_{n,p})$$

$$= \sum_{q,p=1}^{\infty} q \sum_{r=1}^{p \wedge q} (r-1)! \binom{p-1}{r-1} \binom{q-1}{r-1} I_{p+q-2r}(f_{n,q} \widetilde{\otimes}_r f_{n,p}). \tag{6.3.2}$$

Using (6.3.2), one could try to check whether (5.3.1) holds or not. But it is sometimes hard to set up this strategy in practical situations, due to the complexity of the resulting expressions. We shall now prove that Theorem 6.2.3 provides a viable alternative to such heavy computations (Theorem 6.3.1). An explicit application of this result will be our modern proof of the so-called 'Breuer–Major theorem' (see Theorem 7.2.4), as detailed in the next chapter.

Theorem 6.3.1 *Let $(F_n)_{n\geq 1}$ be a sequence in $L^2(\Omega)$ such that $E[F_n] = 0$ for all n. Consider the chaos expansion (6.3.1) of F_n, and suppose in addition that:*

(a) *for every fixed $q \geq 1$, $q!\|f_{n,q}\|^2_{\mathfrak{H}^{\otimes q}} \to \sigma_q^2$ as $n \to \infty$ (for some $\sigma_q^2 \geq 0$);*
(b) *$\sigma^2 = \sum_{q=1}^{\infty} \sigma_q^2 < \infty$;*
(c) *for all $q \geq 2$ and $r = 1, \ldots, q-1$, $\|f_{n,q} \otimes_r f_{n,q}\|_{\mathfrak{H}^{\otimes 2q-2r}} \to 0$ as $n \to \infty$;*
(d) *$\lim_{N\to\infty} \sup_{n\geq 1} \sum_{q=N+1}^{\infty} q!\|f_{n,q}\|^2_{\mathfrak{H}^{\otimes q}} = 0$.*

Then $F_n \overset{\text{Law}}{\to} \mathcal{N}(0, \sigma^2)$ as $n \to \infty$.

Remark 6.3.2 Of course, condition (c) can be replaced by any of the equivalent assertions (i)–(v) of Theorem 5.2.7.

Proof of Theorem 6.3.1. For $n, N \geq 1$, set $F_{n,N} = \sum_{q=1}^{N} I_q(f_{n,q})$. Also, let $G_N \sim \mathcal{N}(0, \sigma_1^2 + \ldots + \sigma_N^2)$ and $G \sim \mathcal{N}(0, \sigma^2)$. For any $t \in \mathbb{R}$, we have

$$\left|E[e^{itF_n}] - E[e^{itG}]\right| \leq \left|E[e^{itF_n}] - E[e^{itF_{n,N}}]\right| + \left|E[e^{itF_{n,N}}]\right.$$
$$\left. - E[e^{itG_N}]\right| + \left|E[e^{itG_N}] - E[e^{itG}]\right| = a_{n,N} + b_{n,N} + c_N.$$

Fix $\varepsilon > 0$. Thanks to (b), observe that

$$c_N = \left|e^{-\frac{t^2(\sigma_1^2+\ldots+\sigma_q^2)}{2}} - e^{-\frac{t^2\sigma^2}{2}}\right| \leq \frac{t^2}{2} \sum_{q=N+1}^{\infty} \sigma_q^2 \to 0 \quad \text{as } N \to \infty.$$

On the other hand, due to (d),

$$\sup_{n\geq 1} a_{n,N} \leq |t| \sup_{n\geq 1} E[|F_n - F_{n,N}|] \leq |t| \sqrt{\sup_{n\geq 1} E[(F_n - F_{n,N})^2]}$$

$$= |t| \sqrt{\sup_{n\geq 1} \sum_{q=N+1}^{\infty} q!\|f_{n,q}\|^2_{\mathfrak{H}^{\otimes q}}} \to 0 \quad \text{as } N \to \infty.$$

Therefore, we can choose N large enough so that $\sup_{n\geq 1} a_{n,N} \leq \varepsilon/3$ and $c_N \leq \varepsilon/3$. Due to (a) and (c), we deduce from Theorem 5.2.7 that, for every fixed $q \geq 2$, $I_q(f_{n,q}) \to \mathcal{N}(0, \sigma_q^2)$ as $n \to \infty$. Due to (a), it is also evident that $I_1(f_{n,1}) \to \mathcal{N}(0, \sigma_1^2)$ as $n \to \infty$. Recalling that $E[I_q(f_{n,q})I_p(f_{n,p})] = 0$ if $p \neq q$ and $E[I_q(f_{n,q})^2] = q!\|f_{n,q}\|^2_{\mathfrak{H}^{\otimes q}}$, we deduce from Theorem 6.2.3 that

$$(I_1(f_{n,1}), \ldots, I_N(f_{n,N})) \overset{\text{Law}}{\to} \mathcal{N}_N\left(0, \text{diag}\left(\sigma_1^2, \ldots, \sigma_N^2\right)\right) \quad \text{as } n \to \infty.$$

In particular, $F_{n,N} \overset{\text{Law}}{\to} \mathcal{N}(0, \sigma_1^2+\ldots+\sigma_N^2)$ as $n \to \infty$, so that $b_{n,N} \leq \varepsilon/3$ if n is large enough. Summarizing, we have shown that $\left|E[e^{itF_n}] - E[e^{itG}]\right| \leq \varepsilon$ if n is large enough, which is the desired conclusion. $\square$

6.4 Exercises

6.4.1 Let $N \sim \mathcal{N}_d(0, C)$ and $N' \sim \mathcal{N}_d(0, D)$ be two d-dimensional centered Gaussian vectors, with covariance matrices C, D.

1. Deduce from (6.1.3) that, for any $h : \mathbb{R}^d \to \mathbb{R}$ belonging to $\mathcal{C}^2$ such that $\|h''\|_\infty < \infty$,

$$\left| E[h(N)] - E[h(N')] \right| \le \frac{1}{2} \|h''\|_\infty \|C - D\|_{HS}. \qquad (6.4.1)$$

2. Use (6.1.1) to prove the following estimate: if both C and D are positive definite, then

$$d_W(N, N') \le M(C, D) \|C - D\|_{HS}, \qquad (6.4.2)$$

where $M(C, D) := \sqrt{d} \, \min\{\|C^{-1}\|_{op} \|C\|_{op}^{1/2} \, ; \, \|D^{-1}\|_{op} \|D\|_{op}^{1/2}\}$.

6.4.2 (Weighted quadratic variations of (fractional) Brownian motion) Let B be a standard Brownian motion, and $f : \mathbb{R} \to \mathbb{R}$ be a $\mathcal{C}^1$ function with a bounded derivative. For any $n \ge 1$, define $S_n(f)$ as

$$S_n(f) = 2^{-n/2} \sum_{k=0}^{2^n - 1} f(B_{k2^{-n}}) \left[n \left(B_{(k+1)2^{-n}} - B_{k2^{-n}} \right)^2 - 1 \right].$$

Our goal is to prove that $S_n(f) \xrightarrow{\text{Law}} \sqrt{2} \int_0^1 f(B_s) dW_s$ as $n \to \infty$, where W stands for an independent Brownian motion.

1. Prove that, for any $m \le n$,

$$S_n(f) = 2^{-n/2} \sum_{l=0}^{2^m - 1} \sum_{k=l2^{n-m}}^{(l+1)2^{n-m} - 1} f(B_{k2^{-n}}) \left[n \left(B_{(k+1)2^{-n}} - B_{k2^{-n}} \right)^2 - 1 \right].$$

2. For $m \le n$, we set

$$T_{n,m}(f) = 2^{-n/2} \sum_{l=0}^{2^m - 1} f(B_{l2^{-m}}) \sum_{k=l2^{n-m}}^{(l+1)2^{n-m} - 1} \left[n \left(B_{(k+1)2^{-n}} - B_{k2^{-n}} \right)^2 - 1 \right].$$

   Prove that $\lim_{m \to \infty} \sup_{n \ge m} E[(S_n - T_{n,m})^2] = 0$.

3. Fix $m \ge 1$. By using Theorem 6.2.3, show that the random vector

$$\left\{ B_{l2^{-m}} \, ; \, 2^{-n/2} \sum_{k=l2^{n-m}}^{(l+1)2^{n-m} - 1} \left[n \left(B_{(k+1)2^{-n}} - B_{k2^{-n}} \right)^2 - 1 \right], \, l = 0, \ldots, 2^m - 1 \right\}$$

converges in law to $\left\{ B_{l2^{-m}}; \sqrt{2}\big(W_{(l+1)2^{-m}} - W_{l2^{-m}}\big), \; l = 0, \ldots, \right.$

$\left. 2^m - 1 \right\}$ as $n \to \infty$. Deduce that $T_{n,m} \overset{\text{Law}}{\longrightarrow} \sqrt{2} \sum_{l=0}^{2^m-1} f(B_{l2^{-m}})$
$\left[W_{(l+1)2^{-m}} - W_{l2^{-m}} \right]$ as $n \to \infty$.

4. Prove that $S_n(f) \overset{\text{Law}}{\longrightarrow} \sqrt{2} \int_0^1 f(B_s) dW_s$ as $n \to \infty$.

5. Using similar arguments, show that $S_n(f)$ continues to converge in distribution to $\sigma_H \int_0^1 f(B_s) dW_s$, with $\sigma_H > 0$ and W a standard Brownian motion independent of B when, more generally, B is a fractional Brownian motion of Hurst index $H \in \left(\frac{1}{4}, \frac{3}{4}\right)$. (See [86] for the details, as well as for a study in the cases $H \leq \frac{1}{4}$ and $H \geq \frac{3}{4}$.)

6.5 Bibliographic comments

Theorem 6.1.1 is taken from Nourdin, Peccati and Réveillac [95], whereas the estimate in Theorem 6.1.2 can be found in Nourdin, Peccati and Reinert [94]. Theorem 6.2.2 was established by Noreddine and Nourdin in [82]. Theorem 6.2.3 was first proved by Peccati and Tudor in [111], by means of stochastic calculus techniques; see also [105]. See Nualart and Ortiz-Latorre [100] for an alternative proof based on Malliavin calculus and on the use of characteristic functions. Some related results are discussed by Airault, Malliavin and Viens in [2]. Hu and Nualart proved Theorem 6.3.1 in [52], in the context of limit theorems for the local times of a fractional Brownian motion. Other distinguished applications of Theorem 6.2.3 are developed, for example, in the papers by Barndorff-Nielsen, Corcuera and Podolskij [8, 9], Barndorff-Nielsen, Corcuera, Podolskij and Woerner [10], and Corcuera, Nualart and Woerner [27]. See Peccati and Zheng [114] for several extensions of the results of this section to the multivariate normal approximation of the functionals of a random Poisson measure.

7

Exploring the Breuer–Major theorem

7.1 Motivation

Let $Y = \{Y_k : k \geq 1\}$ be a sequence of i.i.d. random variables, such that $E[Y_1^2] = 1$ and $E[Y_1] = 0$, and define

$$V_n = \frac{1}{\sqrt{n}} \sum_{k=1}^{n} Y_k, \quad n \geq 1,$$

to be the associated sequence of normalized partial sums. Recall the following statement from Theorem 3.7.1, which contains two of the most fundamental results in the theory of probability.

Theorem 7.1.1 (CLT and Berry–Esseen inequality) *As $n \to \infty$,*

$$V_n \overset{\text{Law}}{\longrightarrow} N \sim \mathcal{N}(0, 1). \tag{7.1.1}$$

Moreover,

$$d_{\text{Kol}}(V_n, N) \leq \frac{0.4785\, E[|Y_1|^3]}{\sqrt{n}}, \quad n \geq 1. \tag{7.1.2}$$

Our aim is now to use the theory developed in the previous chapters to (partially) answer the following question:

Is it possible to prove results analogous to (7.1.1) and (7.1.2) when the random variables $Y_1, Y_2, \ldots$ are no longer independent?

In particular, we shall focus on the stochastic dependence associated with *Gaussian subordinated processes*, that is, with random sequences that can be written as a deterministic function of a Gaussian sequence with a non-trivial covariance structure. A crucial ingredient of our approach is the notion of *Hermite rank* (which roughly measures the degree of polynomial complexity of a given function), as well as a detailed study of some infinite series of covariance

128

coefficients. Note that any central limit result involving Hermite ranks and covariances is customarily called a *Breuer–Major theorem*, in honor of the seminal paper [18]. See the bibliographic comments for a wider historical perspective.

7.2 A general statement

Let $d\gamma(x) = (2\pi)^{-1/2}e^{-x^2/2}dx$ be the standard Gaussian measure on the real line, and let $f \in L^2(\gamma)$ be such that $\int_{\mathbb{R}} f(x)d\gamma(x) = 0$. According to Proposition 1.4.2(iv), the function f can be expanded into a series of Hermite polynomials as follows:

$$f(x) = \sum_{q=1}^{\infty} a_q H_q(x). \tag{7.2.1}$$

Definition 7.2.1 The **Hermite rank** of the function f in (7.2.1) is the integer $d \geq 1$ uniquely defined as

$$d = \inf\{q \geq 1 : a_1 = \ldots = a_{q-1} = 0 \text{ and } a_q \neq 0\},$$

that is, d is the order of the first non-trivial element in the Hermite expansion of f.

Example 7.2.2 Let $f(x) = |x|^p - \int_{\mathbb{R}} |x|^p d\gamma(x)$ for some $p > 0$. Then,

$$\int_{\mathbb{R}} f(x)d\gamma(x) = \int_{\mathbb{R}} xf(x)d\gamma(x) = 0,$$

whereas

$$\int_{\mathbb{R}} f(x)H_2(x)d\gamma(x) = p\int_{\mathbb{R}} |x|^p d\gamma(x) \neq 0.$$

It follows that f has Hermite rank $d = 2$.

Now consider the following elements:

- $X = \{X_k : k \in \mathbb{Z}\}$ is a centered stationary Gaussian sequence with unit variance. For all $v \in \mathbb{Z}$, we set $\rho(v) = E[X_0 X_v]$, so that in particular $\rho(0) = E[X_0^2] = 1$ and therefore, by Cauchy–Schwarz, $|\rho(v)| \leq 1$ for every v. Recall that, by definition of stationary Gaussian sequence (see Definition A.1.3), $\rho(v) = \rho(-v)$.
- $f \in L^2(\gamma)$ is a fixed deterministic function such that $\int_{\mathbb{R}} f(x)d\gamma(x) = E[f(X_1)] = 0$ and f has Hermite rank $d \geq 1$; this implies in particular that f admits the Hermite expansion

$$f(x) = \sum_{q=d}^{\infty} a_q H_q(x). \tag{7.2.2}$$

– The sequence

$$V_n = \frac{1}{\sqrt{n}} \sum_{k=1}^{n} f(X_k), \quad n \geq 1, \tag{7.2.3}$$

is the sequence of the renormalized partial sums associated with the *Gaussian subordinated process* $\{f(X_k) : k \geq 1\}$.

We now wish to establish explicit sufficient conditions for the sequence $(V_n)_{n \geq 1}$ defined above to satisfy a CLT. Note that, if $\rho(k) \neq 0$ for some $k \neq 0$, then the sequence $(f(X_k))_{k \geq 1}$ is not composed of independent random variables, and Theorem 7.1.1 can no longer be applied.

The next result allows us to express the framework of this section in terms of some underlying isonormal Gaussian process.

Proposition 7.2.3 *There exists a real separable Hilbert space $\mathfrak{H}$, as well as an isonormal Gaussian process over $\mathfrak{H}$, written $\{X(h) : h \in \mathfrak{H}\}$, with the property that there exists a set $E = \{\varepsilon_k : k \in \mathbb{Z}\} \subset \mathfrak{H}$ such that (i) E generates $\mathfrak{H}$; (ii) $\langle \varepsilon_k, \varepsilon_l \rangle_{\mathfrak{H}} = \rho(k - l)$ for every $k, l \in \mathbb{Z}$; and (iii) $X_k = X(\varepsilon_k)$ for every $k \in \mathbb{Z}$.*

Proof Denote by $\mathcal{E}$ the set of all real-valued sequences of the type $h = \{h_l : l \in \mathbb{Z}\}$ such that $h_l \neq 0$ only for a finite number of integers l. We define $\mathfrak{H}$ to be the real separable Hilbert space obtained by closing $\mathcal{E}$ with respect to the scalar product

$$\langle g, h \rangle_{\mathfrak{H}} = \sum_{k,l \in \mathbb{Z}} g_k h_l \rho(k - l) = \sum_{k,l \in \mathbb{Z}} g_k h_l E[X_k X_l]. \tag{7.2.4}$$

If $h \in \mathfrak{H}$, set $X(h)$ to be the $L^2(\Omega)$- limit of any sequence of the type $\{X(h_n)\}$, where $\{h_n\} \subset \mathcal{E}$ converges to h in $\mathfrak{H}$. (Note that such a sequence $\{h_n\}$ necessarily exists and may not be unique; however, the definition of $X(h)$ does not depend on the choice of $\{h_n\}$.) Then by construction, the centered Gaussian family $\{X(h) : h \in \mathfrak{H}\}$ is an isonormal Gaussian process over $\mathfrak{H}$. Now define the class $E = \{\varepsilon_k : k \in \mathbb{Z}\}$ by setting $\varepsilon_k = \{\delta_{kl} : l \in \mathbb{Z}\}$, where $\delta_{kl} = 1$ when $l = k$ and $\delta_{kl} = 0$ otherwise. Since $X_k = X(\varepsilon_k)$ by construction and E generates $\mathfrak{H}$, the proof is complete. $\square$

The following statement contains the most important result of this chapter. Recall that V_n is defined by (7.2.3), and that f has Hermite rank $d \geq 1$.

Theorem 7.2.4 (Breuer–Major theorem) *Assume that $\sum_{v \in \mathbb{Z}} |\rho(v)|^d < \infty$, and set*

$$\sigma^2 = \sum_{q=d}^{\infty} q! a_q^2 \sum_{v \in \mathbb{Z}} \rho(v)^q \in [0, \infty).$$

Then

$$V_n \xrightarrow{\text{Law}} \mathcal{N}(0, \sigma^2) \quad \text{as } n \to \infty. \tag{7.2.5}$$

Proof Our main tool will be Theorem 6.3.1. By Proposition 7.2.3, we can assume, without loss of generality, that $X_k = X(\varepsilon_k)$, where $\{X(h) : h \in \mathfrak{H}\}$ is an adequate isonormal Gaussian process and $\langle \varepsilon_k, \varepsilon_l \rangle_{\mathfrak{H}} = \rho(k - l)$ for every $k, l \in \mathbb{Z}$. Since $E[X_k^2] = \|\varepsilon_k\|_{\mathfrak{H}}^2 = 1$ for all k, we have by (2.7.7) that

$$\frac{1}{\sqrt{n}} \sum_{k=1}^{n} f(X_k) = \frac{1}{\sqrt{n}} \sum_{k=1}^{n} \sum_{q=d}^{\infty} a_q H_q(X_k) = \sum_{q=d}^{\infty} I_q(f_{n,q}),$$

where the kernels $f_{n,q} \in \mathfrak{H}^{\odot q}$, $n \geq 1$, are given by

$$f_{n,q} = \frac{a_q}{\sqrt{n}} \sum_{k=1}^{n} \varepsilon_k^{\otimes q}.$$

To conclude the proof, we shall now check that the four conditions (a)–(d) in Theorem 6.3.1 hold.

Condition (a). Fix $q \geq d$. We have

$$q! \|f_{n,q}\|_{\mathfrak{H}^{\otimes q}}^2 = \frac{q! a_q^2}{n} \sum_{k,l=1}^{n} \rho(k-l)^q = q! a_q^2 \sum_{v \in \mathbb{Z}} \rho(v)^q \left(1 - \frac{|v|}{n}\right) \mathbf{1}_{\{|v|<n\}}.$$

It follows from the dominated convergence theorem that

$$q! \|f_{n,q}\|_{\mathfrak{H}^{\otimes q}}^2 \longrightarrow \sigma_q^2 := q! a_q^2 \sum_{v \in \mathbb{Z}} \rho(v)^q \quad \text{as } n \to \infty. \tag{7.2.6}$$

Condition (b). Since $E[X_v^2] = 1$ for all v, we have by Cauchy–Schwarz that $|\rho(v)| \leq 1$ for all v. We can thus write

$$\sum_{q=d}^{\infty} q! a_q^2 \sum_{v \in \mathbb{Z}} \rho(v)^q \leq \sum_{q=d}^{\infty} q! a_q^2 \times \sum_{v \in \mathbb{Z}} |\rho(v)|^d = E[f^2(X_1)] \sum_{v \in \mathbb{Z}} |\rho(v)|^d < \infty.$$

Condition (c). Fix $q \geq d, q \neq 1$. For all $n \geq 1$ and $r = 1, \ldots, q-1$, we have

$$f_{n,q} \otimes_r f_{n,q} = \frac{a_q^2}{n} \sum_{k,l=1}^{n} \rho(k-l)^r \varepsilon_k^{\otimes(q-r)} \otimes \varepsilon_l^{\otimes(q-r)}.$$

We deduce that

$$\|f_{n,q} \otimes_r f_{n,q}\|_{\mathfrak{H}^{\otimes(2q-2r)}}^2 = \frac{a_q^4}{n^2} \sum_{i,j,k,l=1}^{n} \rho(k-l)^r \rho(i-j)^r \rho(k-i)^{q-r} \rho(l-j)^{q-r}.$$

Consequently, using $\left| \rho(k-l)^r \rho(k-i)^{q-r} \right| \leq |\rho(k-l)|^q + |\rho(k-i)|^q$, we obtain that

$$\|f_{n,q} \otimes_r f_{n,q}\|_{\mathfrak{H}^{\otimes(2q-2r)}}^2$$

$$\leq \frac{a_q^4}{n^2} \sum_{i,j,k,l=1}^{n} |\rho(k-l)|^q \left(|\rho(i-j)|^r |\rho(l-j)|^{q-r} + |\rho(i-j)|^{q-r} |\rho(l-j)|^r \right)$$

$$\leq \frac{a_q^4}{n^2} \sum_{k\in\mathbb{Z}} |\rho(k)|^q \sum_{i,j,l=1}^{n} \left(|\rho(i-j)|^r |\rho(l-j)|^{q-r} + |\rho(i-j)|^{q-r} |\rho(l-j)|^r \right)$$

$$\leq \frac{2a_q^4}{n} \sum_{k\in\mathbb{Z}} |\rho(k)|^d \sum_{|i|<n} |\rho(i)|^r \sum_{|j|<n} |\rho(j)|^{q-r}$$

$$= 2a_q^4 \sum_{k\in\mathbb{Z}} |\rho(k)|^d \times n^{-1+\frac{r}{q}} \sum_{|i|<n} |\rho(i)|^r \times n^{-1+\frac{q-r}{q}} \sum_{|j|<n} |\rho(j)|^{q-r}.$$

Therefore, to conclude that $\|f_{n,q} \otimes_r f_{n,q}\|_{\mathfrak{H}^{\otimes(2q-2r)}} \to 0$, it remains to prove that, for any $r = 1, \ldots, q-1$,

$$n^{-1+\frac{r}{q}} \sum_{|j|<n} |\rho(j)|^r \to 0. \tag{7.2.7}$$

To do so, fix $\delta \in (0,1)$, and decompose the sum as $\sum_{|j|<n} = \sum_{|j|\leq[n\delta]} + \sum_{[n\delta]<|j|<n}$. By the Hölder inequality we obtain (recall that $\sum_{j\in\mathbb{Z}} |\rho(j)|^q \leq \sum_{j\in\mathbb{Z}} |\rho(j)|^d < \infty$)

$$n^{-1+r/q} \sum_{|j|\leq[n\delta]} |\rho(j)|^r \leq n^{-1+r/q} (2[n\delta]+1)^{1-r/q} \left(\sum_{j\in\mathbb{Z}} |\rho(j)|^q \right)^{r/q} \leq c\delta^{1-r/q},$$

where c is some constant, as well as

$$n^{-1+r/q} \sum_{[n\delta]<|j|<n} |\rho(j)|^r \leq \left(\sum_{[n\delta]<|j|<n} |\rho(j)|^q \right)^{r/q}.$$

The first term converges to 0 as δ goes to zero (because $1 \le r \le q - 1$), and the second also converges to 0 for fixed δ and $n \to \infty$. This proves that (7.2.7) holds.

Condition (d). For $N \ge d$, we have

$$\sum_{q=N+1}^{\infty} q! \|f_{n,q}\|_{\mathfrak{H}^{\otimes q}}^2 = \frac{1}{n} \sum_{q=N+1}^{\infty} a_q^2 q! \sum_{k,l=1}^{n} \rho(k-l)^q \le \sum_{q=N+1}^{\infty} a_q^2 q! \sum_{v \in \mathbb{Z}} |\rho(v)|^q$$

$$\le \sum_{v \in \mathbb{Z}} |\rho(v)|^d \times \sum_{q=N+1}^{\infty} a_q^2 q!,$$

so that $\lim_{N \to \infty} \sup_{n \ge 1} \sum_{q=N+1}^{\infty} q! \|f_{n,q}\|_{\mathfrak{H}^{\otimes q}}^2 = 0$. The proof of Theorem 7.2.4 is complete. $\qquad\qquad\square$

7.3 Quadratic case

Here we restrict ourselves to the quadratic case, that is, we assume throughout this section that $f(x) = H_2(x) = x^2 - 1$. We continue to deal with a general centered stationary Gaussian sequence $\{X_k : k \in \mathbb{Z}\}$ with unit variance and covariance function given by $E[X_v X_0] = \rho(v)$, $v \in \mathbb{Z}$ (with $\rho(0) = 1$) so that, given our choice of f,

$$V_n = \frac{1}{\sqrt{n}} \sum_{k=0}^{n-1} [X_k^2 - 1], \quad n \ge 1.$$

We also define $v_n > 0$ as $E[V_n^2] = v_n^2$. Our first result is the following.

Theorem 7.3.1 *Let $N \sim \mathcal{N}(0, 1)$. Then, for all $n \ge 1$,*

$$d_{\mathrm{TV}}(V_n/v_n, N) \le \frac{4\sqrt{2}}{v_n^2 \sqrt{n}} \left(\sum_{k=-n+1}^{n-1} |\rho(k)|^{\frac{4}{3}} \right)^{\frac{3}{2}}. \tag{7.3.1}$$

Exercise 7.3.2 1. Let $(z(k))_{k \in \mathbb{Z}}$ be a sequence of positive numbers such that $\sum_{k \in \mathbb{Z}} z(k)^2 < \infty$, and let $e \in [1, 2)$. Prove that, as $n \to \infty$,

$$n^{\frac{e}{2}-1} \sum_{|k| \le n} z(k)^e \to 0.$$

2. Combine the conclusion of part 1 with the estimate (7.3.1) in order to show that, if $\sum_{k \in \mathbb{Z}} \rho(k)^2 < \infty$, then

$$\left(\sum_{k=-n+1}^{n-1} |\rho(k)|^{\frac{4}{3}} \right)^{\frac{3}{2}} = o\left(\sqrt{n}\right),$$

and therefore that V_n/v_n converges to $\mathcal{N}(0, 1)$ in total variation. This shows that Theorem 7.3.1 is indeed a partial refinement of Theorem 7.2.4 (see also Corollary 5.2.8).

Proof of Theorem 7.3.1 As in the proof of Theorem 7.2.4, and thanks to Proposition 7.2.3, we may assume without loss of generality that $X_k = X(\varepsilon_k)$, where $X = \{X(h) : h \in \mathfrak{H}\}$ is some isonormal Gaussian process and $\langle \varepsilon_k, \varepsilon_l \rangle_{\mathfrak{H}} = \rho(k - l)$ for every $k, l \in \mathbb{Z}$. We then have $V_n/v_n = I_2(f_n)$ with

$$f_n = \frac{1}{v_n \sqrt{n}} \sum_{k=0}^{n-1} \varepsilon_k^{\otimes 2}.$$

On the other hand, recall that the convolution of two sequences $\{u(n)\}_{n \in \mathbb{Z}}$ and $\{v(n)\}_{n \in \mathbb{Z}}$ is the sequence $u * v$ defined as $(u * v)(j) = \sum_{n \in \mathbb{Z}} u(n) v(j - n)$, and observe that $(u * v)(l - i) = \sum_{k \in \mathbb{Z}} u(k - l) v(k - i)$ whenever $u(n) = u(-n)$ and $v(n) = v(-n)$ for all $n \in \mathbb{Z}$. Set

$$\rho_n(k) = |\rho(k)| \mathbf{1}_{\{|k| \le n-1\}}, \quad k \in \mathbb{Z}, n \ge 1.$$

We then have (using (5.2.3) for the first equality, and noticing that $f_n \otimes_1 f_n = f_n \widetilde{\otimes}_1 f_n$),

$$\text{Var}\left(\frac{1}{2} \| D[I_2(f_n)] \|_{\mathfrak{H}}^2 \right) = 8 \| f_n \otimes_1 f_n \|_{\mathfrak{H}^{\otimes 2}}^2$$

$$= \frac{8}{v_n^4 n^2} \sum_{i,j,k,l=0}^{n-1} \rho(k - l) \rho(i - j) \rho(k - i) \rho(l - j)$$

$$\le \frac{8}{v_n^4 n^2} \sum_{i,l=0}^{n-1} \sum_{j,k \in \mathbb{Z}} \rho_n(k - l) \rho_n(i - j) \rho_n(k - i) \rho_n(l - j)$$

$$= \frac{8}{v_n^4 n^2} \sum_{i,l=0}^{n-1} (\rho_n * \rho_n)(l - i)^2$$

$$\le \frac{8}{v_n^4 n} \sum_{k \in \mathbb{Z}} (\rho_n * \rho_n)(k)^2 = \frac{8}{v_n^4 n} \| \rho_n * \rho_n \|_{\ell^2(\mathbb{Z})}^2.$$

Recall Young's inequality: if $s, p, q \geq 1$ are such that $\frac{1}{p} + \frac{1}{q} = 1 + \frac{1}{s}$, then

$$\|u * v\|_{\ell^s(\mathbb{Z})} \leq \|u\|_{\ell^p(\mathbb{Z})} \|v\|_{\ell^q(\mathbb{Z})}. \tag{7.3.2}$$

Let us apply (7.3.2) with $u = v = \rho_n$, $s = 2$ and $p = \frac{4}{3}$. We get

$$\|\rho_n * \rho_n\|^2_{\ell^2(\mathbb{Z})} \leq \|\rho_n\|^4_{\ell^{\frac{4}{3}}(\mathbb{Z})},$$

so that

$$\mathrm{Var}\left(\frac{1}{2}\|D[I_2(f_n)]\|^2_{\mathfrak{H}}\right) \leq \frac{8}{v_n^4 n} \left(\sum_{k=-n+1}^{n-1} |\rho(k)|^{\frac{4}{3}}\right)^3.$$

We then recover the bound (7.3.1) from the general estimate (5.2.13). $\qquad\square$

The following result characterizes the asymptotic behavior of the cumulants of V_n under a suitable integrability condition on ρ.

Proposition 7.3.3 *Fix an integer $s \geq 2$, and assume that $\rho \in \ell^{\frac{s}{s-1}}(\mathbb{Z})$. Then the function*

$$h(x) = \sum_{k \in \mathbb{Z}} \rho(k)e^{ikx}, \quad x \in [-\pi, \pi]$$

(the sum being understood in the $L^2([-\pi, \pi])$ sense), is even, real-valued, positive, and belongs to $L^s([-\pi, \pi])$. Moreover, the sth cumulant of V_n behaves asymptotically as

$$\kappa_s(V_n) \sim n^{1-s/2} \frac{2^{s-1}(s-1)!}{2\pi} \int_{-\pi}^{\pi} h^s(x)dx, \quad \text{as } n \to \infty. \tag{7.3.3}$$

Proof As in the proof of Theorem 7.2.4, we assume that $X_k = X(\varepsilon_k)$, where $X = \{X(h) : h \in \mathfrak{H}\}$ is some isonormal Gaussian process and $\langle \varepsilon_k, \varepsilon_l \rangle_{\mathfrak{H}} = \rho(k-l)$ for every $k, l \in \mathbb{Z}$. We then have $V_n = I_2(f_n)$ with $f_n = \frac{1}{\sqrt{n}} \sum_{k=0}^{n-1} \varepsilon_k^{\otimes 2}$. Due to our assumption, we automatically have that $\rho \in \ell^r(\mathbb{Z})$ for all $r \geq \frac{s}{s-1}$, in particular for $r = 2$. Now let us proceed with the proof. It is divided into several steps.

First step Using the formula (2.7.17) giving the cumulants of $V_n = I_2(f_n)$ (see also (8.4.3)) as well as the very definition of the contraction $\otimes_1$, we can immediately see that

$$\kappa_s(V_n) = \frac{2^{s-1}(s-1)!}{n^{s/2}} \sum_{k_1,\dots,k_s=0}^{n-1} \rho(k_s - k_{s-1})\dots\rho(k_2 - k_1)\rho(k_1 - k_s).$$

Second step Let us prove that $0 \leq \langle |\rho|^{*(s-1)}, |\rho| \rangle_{\ell^2(\mathbb{Z})} < \infty$. Thanks to the Hölder inequality, we are left to show that $|\rho|^{*(s-1)} \in \ell^s(\mathbb{Z})$. By repeatedly applying Young inequality, we have

$$\| \, |\rho|^{*(s-1)} \|_{\ell^s(\mathbb{Z})} \leq \|\rho\|_{\ell^{\frac{s}{s-1}}(\mathbb{Z})} \| \, |\rho|^{*(s-2)} \|_{\ell^{\frac{s}{2}}(\mathbb{Z})} \leq \|\rho\|^2_{\ell^{\frac{s}{s-1}}(\mathbb{Z})} \| \, |\rho|^{*(s-3)} \|_{\ell^{\frac{s}{3}}(\mathbb{Z})}$$

$$\leq \ldots \leq \|\rho\|^{s-1}_{\ell^{\frac{s}{s-1}}(\mathbb{Z})} < \infty \quad \text{(because } \rho \in \ell^{\frac{s}{s-1}} \text{ by assumption)},$$

so that the desired conclusion follows.

Third step Thanks to the result shown in the previous step, observe first that

$$\sum_{k_2,\ldots,k_s \in \mathbb{Z}} |\rho(k_2)\rho(k_2 - k_3)\rho(k_3 - k_4)\ldots\rho(k_{s-1} - k_s)\rho(k_s)|$$

$$= \langle |\rho|^{*(s-1)}, |\rho| \rangle_{\ell^2(\mathbb{Z})} < \infty.$$

Hence, one can apply dominated convergence to get, as $n \to \infty$, that

$$\frac{n^{s/2-1}}{2^{s-1}(s-1)!} \kappa_s(V_n)$$

$$= \frac{1}{n} \sum_{k_1=0}^{n-1} \sum_{k_2,\ldots,k_s=-k_1}^{n-1-k_1} \rho(k_2)\rho(k_2 - k_3)\rho(k_3 - k_4)\ldots\rho(k_{s-1} - k_s)\rho(k_s)$$

$$= \sum_{k_2,\ldots,k_s \in \mathbb{Z}} \rho(k_2)\rho(k_2 - k_3)\rho(k_3 - k_4)\ldots\rho(k_{s-1} - k_s)\rho(k_s)$$

$$\times \left[1 \wedge \left(1 - \frac{\max\{k_2,\ldots,k_s\}}{n} \right) - 0 \vee \left(\frac{\min\{k_2,\ldots,k_s\}}{n} \right) \right] \mathbf{1}_{\{|k_2|<n,\ldots,|k_s|<n\}}$$

$$\to \sum_{k_2,\ldots,k_s \in \mathbb{Z}} \rho(k_2)\rho(k_2 - k_3)\rho(k_3 - k_4)\ldots\rho(k_{s-1} - k_s)\rho(k_s) = \langle \rho^{*(s-1)}, \rho \rangle_{\ell^2(\mathbb{Z})}.$$

$$(7.3.4)$$

Fourth step That h is a well-defined function of $L^2([-\pi,\pi])$ is a consequence of the Parseval identity and $\rho \in \ell^2(\mathbb{Z})$, whereas that h is an even real-valued function is an immediate consequence of $\rho(k) = \rho(-k) \in \mathbb{R}$ for all $k \in \mathbb{Z}$. Let us now focus on the positivity of h. For all $x \in \mathbb{R}$,

$$h_n(x) := \sum_{|k|<n} e^{ikx} \rho(k) \left(1 - \frac{|k|}{n} \right) = \frac{1}{n} \sum_{k,l=0}^{n-1} e^{i(k-l)x} \rho(k-l)$$

$$= \frac{1}{n} E \left| \sum_{k=0}^{n-1} e^{ikx} X_k \right|^2 \geq 0.$$

Moreover, for all $m \geq 1$ and all $n \geq m$, we can write

$$\|h_n - h\|^2_{L^2([-\pi,\pi])} = \frac{1}{n^2} \sum_{|k|<n} k^2 \rho^2(k) + \sum_{|k|\geq n} \rho^2(k)$$

$$\leq \frac{1}{n^2} \sum_{|k|<m} k^2 \rho^2(k) + 2 \sum_{|k|\geq m} \rho^2(k),$$

so that, for all m,

$$\limsup_{n \to \infty} \|h_n - h\|^2_{L^2([-\pi,\pi])} \leq 2 \sum_{|k| \geq m} \rho^2(k),$$

and then $\lim_{n \to \infty} \|h_n - h\|^2_{L^2([-\pi,\pi])} = 0$ by letting $m \to \infty$. Consequently, the function h is a.e. positive.

Fifth and final step Using Cauchy products, we can immediately see that $h^{s-1}(x) = \sum_{k \in \mathbb{Z}} \rho^{*(s-1)}(k)e^{ikx}$. Hence, using the Parseval identity (and because $\rho(k) = \rho(-k)$), we get that

$$\langle \rho^{*(s-1)}, \rho \rangle_{\ell^2(\mathbb{Z})} = \frac{1}{2\pi} \int_{-\pi}^{\pi} h^s(x)dx,$$

which, according to the third step, shows that $h \in L^s([-\pi, \pi])$ and that (7.3.3) holds. $\qquad\square$

We conclude this section by proving a result that will be useful in Chapter 9.

Proposition 7.3.4 *For any fixed $n \geq 1$, there exists an integer $m(n) \geq 1$, a sequence $(\lambda_{n,i})_{0 \leq i \leq m(n)-1}$ of non-negative real numbers as well as a sequence $(N_i)_{0 \leq i \leq m(n)-1}$ of independent $\mathcal{N}(0, 1)$ random variables, such that*

$$\sqrt{n} \, V_n = \sum_{i=0}^{n-1} \left(X_i^2 - 1\right) = \sum_{i=0}^{m(n)-1} \lambda_{n,i}\left(N_i^2 - 1\right).$$

In particular, by (2.7.17), all the cumulants of V_n are non-negative.

Proof By Proposition 7.2.3, we can assume, without loss of generality, that $X_k = X(\varepsilon_k)$, for some isonormal Gaussian process $X = \{X(h) : h \in \mathfrak{H}\}$. We have $\|\varepsilon_k\|_{\mathfrak{H}} = 1$ for each k. Let $\{\eta_j : 0 \leq j \leq m(n) - 1\}$ be an orthonormal family of $\mathfrak{H}$ such that $\text{Vect}\{\eta_j : 0 \leq j \leq m(n) - 1\} = \text{Vect}\{\varepsilon_k : 0 \leq k \leq n - 1\}$. Set $Y_j = X(\eta_j)$, $j = 0, \ldots, m(n) - 1$. Let B be the matrix defined as $B_{i,j} = \langle \varepsilon_i, \eta_j \rangle_{\mathfrak{H}}$, $i = 0, \ldots, n - 1$; $j = 0, \ldots, m(n) - 1$. The matrix $B^T B$ being real symmetric and positive, we have $B^T B = P^T D P$ for some orthogonal matrix $P \in O_n(\mathbb{R})$ and some diagonal matrix $D = [\lambda_{n,0}, \ldots, \lambda_{n,m(n)-1}]$ having non-negative entries. Moreover,

$$\text{Tr}(B^T B) = \sum_{k=0}^{n-1} \sum_{i=0}^{m(n)-1} \langle \varepsilon_k, \eta_i \rangle^2_{\mathfrak{H}} = \sum_{k=0}^{n-1} \|\varepsilon_k\|^2_{\mathfrak{H}} = n.$$

Set $N^T = PY^T$. Because P is an orthogonal matrix and $\{Y_0, \ldots, Y_{m(n)-1}\}$ is a family of independent $\mathcal{N}(0, 1)$ random variables, the family $\{N_0, \ldots, N_{m(n)-1}\}$ is formed by independent $\mathcal{N}(0, 1)$ random variables as well. For any $i = 0, \ldots, n-1$, we have $\varepsilon_i = \sum_{j=0}^{m(n)-1} B_{i,j}\, \eta_j$, leading to $(X_0, \ldots, X_{n-1})^T = B(Y_0, \ldots, Y_{m(n)-1})^T$. Therefore

$$\sum_{i=0}^{n-1} X_i^2 = \|X^T\|_{\mathbb{R}^n}^2 = \|BY^T\|_{\mathbb{R}^m}^2 = \langle N^T, DN^T\rangle_{\mathbb{R}^m} = \sum_{i=0}^{m(n)-1} \lambda_{n,i}\, N_i^2,$$

so that, because $\sum_{i=0}^{m(n)-1} \lambda_{n,i} = \mathrm{Tr}(B^T B) = n$,

$$\sum_{i=0}^{n-1} (X_i^2 - 1) = \sum_{i=0}^{m(n)-1} \lambda_{n,i}(N_i^2 - 1). \qquad \square$$

7.4 The increments of a fractional Brownian motion

Fix $H \in (0, 1)$. We recall from Appendix D that a *fractional Brownian motion* of Hurst index H is a centered Gaussian process of type $B^H = (B_t^H)_{t \in \mathbb{R}}$, such that

$$E[B_t^H B_s^H] = \frac{1}{2}\Big(|t|^{2H} + |s|^{2H} - |t-s|^{2H}\Big), \quad s, t \in \mathbb{R}.$$

Its increments

$$X_k = B_{k+1}^H - B_k^H, \quad k \in \mathbb{Z}, \qquad (7.4.1)$$

customarily called 'fractional Gaussian noise', constitute a stationary Gaussian sequence with covariance given by

$$\rho(v) = E[X_k X_{k+v}] = \frac{1}{2}\big(|v+1|^{2H} + |v-1|^{2H} - 2|v|^{2H}\big), \quad v \in \mathbb{Z}. \;\; (7.4.2)$$

This covariance behaves asymptotically as

$$\rho(v) = H(2H-1)|v|^{2H-2} + o(|v|^{2H-2}), \quad \text{as } |v| \to \infty. \qquad (7.4.3)$$

Observe that $\rho(0) = 1$ and that:

(1) for $0 < H < 1/2$, we have $\rho(v) < 0$ for $v \neq 0$, $\sum_{v \in \mathbb{Z}} |\rho(v)| < \infty$ and $\sum_{v \in \mathbb{Z}} \rho(v) = 0$;
(2) for $H = 1/2$, we have $\rho(v) = 0$ if $v \neq 0$;
(3) for $1/2 < H < 1$, we have $\sum_{v \in \mathbb{Z}} |\rho(v)| = \infty$.

When $H \geq 1/2$, the Hurst index H measures the strength of the dependence between the increments of B^H: the larger is H, the stronger is the dependence.

In this section, we are interested in the asymptotic behavior of V_n defined by (7.2.3) in the particular case where f is a Hermite polynomial and the sequence $(X_k)_{k\in\mathbb{Z}}$ is given by (7.4.1). More specifically, suppose that

$$V_n = \frac{1}{\sqrt{n}} \sum_{k=0}^{n-1} H_q(B_{k+1}^H - B_k^H), \quad n \geq 1, \qquad (7.4.4)$$

where H_q stands for a Hermite polynomial of degree $q \geq 2$ (observe that the case where $q = 1$ is indeed trivial). Our main result is the following.

Theorem 7.4.1 *As $n \to \infty$, the following asymptotic relations hold:*

− *If $0 < H < 1 - \frac{1}{2q}$, then*

$$V_n \xrightarrow{\text{Law}} \mathcal{N}(0, \sigma_{H,q}^2), \qquad (7.4.5)$$

with $\sigma_{H,q}^2 = \frac{q!}{2^q} \sum_{r\in\mathbb{Z}} \left(|r+1|^{2H} + |r-1|^{2H} - 2|r|^{2H}\right)^q \in (0, \infty)$.
− *If $H = 1 - \frac{1}{2q}$, then*

$$\frac{V_n}{\sqrt{\log n}} \xrightarrow{\text{Law}} \mathcal{N}(0, \sigma_{1-1/(2q),q}^2), \qquad (7.4.6)$$

with $\sigma_{1-1/(2q),q}^2 = 2q! \left(1 - \frac{1}{q}\right)^q \left(1 - \frac{1}{2q}\right)^q$.
− *If $H > 1 - \frac{1}{2q}$, then*

$$n^{q(1-H)-\frac{1}{2}} V_n \xrightarrow{\text{Law}} F_\infty \qquad (7.4.7)$$

where F_∞ has a so-called 'Hermite distribution' (see Proposition 7.4.2).

We present three separate proofs for (7.4.5)–(7.4.7).

Proof of (7.4.5) This convergence follows directly from Theorem 7.2.4. Indeed, (7.4.3) implies that $\sum_{v\in\mathbb{Z}} |\rho(v)|^q < \infty$ if (and only if) $H < 1 - \frac{1}{2q}$.
$\square$

Proof of (7.4.6) The proof is straightforward but involves tedious calculations. The interested reader is refered to [17] or to the proof of Theorem 7.4.5 for the quadratic case.
$\square$

Proof of (7.4.7) Our proof of (7.4.7) is very simple. It is based on the fact that, for *fixed n*, the random variables F_n, as defined in formula (7.4.8) below, and $n^{q(1-H)-\frac{1}{2}} V_n$ share the same law, because of the self-similarity property of fractional Brownian motion (show this as an easy exercise!). Moreover, we have the following result that allows us to conclude our result.

Proposition 7.4.2 *Assume $H > 1 - \frac{1}{2q}$, and define F_n by*

$$F_n = n^{q(1-H)-1} \sum_{k=0}^{n-1} H_q \big(n^H (B_{(k+1)/n}^H - B_{k/n}^H) \big), \quad n \geq 1. \tag{7.4.8}$$

Then, as $n \to \infty$, F_n converges in $L^2(\Omega)$ to some random variable, written F_∞, which belongs to the qth chaos of B^H. The law of F_∞ (which is therefore not Gaussian, since $q \geq 2$) is customarily called a 'Hermite distribution'.

Proof For $n, m \geq 1$, we have

$$E[F_n F_m] = q!(nm)^{q-1} \sum_{k=0}^{n-1} \sum_{l=0}^{m-1} \Big(E\big[(B_{(k+1)/n}^H - B_{k/n}^H)(B_{(l+1)/m}^H - B_{l/m}^H) \big] \Big)^q.$$

Furthermore, since $H > 1/2$, we have for all $s, t \geq 0$,

$$E[B_s^H B_t^H] = H(2H - 1) \int_0^t du \int_0^s dv |u - v|^{2H-2}$$

(see also (D.1.2)). Hence,

$$E[F_n F_m] = q! H^q (2H - 1)^q$$

$$\times \frac{1}{nm} \sum_{k=0}^{n-1} \sum_{l=0}^{m-1} \Big(nm \int_{k/n}^{(k+1)/n} du \int_{l/m}^{(l+1)/m} dv |v - u|^{2H-2} \Big)^q.$$

Therefore, as $n, m \to \infty$, we have by a Riemann sums argument that

$$E[F_n F_m] \to q! H^q (2H - 1)^q \int_{[0,1]^2} |u - v|^{(2H-2)q} du dv,$$

and the limit is finite since $H > 1 - \frac{1}{2q}$. This result implies that the sequence $\{F_n : n \geq 1\}$ is Cauchy in $L^2(\Omega)$, and hence converges in $L^2(\Omega)$ to some F_∞. $\qquad\square$

We conclude this section by specializing our findings to the quadratic case (for the general case $q \geq 3$, see Remark 7.4.4(2) below), that is, from now on we set

$$V_n = \frac{1}{\sqrt{n}} \sum_{k=0}^{n-1} \big[(B_{k+1}^H - B_k^H)^2 - 1 \big], \quad n \geq 1.$$

Corollary 7.4.3 *Assume $H \leq \frac{3}{4}$, let $N \sim \mathcal{N}(0, 1)$ and define $v_n > 0$ by $E[V_n^2] = v_n^2$. Then there exists a constant $c_H > 0$ (depending only on H) such that, for all $n \geq 2$:*

$$d_{\mathrm{TV}}(V_n/v_n, N) \le c_H \times \begin{cases} \dfrac{1}{\sqrt{n}} & \text{if } H \in \left(0, \dfrac{5}{8}\right) \\[2ex] \dfrac{(\log n)^{\frac{3}{2}}}{\sqrt{n}} & \text{if } H = \dfrac{5}{8} \\[2ex] n^{4H-3} & \text{if } H \in \left(\dfrac{5}{8}, \dfrac{3}{4}\right) \\[2ex] \dfrac{1}{\log n} & \text{if } H = \dfrac{3}{4}. \end{cases} \qquad (7.4.9)$$

Proof Recall (7.4.3), the asymptotic behavior of $\rho(k)$ as $|k| \to \infty$. Hence

$$\sum_{k=-n+1}^{n-1} |\rho(k)|^{\frac{4}{3}} = \begin{cases} O(1) & \text{if } H \in (0, \frac{5}{8}) \\ O(\log n) & \text{if } H = \frac{5}{8} \\ O(n^{(8H-5)/3}) & \text{if } H \in (\frac{5}{8}, 1). \end{cases} \qquad (7.4.10)$$

Assume first that $H < \frac{3}{4}$. From (7.2.6) we have that

$$\lim_{n \to \infty} v_n^2 = 2 \sum_{v \in \mathbb{Z}} \rho^2(v) \in (0, \infty). \qquad (7.4.11)$$

This, together with (7.4.10) and (7.3.1), implies the desired conclusion for $H \in (0, \frac{3}{4})$. Assume now that $H = \frac{3}{4}$. In this case, elementary calculations lead to

$$\lim_{n \to \infty} \frac{v_n^2}{\log n} = \frac{9}{16}.$$

This, together with (7.4.10) and (7.3.1), implies the desired conclusion for $H = \frac{3}{4}$. The proof of the corollary is complete. $\qquad \square$

Remark 7.4.4 1. When $H < 5/8$ (that is, when $\rho \in \ell^{4/3}(\mathbb{Z})$), Theorem 9.5.1 implies that $\liminf_{n \to \infty} \sqrt{n} \, d_{\mathrm{TV}}(V_n/v_n, N) > 0$. Therefore, the bound $O(n^{-1/2})$ in (7.4.9) is sharp. Whether the bounds are sharp for $H > 5/8$ remains an open problem.

2. More generally, one can associate bounds to the convergence of

$$V_n = \frac{1}{\sqrt{n}} \sum_{k=0}^{n-1} H_q(B_{k+1}^H - B_k^H),$$

where H_q ($q \ge 3$) denotes the qth Hermite polynomial, see Definition 1.4.1. We refer to Exercise 7.5.1 or to [17, 88] for the details.

The next statement deals with the case $H > 3/4$. The proof makes use (without proof) of the main estimate in a remarkable paper by Davydov and Martynova [28].

Theorem 7.4.5 *Let the above assumptions and notation prevail, assume $H >$
$3/4$ and let F_∞ be as in Proposition 7.4.2. Then there exists a constant $c_H > 0$
(depending only on H) such that, for every $n \geq 1$:*

$$d_{\mathrm{TV}}(n^{\frac{3}{2}-2H} V_n, F_\infty) \leq c_H \times n^{\frac{3}{4}-H}. \tag{7.4.12}$$

Proof Since they have the same law for any fixed n, without loss of generality
we can replace $n^{\frac{3}{2}-2H} V_n$ by

$$F_n = n^{1-2H} \sum_{k=0}^{n-1} \left[n^{2H} \left(B_{(k+1)/n}^H - B_{k/n}^H \right)^2 - 1 \right].$$

In Proposition 7.4.2, it is shown that the sequence $\{F_n\}$ converges in L^2
towards F_∞. Let us specify the rate of this convergence, by showing that

$$E[(F_n - F_\infty)^2] = O(n^{3-4H}), \quad \text{as } n \to \infty. \tag{7.4.13}$$

By Proposition 2.2.1, we can write

$$E[F_n^2] = 2H^2(2H-1)^2 n^2 \sum_{k,l=0}^{n-1} \left(\int_{k/n}^{(k+1)/n} du \int_{l/n}^{(l+1)/n} dv |u-v|^{2H-2} \right)^2. \tag{7.4.14}$$

Note that we also use a straightforward equality: for $b > a \geq 0$ and $d > c \geq 0$,

$$E[(B_b - B_a)(B_d - B_c)] = H(2H-1) \int_a^b du \int_c^d dv |u-v|^{2H-2}. \tag{7.4.15}$$

By letting n go to infinity in (7.4.14), we obtain through a Riemann sum
argument that

$$E[F_\infty^2] = 2H^2(2H-1)^2 \int_{[0,1]^2} |u-v|^{4H-4} du dv$$

$$= 2H^2(2H-1)^2 \sum_{k,l=0}^{n-1} \int_{k/n}^{(k+1)/n} du \int_{l/n}^{(l+1)/n} dv |u-v|^{4H-4}. \tag{7.4.16}$$

Now, let a, b be such that $b > a \geq 0$. Still using Proposition 2.2.1 and (7.4.15),
we have

$$E\left[\left((B_b^H - B_a^H)^2 - (b-a)^{2H} \right) F_n \right]$$

$$= 2H^2(2H-1)^2 n \sum_{l=0}^{n-1} \left(\int_{l/n}^{(l+1)/n} dv \int_0^1 du \, \mathbf{1}_{[a,b]}(u) |u-v|^{2H-2} \right)^2.$$

By letting n go to infinity, we obtain

$$E\left[\left((B_b^H - B_a^H)^2 - (b-a)^{2H}\right) F_\infty\right]$$

$$= 2H^2(2H-1)^2 \int_0^1 dv \left(\int_0^1 du \, \mathbf{1}_{[a,b]}(u)|u-v|^{2H-2}\right)^2.$$

Hence,

$$E[F_n F_\infty] = 2H^2(2H-1)^2 n \sum_{k=0}^{n-1} \int_0^1 dv \left(\int_{k/n}^{(k+1)/n} du |u-v|^{2H-2}\right)^2$$

$$= 2H^2(2H-1)^2 n \sum_{k,l=0}^{n-1} \int_{l/n}^{(l+1)/n} dv \left(\int_{k/n}^{(k+1)/n} du |u-v|^{2H-2}\right)^2.$$

$$(7.4.17)$$

Finally, by combining (7.4.14), (7.4.16) and (7.4.17), and by using elementary changes of variables as well, we can write:

$$E[(F_n - F_\infty)^2]$$

$$= 2H(2H-1)^2 \sum_{k,l=0}^{n-1} \left\{ n^2 \left(\int_{k/n}^{(k+1)/n} du \int_{l/n}^{(l+1)/n} dv |u-v|^{2H-2}\right)^2 \right.$$

$$- 2n \int_{l/n}^{(l+1)/n} dv \left(\int_{k/n}^{(k+1)/n} du |u-v|^{2H-2}\right)^2$$

$$\left. + \int_{k/n}^{(k+1)/n} dw \int_{l/n}^{(l+1)/n} dz |w-z|^{4H-4}\right\}$$

$$= 2H^2(2H-1)^2 n^{2-4H} \sum_{k,l=0}^{n-1} \left\{ \left(\int_0^1 du \int_0^1 dv |k-l+u-v|^{2H-2}\right)^2 \right.$$

$$\left. -2\int_0^1 dv \left(\int_0^1 du |k-l+u-v|^{2H-2}\right)^2 + \int_0^1 du \int_0^1 dv |k-l+u-v|^{4H-4}\right\}$$

$$\le 2H^2(2H-1)^2 n^{3-4H} \sum_{r\in\mathbb{Z}} \left|\left(\int_0^1 du \int_0^1 dv |r+u-v|^{2H-2}\right)^2\right.$$

$$\left. -2\int_0^1 dv \left(\int_0^1 du |r+u-v|^{2H-2}\right)^2 + \int_0^1 du \int_0^1 dv |r+u-v|^{4H-4}\right|.$$

$$(7.4.18)$$

To complete the proof of (7.4.13), it remains to ensure that the sum over $\mathbb{Z}$ in (7.4.18) is finite. For $r > 1$, elementary computations give

$$\left(\int_0^1 du \int_0^1 dv |r + u - v|^{2H-2} \right)^2$$

$$= (2H(2H - 1))^{-2} ((r + 1)^{2H} - 2r^{2H} + (r - 1)^{2H})^2$$

$$= \left(r^{2H-2} + O(r^{2H-4}) \right)^2 = r^{4H-4} + O(r^{4H-6}) \qquad (7.4.19)$$

and

$$\int_0^1 du \int_0^1 dv |r + u - v|^{4H-4} = \frac{(r + 1)^{4H-2} - 2r^{4H-2} + (r - 1)^{4H-2}}{(4H - 3)(4H - 2)}$$

$$= r^{4H-4} + O(r^{4H-6}). \qquad (7.4.20)$$

Moreover, using the inequality

$$\left| (1 + x)^{2H-1} - 1 - (2H - 1)x \right|$$

$$= (2H - 1)(2 - 2H) \int_0^x du \int_0^u \frac{dv}{(1 + v)^{3-2H}}$$

$$\leq (2H - 1)(2 - 2H) \int_0^x du \int_0^u dv = (2H - 1)(1 - H)x^2,$$

for all $x \geq 0$, we can write

$$\int_0^1 dv \left(\int_0^1 du |r + u - v|^{2H-2} \right)^2$$

$$= (2H - 1)^{-2} \int_0^1 \left((r + 1 - v)^{2H-1} - (r - v)^{2H-1} \right)^2 dv$$

$$= (2H - 1)^{-2} \int_0^1 (r - v)^{4H-2} \left(\left(1 + \frac{1}{r - v} \right)^{2H-1} - 1 \right)^2 dv$$

$$= \int_0^1 (r - v)^{4H-2} \left(\frac{1}{r - v} + R \left(\frac{1}{r - v} \right) \right)^2 dv$$

where the remainder term R satisfies $|R(u)| \leq (1 - H)u^2$. In particular, for any $v \in [0, 1]$, we have

$$(r - v) \left| R \left(\frac{1}{r - v} \right) \right| \leq \frac{1 - H}{r - 1}.$$

Hence, we deduce that

$$\int_0^1 dv \left(\int_0^1 du |r + u - v|^{2H-2} \right)^2 = \int_0^1 (r - v)^{4H-4} (1 + O(1/r))^2 dv$$

$$= r^{4H-3} \frac{1 - (1 - 1/r)^{4H-3}}{4H - 3} (1 + O(1/r))$$

$$= r^{4H-4} + O(r^{4H-5}). \qquad (7.4.21)$$

By combining (7.4.19), (7.4.20) and (7.4.21), we obtain (since similar arguments also apply to the case $r < -1$) that

$$\left(\int_0^1 du \int_0^1 dv |r + u - v|^{2H-2} \right)^2 - 2 \int_0^1 dv \left(\int_0^1 du |r + u - v|^{2H-2} \right)^2$$

$$+ \int_0^1 du \int_0^1 dv |r + u - v|^{4H-4}$$

is $O(|r|^{4H-5})$, so that the sum over $\mathbb{Z}$ in (7.4.18) is finite. The proof of (7.4.13) is done. To conclude the proof of (7.4.12), it remains to combine (7.4.13) with the already quoted result due to Davydov and Martynova [28] that states that there exists a constant $c(f)$, depending only on f, such that:

$$d_{\mathrm{TV}}(F_n, F_\infty) \leq c(f) \times \left(E[(F_n - F_\infty)^2] \right)^{1/4}. \qquad (7.4.22)$$

$\square$

7.5 Exercises

7.5.1 Fix an integer $q \geq 3$, let H_q be the qth Hermite polynomial as in Definition 1.4.1, and let B^H be a fractional Brownian motion of Hurst index $H \in (0, 1)$. Set

$$V_n = \frac{1}{\sqrt{n}} \sum_{k=0}^{n-1} H_q(B_{k+1}^H - B_k^H).$$

1. Assume $H \leq 1 - \frac{1}{2q}$, let $N \sim \mathcal{N}(0, 1)$ and define $v_n > 0$ by $E[V_n^2] = v_n^2$. Show the existence of a constant $c_{q,H} > 0$ (depending only on q and H) such that, for every $n \geq 2$,

$$d_{\mathrm{TV}}(V_n/v_n, N) \le c_{q,H} \times \begin{cases} \dfrac{1}{\sqrt{n}} & \text{if } H \in \left(0, \dfrac{3}{4}\right) \\[2ex] n^{2H-2} & \text{if } H \in \left(\dfrac{3}{4}, \dfrac{2q-3}{2q-2}\right) \\[2ex] n^{2qH-2q+1} & \text{if } H \in \left(\dfrac{2q-3}{2q-2}, 1 - \dfrac{1}{2q}\right) \\[2ex] \dfrac{1}{\log n} & \text{if } H = 1 - \dfrac{1}{2q}. \end{cases}$$

2. Assume $H > 1 - \frac{1}{2q}$ and let F_∞ be as in Proposition 7.4.2. Show the existence of a constant $c_{q,H} > 0$ (depending only on q and H) such that, for every $n \ge 1$:

$$d_{\mathrm{TV}}(n^{q(1-H)-1} V_n, F_\infty) \le c_{q,H} \times n^{1 - \frac{1}{2q} - H}.$$

(Hint: In this case, the crucial inequality (7.4.22) due to Davydov and Martynova [28] becomes $d_{\mathrm{TV}}\big(F_n, F_\infty\big) \le c(q, f) \times \big(E[(F_n - F_\infty)^2]\big)^{1/2q}$.)

7.5.2 Let X be an isonormal Gaussian process, and let $\mathscr{H}_2$ denote the second Wiener chaos of X. Using (7.4.13), the Borel–Cantelli lemma and Theorem 2.7.2, show that the sequence $\{F_n : n \ge 1\}$ defined in Proposition 7.4.2 converges almost surely towards F_∞.

7.6 Bibliographic comments

The notion of 'Hermite rank' was introduced by Taqqu in [143]. The fundamental Theorem 7.2.4 was proved by Breuer and Major in [18] (see also Sun [138] for an early statement with some additional assumptions). Important generalizations can be found in Arcones [4] (multidimensional case), Chambers and Slud [19] and Giraitis and Surgailis [41] (continuous-time case); Surgailis [139] provides a nice survey of further generalizations and applications (up to the year 2003). It should be noted that, in all these references, Breuer–Major results are proved by means of combinatorial moments/cumulants computations. Our proof, which is indeed based on the Malliavin/Stein techniques developed in this book, is close to that given by Nourdin, Peccati and Podolskij in [92], where one can also find explicit bounds in the total variation and Wasserstein distances. The Hermite distribution appearing in (7.4.7) and (7.4.8) (as well as its stochastic process counterpart, called the *Rosenblatt process*) plays a marginal role in our monograph, so we do not provide a full description of its structure. Hermite distributions first appeared as limits of partial sums of Gaussian subordinated sequences in the landmark paper by Rosenblatt [121]. In [142], Taqqu provided a general result for subordinated sequences with Hermite rank that is either 1 (yielding convergence to the fractional Brownian motion) or 2 (with convergence to the Rosenblatt process of order 2). Finally, Taqqu [143] and Dobrushin and Major [30] obtained the general picture for functions of arbitrary Hermite rank in the discrete-time and continuous-time case, respectively. Several refinements of these results are obtained by Breton and Nourdin [17], where explicit bounds in total variation are deduced by means of the 'stratification techniques' by Davydov and Martynova [28]. In particular, the estimates of Theorem 7.4.5 are special instances of the results contained in [17]. Another modern reference for Rosenblatt processes is the paper by Tudor [146]. The bounds of Corollary 7.4.3 and Exercise 7.5.1 are taken from Biermé, Bonami and Léon [12], improving analogous bounds by Nourdin and Peccati [88]. For some examples of the statistical use of the Breuer–Major theorem in the quadratic case, see, for instance, Coeurjolly [26] and the references therein.

8

Computation of cumulants

In this chapter, we apply the integration by parts techniques of Malliavin calculus to the explicit computations of the cumulants associated with regular functionals of Gaussian fields. The necessary definitions and basic result concerning cumulants are gathered together in Section A.2 of Appendix A. One should observe that the formulae obtained below represent an effective alternative to the usual expressions based on 'diagram formulae' (see e.g. [68, 110]). See the bibliographic remarks for more details on this point.

8.1 Decomposing multi-indices

We adopt the multi-index notation detailed in Section A.2.3 of Appendix A. For any fixed integer $d \geq 1$, we shall consider multi-indices of the form $m = (m_1, \ldots, m_d)$, where each m_i is a non-negative integer. For every $i = 1, \ldots, d$, we denote by e_i the multi-index of length d whose entries are all zero, except for the ith, which is equal to one. For instance, for $d = 4$,

$$e_1 = (1, 0, 0, 0), \ e_2 = (0, 1, 0, 0), \ e_3 = (0, 0, 1, 0) \text{ and } e_4 = (0, 0, 0, 1).$$

The following elementary lemma proves that the class $\{e_1, \ldots, e_d\}$ is a basis for the set of all multi-indices of length d. The proof is immediate and left to the reader.

Lemma 8.1.1 *Let $m = (m_1, \ldots, m_d)$ be a multi-index, and write $|m| = \sum_{i=1}^{d} m_i$. Then there exists a sequence $(l_1, \ldots, l_{|m|}) \in \{e_1, \ldots, e_d\}^{|m|}$ decomposing m, that is, such that*

$$l_1 + l_2 + \ldots + l_{|m|} = m.$$

The sequence $(l_1, \ldots, l_{|m|})$ is unique in the following sense: if $(l'_1, \ldots, l'_{|m|})$ is another sequence of length $|m|$ decomposing m, then there exists a permutation

σ *of* $\{1, \ldots, |m|\}$ *such that*

$$(l'_1, \ldots, l'_{|m|}) = (l_{\sigma(1)}, \ldots, l_{\sigma(|m|)}).$$

Example 8.1.2 The following sequences of length 4 decompose the multi-index $(3, 1, 0, 0)$:

$$(e_1, e_1, e_1, e_2), \ (e_1, e_1, e_2, e_1), \ (e_1, e_2, e_1, e_1), \ \text{and} \ (e_2, e_1, e_1, e_1).$$

Note that each sequence is a permutation of the others.

8.2 General formulae

For the rest of this chapter, we fix an isonormal Gaussian process $X = \{X(h), \ h \in \mathfrak{H}\}$, defined on some probability space $(\Omega, \mathscr{F}, P)$. We use the Malliavin calculus notation introduced in Chapter 2.

The following recursive definition is crucial for the whole chapter.

Definition 8.2.1 Let $F = (F_1, \ldots, F_d)$ be an $\mathbb{R}^d$-valued random vector with $F_i \in \mathbb{D}^{1,2}$ for each i. Let $l_1, l_2, \ldots$ be a sequence taking values in the multi-index set $\{e_1, \ldots, e_d\}$. We set $\Gamma_{l_1}(F) = F^{l_1} = F_j$, where j is such that $l_1 = e_j$. If the random variable $\Gamma_{l_1, \ldots, l_k}(F)$ is a well-defined element of $L^2(\Omega)$ for some $k \geq 1$, we set

$$\Gamma_{l_1, \ldots, l_{k+1}}(F) = \langle DF^{l_{k+1}}, -DL^{-1}\Gamma_{l_1, \ldots, l_k}(F) \rangle_{\mathfrak{H}}.$$

Remark 8.2.2 1. Since the square-integrability of $\Gamma_{l_1, \ldots, l_k}(F)$ implies that $L^{-1}\Gamma_{l_1, \ldots, l_k}(F) \in \text{Dom} \, L \subset \mathbb{D}^{1,2}$, $\Gamma_{l_1, \ldots, l_{k+1}}(F)$ is well defined.
2. According to the multi-index conventions adopted in this book (see formula (A.2.6)), if $l = e_i$, then $F^l = F_i$.

Example 8.2.3 If $l_1 = e_1$ and $l_2 = e_2$, then $\Gamma_{l_1} = F^{e_1} = F_1$ and

$$\Gamma_{l_1, l_2} = \langle DF_2, -DL^{-1}F_1 \rangle_{\mathfrak{H}}.$$

The next lemma gives sufficient conditions on F ensuring that the random variable $\Gamma_{l_1, \ldots, l_k}(F)$ is a well-defined element of $L^2(\Omega)$.

Lemma 8.2.4 (i) *Fix an integer* $j \geq 1$, *and assume that* $F = (F_1, \ldots, F_d)$ *is such that* $F_i \in \mathbb{D}^{j,2^j}$ *for each* i. *Let* $l_1, l_2, \ldots, l_j$ *be elements of the multi-index set* $\{e_1, \ldots, e_d\}$. *Then, for all* $k = 1, \ldots, j$, *we have that* $\Gamma_{l_1, \ldots, l_k}(F)$ *is a well-defined element of* $\mathbb{D}^{j-k+1, 2^{j-k+1}}$; *in particular,* $\Gamma_{l_1, \ldots, l_j}(F) \in \mathbb{D}^{1,2} \subset L^2(\Omega)$ *and the quantity* $E[\Gamma_{l_1, \ldots, l_j}(F)]$ *is well defined and finite.*

(ii) *Assume that $F = (F_1, \ldots, F_d)$ is such that $F_i \in \cap_{q \geq 1} \mathbb{D}^{\infty, q}$ for each i. Let $l_1, l_2, \ldots$ be a sequence taking values in $\{e_1, \ldots, e_d\}$. Then, for all $k \geq 1$, the random variable $\Gamma_{l_1, \ldots, l_k}(F)$ is a well-defined element of $\cap_{q \geq 1} \mathbb{D}^{\infty, q}$.*

Proof Without loss of generality, we may assume throughout the proof that $\mathfrak{H}$ has the form $L^2(A, \mathscr{A}, \mu)$, where $(A, \mathscr{A})$ is a measurable space and μ is a σ-finite measure with no atoms.

(i) The proof is by induction on $k \in \{1, \ldots, j\}$. For $k = 1$, we have that $\Gamma_{l_1}(F) = F^{l_1} \in \mathbb{D}^{j, 2^j}$. Assume now that $k \geq 2$ and that $\Gamma_{l_1, \ldots, l_{k-1}}(F) \in \mathbb{D}^{j-k+2, 2^{j-k+2}}$, and let us prove that $\Gamma_{l_1, \ldots, l_k}(F) \in \mathbb{D}^{j-k+1, 2^{j-k+1}}$. Write G instead of $\Gamma_{l_1, \ldots, l_{k-1}}(F)$ to simplify the discussion, and fix $r = 1, \ldots, j-k+1$. Since $\Gamma_{l_1, \ldots, l_k}(F)$ is an element of $\mathbb{D}^{r, 2} \cap L^{2^{j-k+1}}(\Omega)$ for $r \leq j - k + 1$ (see Exercise 8.5.1), the derivative $D^r \Gamma_{l_1, \ldots, l_k}(F)$ is well defined, and we can apply the Leibniz rule for D (see Exercise 2.3.10), together with an approximation argument in order to differentiate under the integral sign, thus yielding

$$-D^r \Gamma_{l_1, \ldots, l_k}(F) = -D^r \langle DF^{l_k}, -DL^{-1}G \rangle_{\mathfrak{H}} = D^r \int_A D_a F^{l_k} D_a L^{-1} G \, \mu(da)$$

$$= \sum_{l=0}^{r} \binom{r}{l} \int_A D^{r-l}(D_a F^{l_k}) \widetilde{\otimes} D^l(D_a L^{-1}G) \, \mu(da),$$

with $\widetilde{\otimes}$ the usual symmetric tensor product. We deduce that

$$\|D^r \Gamma_{l_1, \ldots, l_k}(F)\|_{\mathfrak{H}^{\otimes k}}$$

$$\leq \sum_{l=0}^{r} \binom{r}{l} \left\| \int_A D^{r-l}(D_a F^{l_k}) \widetilde{\otimes} D^l(D_a L^{-1}G) \, \mu(da) \right\|_{\mathfrak{H}^{\otimes k}}$$

$$\leq \sum_{l=0}^{r} \binom{r}{l} \int_A \|D^{r-l}(D_a F^{l_k})\|_{\mathfrak{H}^{\otimes(k-l)}} \|D^l(D_a L^{-1}G)\|_{\mathfrak{H}^{\otimes l}} \, \mu(da)$$

$$\leq \sum_{l=0}^{r} \binom{r}{l} \sqrt{\int_A \|D^{r-l}(D_a F^{l_k})\|^2_{\mathfrak{H}^{\otimes(k-l)}} \mu(da)} \sqrt{\int_A \|D^l(D_a L^{-1}G)\|^2_{\mathfrak{H}^{\otimes l}} \mu(da)}$$

$$= \sum_{l=0}^{r} \binom{r}{l} \|D^{r+1-l} F^{l_k}\|_{\mathfrak{H}^{\otimes(k-l+1)}} \|D^{l+1} L^{-1} G\|_{\mathfrak{H}^{\otimes(l+1)}}. \tag{8.2.1}$$

By mimicking the arguments used in the proof of (2.9.2), we get

$$-D^{l+1} L^{-1} G = \int_0^\infty e^{-(l+1)t} P_t D^{l+1} G \, dt.$$

Consequently, for any real $p \geq 1$,

$$E\left[\|D^{l+1}L^{-1}G\|_{\mathfrak{H}^{\otimes(l+1)}}^{p}\right] \leq E\left[\left(\int_{0}^{\infty} e^{-(l+1)t}\|P_{t}D^{l+1}G\|_{\mathfrak{H}^{\otimes(l+1)}} dt\right)^{p}\right]$$

$$\leq \frac{1}{(l+1)^{p-1}} \int_{0}^{\infty} e^{-(l+1)t} E\left[\|P_{t}D^{l+1}G\|_{\mathfrak{H}^{\otimes(l+1)}}^{p}\right] dt$$

$$\leq \frac{1}{(l+1)^{p-1}} E\left[\|D^{l+1}G\|_{\mathfrak{H}^{\otimes(l+1)}}^{p}\right] \int_{0}^{\infty} e^{-(l+1)t} dt$$

$$= \frac{1}{(l+1)^{p}} E\left[\|D^{l+1}G\|_{\mathfrak{H}^{\otimes(l+1)}}^{p}\right], \qquad (8.2.2)$$

where the last inequality is obtained by using the contraction property of P_t on $L^p(\Omega)$. Finally, by combining (8.2.1) with (8.2.2) and the Cauchy–Schwarz inequality on the one hand, and by virtue of the assumptions on F^{l_k} and G on the other hand, we immediately infer that $\|D^r \Gamma_{l_1,\ldots,l_k}(F)\|_{\mathfrak{H}^{\otimes k}}$ is an element $L^{2^{j-k+1}}(\Omega)$ for each $r = 1,\ldots, j-k+1$, and the result is proved.

(ii) The proof is immediately obtained by a repeated application of (i). $\square$

We are now ready to state and prove the main result of this section, which is an abstract formula for the cumulants of any random vector F whose components are sufficiently regular in the sense of Malliavin calculus. In the following statement, $d \geq 1$ is an integer; also, we use the joint cumulant notation $\kappa_m(\cdot)$, where m is a multi-index, as described in Definition A.2.1 and the subsequent discussion.

Theorem 8.2.5 *Let* $m = (m_1,\ldots,m_d) \in \mathbb{N}^d \setminus \{0\}$ *be a multi-index. Write* $m = l_1 + \ldots + l_{|m|}$ *where the multi-indices* $l_i \in \{e_1,\ldots,e_d\}$, $i = 1,\ldots,|m|$, *are unique in the sense of Lemma 8.1.1. Suppose that the random vector* $F = (F_1,\ldots,F_d)$ *is such that* $F_i \in \mathbb{D}^{|m|,2^{|m|}}$ *for each* i. *Then*

$$\kappa_m(F) = \sum_{\sigma \in \mathfrak{S}_{\{2,\ldots,|m|\}}} E\left[\Gamma_{l_1,l_{\sigma(2)},\ldots,l_{\sigma(|m|)}}(F)\right]. \qquad (8.2.3)$$

Remark 8.2.6 According to Lemma 8.1.1, in the statement of Theorem 8.2.5 one could replace the set $\{l_1,\ldots,l_{|m|}\}$ with any of its permutations $\{l'_1,\ldots,l'_{|m|}\} = \{l_{\rho(1)},\ldots,l_{\rho(|m|)}\}$, where $\rho \in \mathfrak{S}_{\{1,\ldots,|m|\}}$. It follows that, for every $j = 1,\ldots,m$, the following generalization of (8.2.3) holds:

$$\kappa_m(F) = \sum_{\sigma \in \mathfrak{S}_{\{1,\ldots,|m|\}\setminus\{j\}}} E\left[\Gamma_{l_j,l_{\sigma(x_1)},\ldots,l_{\sigma(x_{|m|-1})}}(F)\right], \qquad (8.2.4)$$

where $\{x_1,\ldots,x_{|m|-1}\} = \{1,\ldots,|m|\}\setminus\{j\}$. For instance, if $m = (1,1,1)$, then

$$\kappa_m(F) = E[\Gamma_{e_1,e_2,e_3}] + E[\Gamma_{e_1,e_3,e_2}] = E[\Gamma_{e_2,e_1,e_3}] + E[\Gamma_{e_2,e_3,e_1}]$$
$$= E[\Gamma_{e_3,e_1,e_2}] + E[\Gamma_{e_3,e_2,e_1}].$$

Proof of Theorem 8.2.5 The proof is by induction on the value of the integer $|m|$. The case $|m| = 1$ is clear because $\kappa_{e_j}(F) = E[F_j] = E[\Gamma_{e_j}(F)]$ for all j. Now assume that (8.2.3) holds for all multi-indices $m \in \mathbb{N}^d$ such that $|m| \leq N$, for some $N \geq 1$ fixed, and let us prove that it continues to hold for all the multi-indices m satisfying $|m| = N + 1$. Let $m \in \mathbb{N}^d$ be such that $|m| \leq N$, and fix $j = 1, \ldots, d$. By following the same route as in the proof of Theorem 2.9.1, we can write (note that, by a standard approximation argument, one can apply the chain rule of Proposition 2.3.7 in this situation, even if monomials do not have bounded derivatives)

$$E[F^{m+e_j}] = E[F^m \Gamma_{e_j}(F)] = \quad E[F^m]E[\Gamma_{e_j}(F)] + E[F^m LL^{-1}\Gamma_{e_j}(F)]$$
$$= E[F^m]E[\Gamma_{e_j}(F)] + E[\langle DF^m, -DL^{-1}\Gamma_{e_j}(F)\rangle_{\mathfrak{H}}]$$
$$= E[F^m]E[\Gamma_{e_j}(F)] + \sum_{1 \leq i_1 \leq |m|} E[F^{m-l_{i_1}}\langle DF^{l_{i_1}}, -DL^{-1}\Gamma_{e_j}(F)\rangle_{\mathfrak{H}}]$$
$$= E[F^m]E[\Gamma_{e_j}(F)] + \sum_{1 \leq i_1 \leq |m|} E[F^{m-l_{i_1}}\Gamma_{e_j,l_{i_1}}(F)]$$
$$= E[F^m]E[\Gamma_{e_j}(F)] + \sum_{1 \leq i_1 \leq |m|} E[F^{m-l_{i_1}}]E[\Gamma_{e_j,l_{i_1}}(F)]$$
$$+ \sum_{\substack{1 \leq i_1,i_2 \leq |m| \\ i_1,i_2 \text{ different}}} E[F^{m-l_{i_1}-l_{i_2}}\Gamma_{e_j,l_{i_1},l_{i_2}}(F)],$$

and therefore, by iterating the same procedure,

$$E[F^{m+e_j}] = E[F^m]E[\Gamma_{e_j}(F)] + \sum_{1 \leq i_1 \leq |m|} E[F^{m-l_{i_1}}]E[\Gamma_{e_j,l_{i_1}}(F)]$$
$$+ \sum_{\substack{1 \leq i_1,i_2 \leq |m| \\ i_1,i_2 \text{ different}}} E[F^{m-l_{i_1}-l_{i_2}}]E[\Gamma_{e_j,l_{i_1},l_{i_2}}(F)]$$
$$+ \ldots + \sum_{\substack{1 \leq i_1,\ldots,i_{|m|-1} \leq |m| \\ i_1,\ldots,i_{|m|-1} \text{ pairwise different}}} E[F^{m-l_{i_1}-\ldots-l_{i_{|m|-1}}}]E[\Gamma_{e_j,l_{i_1},\ldots,l_{i_{|m|-1}}}(F)]$$
$$+ \sum_{\substack{1 \leq i_1,\ldots,i_{|m|} \leq |m| \\ i_1,\ldots,i_{|m|} \text{ pairwise different}}} E[\Gamma_{e_j,l_{i_1},\ldots,l_{i_{|m|}}}(F)]$$

$$= \sum_{s \leq m} E[F^{m-s}] \sum_{\substack{1 \leq i_1,\ldots,i_{|s|} \leq |m| \\ i_1,\ldots,i_{|s|} \text{ pairwise different} \\ l_{i_1}+\ldots+l_{i_{|s|}}=s}} E[\Gamma_{e_j,l_{i_1},\ldots,l_{i_{|s|}}}(F)]$$

$$= \sum_{s \leq m} E[F^{m-s}] \sum_{\substack{1 \leq i_1 < \ldots < i_{|s|} \leq |m| \\ l_{i_1}+\ldots+l_{i_{|s|}}=s}} \sum_{\sigma \in \mathfrak{S}_{|s|}} E[\Gamma_{e_j,l_{\sigma(i_1)},\ldots,l_{\sigma(i_{|s|})}}(F)]. \quad (8.2.5)$$

Recall that $s \leq m$ means that $s = (s_1,\ldots,s_d)$ is a multi-index satisfying $s_i \leq m_i$ for every $i = 1,\ldots,d$. Let B_s stand for the set of vectors $(i_1,\ldots,i_{|s|}) \in \{1,\ldots,|m|\}^{|s|}$ composed of indices such that $i_1 < \ldots < i_{|s|}$ and such that $l_{i_1} + \ldots + l_{i_{|s|}} = s$. For any $s \leq m$, it is easy to check that the cardinality of B_s is given by $\binom{m_1}{s_1} \ldots \binom{m_d}{s_d}$ and that the sum $\sum_{\sigma \in \mathfrak{S}_{|s|}} E[\Gamma_{e_j,l_{\sigma(i_1)},\ldots,l_{\sigma(i_{|s|})}}(F)]$ appearing in (8.2.5) only depends on s. By induction, we consequently deduce that

$$E[F^{m+e_j}] = \sum_{s \leq m} E[F^{m-s}]\binom{m_1}{s_1}\ldots\binom{m_d}{s_d}\kappa_{s+e_j}(F)$$
$$+ \sum_{\sigma \in \mathfrak{S}_{|m|}} E[\Gamma_{e_j,l_{\sigma(i_1)},\ldots,l_{\sigma(i_{|m|})}}(F)] - \kappa_{m+e_j}(F).$$

It follows that

$$E[F^{m+e_j}]$$
$$= \sum_{s \leq m} \binom{m_1}{s_1}\ldots\binom{m_d}{s_d}(-i)^{|m|-|s|}\partial^{m-s}\phi_F(0) \times (-i)^{|s|+1}\partial^{s+e_j}\log\phi_F(0)$$
$$+ \sum_{\sigma \in \mathfrak{S}_{|m|}} E[\Gamma_{e_j,l_{\sigma(i_1)},\ldots,l_{\sigma(i_{|m|})}}(F)] - \kappa_{m+e_j}(F)$$
$$= (-i)^{|m|+1}\partial^m\big(\phi_F\partial^{e_j}\log\phi_F\big)(0) + \sum_{\sigma \in \mathfrak{S}_{|m|}} E[\Gamma_{e_j,l_{\sigma(i_1)},\ldots,l_{\sigma(i_{|m|})}}(F)] - \kappa_{m+e_j}(F)$$
$$= (-i)^{|m|+1}\partial^{m+e_j}\phi_F(0) + \sum_{\sigma \in \mathfrak{S}_{|m|}} E[\Gamma_{e_j,l_{\sigma(i_1)},\ldots,l_{\sigma(i_{|m|})}}(F)] - \kappa_{m+e_j}(F)$$
$$= E[F^{m+e_j}] + \sum_{\sigma \in \mathfrak{S}_{|m|}} E[\Gamma_{e_j,l_{\sigma(i_1)},\ldots,l_{\sigma(i_{|m|})}}(F)] - \kappa_{m+e_j}(F),$$

which is the same as

$$\sum_{\sigma \in \mathfrak{S}_{|m|}} E[\Gamma_{e_j,l_{\sigma(i_1)},\ldots,l_{\sigma(i_{|m|})}}(F)] = \kappa_{m+e_j}(F).$$

Thus that (8.2.3) holds with m replaced by $m + e_j$. The proof by induction is complete. $\square$

8.3 Application to multiple integrals

We shall now focus on the explicit computation of the cumulants of a random vector whose entries are elements of some Wiener chaos.

Theorem 8.3.1 *Let $m \in \mathbb{N}^d \setminus \{0\}$ be a multi-index such that $|m| \geq 3$. Write $m = l_1 + \ldots + l_{|m|}$ with $l_i \in \{e_1, \ldots, e_d\}$ for each i (see Lemma 8.1.1). Consider an $\mathbb{R}^d$-valued random vector of the form*

$$F = (F_1, \ldots, F_d) = \left(I_{q_1}(f_1), \ldots, I_{q_d}(f_d) \right),$$

where each f_i belongs to $\mathfrak{H}^{\odot q_i}$. When $l_k = e_j$, we set $\lambda_k = j$, so that $F^{l_k} = F_{\lambda_k}$ for all $k = 1, \ldots, |m|$. Then

$$\kappa_m(F) = \sum_{\sigma \in \mathfrak{S}_{\{2,\ldots,|m|\}}} (q_{\lambda_{\sigma(|m|)}})! \sum_{*} c_{q,l,\sigma}(r_2, \ldots, r_{|m|-1}) \qquad (8.3.1)$$

$$\times \langle (\ldots ((f_{\lambda_1} \widetilde{\otimes}_{r_2} f_{\lambda_{\sigma(2)}}) \widetilde{\otimes}_{r_3} f_{\lambda_{\sigma(3)}}) \ldots) \widetilde{\otimes}_{r_{|m|-1}} f_{\lambda_{\sigma(|m|-1)}}; f_{\lambda_{\sigma(|m|)}} \rangle_{\mathfrak{H}^{\otimes q_{\lambda_{\sigma(|m|)}}}},$$

where the second sum $\displaystyle\sum_{*}$ runs over all collections of integers $r_2, \ldots, r_{|m|-1}$ such that:

(i) $1 \leq r_i \leq q_{\lambda_{\sigma(i)}}$ for all $i = 2, \ldots, |m| - 1$;

(ii) $r_2 + \ldots + r_{|m|-1} = \dfrac{q_{\lambda_1} + q_{\lambda_{\sigma(2)}} + \ldots + q_{\lambda_{\sigma(|m|-1)}} - q_{\lambda_{\sigma(|m|)}}}{2}$;

(iii) $r_2 < \dfrac{q_{\lambda_1} + q_{\lambda_{\sigma(2)}}}{2}, \ldots, r_2 + \ldots + r_{|m|-2} < \dfrac{q_{\lambda_1} + q_{\lambda_{\sigma(2)}} + \ldots + q_{\lambda_{\sigma(|m|-2)}}}{2}$;

(iv) $r_2 \leq q_{\lambda_1}, r_3 \leq q_{\lambda_1} + q_{\lambda_{\sigma(2)}} - 2r_2, \ldots, r_{|m|-1} \leq q_{\lambda_1} + q_{\lambda_{\sigma(2)}} + \ldots + q_{\lambda_{\sigma(|m|-2)}} - 2r_2 - \ldots - 2r_{|m|-2}$;

and where the combinatorial constants $c_{q,l,\sigma}(r_2, \ldots, r_s)$ are recursively defined by the relations

$$c_{q,l,\sigma}(r_2) = q_{\lambda_{\sigma(2)}}(r_2 - 1)! \binom{q_{\lambda_1} - 1}{r_2 - 1} \binom{q_{\lambda_{\sigma(2)}} - 1}{r_2 - 1},$$

and, for $s \geq 3$,

$$c_{q,l,\sigma}(r_2, \ldots, r_s)$$
$$= q_{\lambda_{\sigma(s)}}(r_s - 1)! \binom{q_{\lambda_1} + q_{\lambda_{\sigma(2)}} + \ldots + q_{\lambda_{\sigma(s)}} - 2r_2 - \ldots - 2r_{s-1} - 1}{r_s - 1}$$
$$\times \binom{q_{\lambda_{\sigma(s)}} - 1}{r_s - 1} c_{q,l,\sigma}(r_2, \ldots, r_{s-1}).$$

Proof To simplify the discussion, in what follows we shall drop the brackets in writing $(\ldots ((f_{\lambda_1} \widetilde{\otimes}_{r_2} f_{\lambda_{\sigma(2)}}) \widetilde{\otimes}_{r_3} f_{\lambda_{\sigma(3)}}) \ldots) \widetilde{\otimes}_{r_{|m|-1}} f_{\lambda_{\sigma(|m|-1)}}$, by implicitly assuming that this quantity is always defined iteratively from left to right. For instance, $f \widetilde{\otimes}_\alpha g \widetilde{\otimes}_\beta h \widetilde{\otimes}_\gamma k$ actually means $((f \widetilde{\otimes}_\alpha g) \widetilde{\otimes}_\beta h) \widetilde{\otimes}_\gamma k$.

To prove (8.3.1), we shall use Theorem 8.2.5. If $f \in \mathfrak{H}^{\odot p}$ and $g \in \mathfrak{H}^{\odot q}$ $(p, q \geq 1)$, the multiplication formula yields

$$
\langle DI_p(f), -DL^{-1}I_q(g) \rangle_{\mathfrak{H}} = p \langle I_{p-1}(f), I_{q-1}(g) \rangle_{\mathfrak{H}}
$$

$$
= p \sum_{r=0}^{p \wedge q - 1} r! \binom{p-1}{r} \binom{q-1}{r} I_{p+q-2-2r}(f \widetilde{\otimes}_{r+1} g)
$$

$$
= p \sum_{r=1}^{p \wedge q} (r-1)! \binom{p-1}{r-1} \binom{q-1}{r-1} I_{p+q-2r}(f \widetilde{\otimes}_r g). \quad (8.3.2)
$$

Let $\sigma \in \mathfrak{S}_{\{2,\dots,|m|\}}$. Thanks to (8.3.2), it is straightforward to prove by induction on $|m|$ that

$$
\Gamma_{l_1, l_{\sigma(2)}, \dots, l_{\sigma(|m|)}}(F) \tag{8.3.3}
$$

$$
= \sum_{r_2=1}^{q_{\lambda_1} \wedge q_{\lambda_{\sigma(2)}}} \cdots \sum_{r_{|m|}=1}^{[q_{\lambda_1}+q_{\lambda_{\sigma(2)}}+\dots+q_{\lambda_{\sigma(|m|-1)}}-2r_2-\dots-2r_{|m|-1}] \wedge q_{\lambda_{\sigma(|m|)}}} c_{q,l}(r_2, \dots, r_{|m|})
$$

$$
\times \mathbf{1}_{\left\{ r_2 < \frac{q_{\lambda_1}+q_{\lambda_{\sigma(2)}}}{2} \right\}} \cdots \mathbf{1}_{\left\{ r_2+\dots+r_{|m|-1} < \frac{q_{\lambda_1}+q_{\lambda_{\sigma(2)}}+\dots+q_{\lambda_{\sigma(|m|-1)}}}{2} \right\}}
$$

$$
\times I_{q_{\lambda_1}+q_{\lambda_{\sigma(2)}}+\dots+q_{\lambda_{\sigma(|m|)}}-2r_2-\dots-2r_{|m|}} \left(f_{\lambda_1} \widetilde{\otimes}_{r_2} f_{\lambda_{\sigma(2)}} \cdots \widetilde{\otimes}_{r_{|m|}} f_{\lambda_{\sigma(|m|)}} \right).
$$

$$
\tag{8.3.4}
$$

Now let us take the expectation on both sides of (8.3.4). We get

$$
\kappa_m(F)
$$

$$
= \sum_{\sigma \in \mathfrak{S}_{\{2,\dots,|m|\}}} E[\Gamma_{l_1, l_{\sigma(2)}, \dots, l_{\sigma(|m|)}}(F)]
$$

$$
= \sum_{\sigma \in \mathfrak{S}_{\{2,\dots,|m|\}}} \sum_{r_2=1}^{q_{\lambda_1} \wedge q_{\lambda_{\sigma(2)}}} \cdots \sum_{r_{|m|}=1}^{[q_{\lambda_1}+q_{\lambda_{\sigma(2)}}+\dots+q_{\lambda_{\sigma(|m|-1)}}-2r_2-\dots-2r_{|m|-1}] \wedge q_{\lambda_{\sigma(|m|)}}} c_{q,l}(r_2, \dots, r_{|m|})
$$

$$
\times \mathbf{1}_{\left\{ r_2 < \frac{q_{\lambda_1}+q_{\lambda_{\sigma(2)}}}{2} \right\}} \cdots \mathbf{1}_{\left\{ r_2+\dots+r_{|m|-1} < \frac{q_{\lambda_1}+q_{\lambda_{\sigma(2)}}+\dots+q_{\lambda_{\sigma(|m|-1)}}}{2} \right\}}
$$

$$
\times \mathbf{1}_{\left\{ r_2+\dots+r_{|m|} = \frac{q_{\lambda_1}+q_{\lambda_{\sigma(2)}}+\dots+q_{\lambda_{\sigma(|m|)}}}{2} \right\}} \times f_{\lambda_1} \widetilde{\otimes}_{r_2} f_{\lambda_{\sigma(2)}} \widetilde{\otimes}_{r_3} \cdots \widetilde{\otimes}_{r_{|m|}} f_{\lambda_{\sigma(|m|)}}.
$$

Observe that if $2r_2 + \dots + 2r_{|m|} = q_{\lambda_1} + q_{\lambda_{\sigma(2)}} + \dots + q_{\lambda_{\sigma(|m|)}}$ and $r_{|m|} \leq q_{\lambda_1} + q_{\lambda_{\sigma(2)}} + \dots + q_{\lambda_{\sigma(|m|-1)}} - 2r_2 - \dots - 2r_{|m|-1}$, then

$$2r_{|m|} = q_{\lambda_{\sigma(|m|)}} + \left(q_{\lambda_1} + q_{\lambda_{\sigma(2)}} + \ldots + q_{\lambda_{\sigma(|m|-1)}} - 2r_2 - \ldots - 2r_{|m|-1}\right)$$
$$\geq q_{\lambda_{\sigma(|m|)}} + r_{|m|},$$

that is, $r_{|m|} \geq q_{\lambda_{\sigma(|m|)}}$, so that $r_{|m|} = q_{\lambda_{\sigma(|m|)}}$. Therefore,

$\kappa_m(F)$

$$= \sum_{\sigma \in \mathfrak{S}_{\{2,\ldots,|m|\}}} \sum_{r_2=1}^{q_{\lambda_1} \wedge q_{\lambda_{\sigma(2)}}} \cdots \sum_{r_{|m|-1}=1}^{[q_{\lambda_1} + q_{\lambda_{\sigma(2)}} + \ldots + q_{\lambda_{\sigma(|m|-2)}} - 2r_2 - \ldots - 2r_{|m|-2}] \wedge q_{\lambda_{\sigma(|m|-1)}}} c_{q,l}(r_2, \ldots, r_{|m|-1}, q_{\lambda_{\sigma(|m|)}})$$

$$\times \mathbf{1}_{\{r_2 < \frac{q_{\lambda_1} + q_{\lambda_{\sigma(2)}}}{2}\}} \cdots \mathbf{1}_{\{r_2 + \ldots + r_{|m|-2} < \frac{q_{\lambda_1} + q_{\lambda_{\sigma(2)}} + \ldots + q_{\lambda_{\sigma(|m|-2)}}}{2}\}}$$

$$\times \mathbf{1}_{\{r_2 + \ldots + r_{|m|-1} = \frac{q_{\lambda_1} + q_{\lambda_{\sigma(2)}} + \ldots + q_{\lambda_{\sigma(|m|-1)}} - q_{\lambda_{\sigma(|m|)}}}{2}\}}$$

$$\times \langle f_{\lambda_1} \widetilde{\otimes}_{r_2} f_{\lambda_{\sigma(2)}} \cdots \widetilde{\otimes}_{r_{|m|-1}} f_{\lambda_{\sigma(|m|-1)}}, f_{\lambda_{\sigma(|m|)}} \rangle_{\mathfrak{H}^{\otimes q_{\lambda_{\sigma(|m|)}}}},$$

which is the desired result, since

$$c_{q,l}(r_2, \ldots, r_{|m|-1}, q_{\lambda_{\sigma(|m|)}}) = (q_{\lambda_{\sigma(|m|)}})! c_{q,l}(r_2, \ldots, r_{|m|-1}). \qquad \square$$

Example 8.3.2 Assume that $|m| \geq 3$ and that $q_1 = \ldots = q_d = 1$. Then there is no collection of integers $r_2, \ldots, r_{|m|-1}$ satisfying conditions (i) and (ii) in Theorem 8.3.1, and consequently formula (8.3.1) implies that $\kappa_m(F) = 0$ (as expected, since the joint cumulants of a Gaussian vector are zero whenever $|m| \geq 3$ – see Appendix A).

Example 8.3.3 As a more challenging illustration of Theorem 8.3.1, we can deduce yet another proof of the 'difficult' implication (b)→(a) in Theorem 6.2.3. Let the notation and assumptions of Theorem 6.2.3 prevail, suppose that (b) holds, and let us prove that (a) holds. Applying the method of moments and cumulants (see Theorem A.3.1), we must prove that the cumulants of F_n satisfy, for all $m \in \mathbb{N}^d \setminus \{0\}$,

$$\kappa_m(F_n) \to \kappa_m(N) = \begin{cases} 0 & \text{if } |m| \neq 2 \\ C(i,j) & \text{if } m = e_i + e_j \end{cases} \quad \text{as } n \to \infty.$$

Let $m \in \mathbb{N}^d \setminus \{0\}$. If $m = e_j$ for some j (which holds if and only if $|m| = 1$), then $\kappa_m(F_n) = E[F_{j,n}] = 0$. If $m = e_i + e_j$ for some i, j (which holds if and only if $|m| = 2$), then $\kappa_m(F_n) = E[F_{i,n} F_{j,n}] \to C(i,j)$ by assumption (6.2.8). If $|m| \geq 3$, we consider expression (8.3.1). Thanks to Theorem 5.2.7, we deduce that $\|f_{i,n} \otimes_r f_{i,n}\|_{\mathfrak{H}^{\otimes q_i}} \to 0$ as $n \to \infty$ for all i and all $r = 1, \ldots, q_i - 1$, whereas, thanks to (6.2.8), we deduce that $q_i! \|f_{i,n}\|_{\mathfrak{H}^{\odot q_i}}^2 = E[F_{i,n}^2] \to C(i,i)$ for all i, so that $\sup_{n \geq 1} \|f_{i,n}\|_{L^2([0,T]^{q_i})} < \infty$ for all i. Let $\sigma \in \mathfrak{S}_{\{2,\ldots,m\}}$ and let $r_2, \ldots, r_{|m|-1}$ be some integers such that (i)–(iv) in Theorem 8.3.1 are satisfied. In particular, $r_2 < \frac{q_{\lambda_1} + q_{\lambda_{\sigma(2)}}}{2}$. From (6.2.4) and (6.2.5), it follows that $\| f_{\lambda_1,n} \widetilde{\otimes}_{r_2} f_{\lambda_2,n} \|_{\mathfrak{H}^{\odot(q_{\lambda_1} + q_{\lambda_{\sigma(2)}} - 2r_2)}} \to 0$ as $n \to \infty$. Hence, combining an iterated use of the Cauchy–Schwarz inequality with the relations

$$\| g \widetilde{\otimes}_r h \|_{\mathfrak{H}^{\otimes(p+q-2r)}} \leq \| g \otimes_r h \|_{\mathfrak{H}^{\otimes(p+q-2r)}} \leq \| g \|_{\mathfrak{H}^{\otimes p}} \| h \|_{\mathfrak{H}^{\otimes q}}$$

valid for $g \in \mathfrak{H}^{\odot p}$, $h \in \mathfrak{H}^{\odot q}$ and $r = 1, \ldots, p \wedge q$, we infer that

$$\langle (\ldots ((f_{\lambda_1} \widetilde{\otimes}_{r_2} f_{\lambda_{\sigma(2)}}) \widetilde{\otimes}_{r_3} f_{\lambda_{\sigma(3)}}) \ldots) \widetilde{\otimes}_{r_{|m|-1}} f_{\lambda_{\sigma(|m|-1)}}; f_{\lambda_{\sigma(|m|)}} \rangle_{\mathfrak{H}^{\otimes q_{\lambda_{\sigma(|m|)}}}} \to 0$$

as $n \to \infty$. As a consequence, $\kappa_m(F_n) \to 0$ as $n \to \infty$ by (8.3.1), and the desired implication holds.

8.4 Formulae in dimension one

In this section, we specialize the previous computations to the one-dimensional case (that is, $d = 1$); proofs (which are elementary consequences of the previous statements) are left to the reader. If F is a one-dimensional random variable having finite moments up to some order m, we denote by $\kappa_j(F)$, $j = 1, \ldots, m$, the jth cumulant of F, as defined in Section A.2.2. We start with a one-dimensional version of Definition 8.2.1.

Definition 8.4.1 Let $F \in \mathbb{D}^{1,2}$. We set $\Gamma_1(F) = F$. If the random variable $\Gamma_j(F)$ is a well-defined element of $L^2(\Omega)$ for some $j \geq 1$, we set $\Gamma_{j+1}(F) = \langle DF, -DL^{-1}\Gamma_j(F) \rangle_{\mathfrak{H}}$.

Remark 8.4.2 In [91], the random variables $\Gamma_j(F)$ are recursively defined as in Definition 8.4.1, but starting by setting $\Gamma_0(F) = F$ (so that the sequence $\{\Gamma_j(F) : j \geq 1\}$ in [91] appears 'shifted' by one step).

Here is how Theorem 8.2.5 can be reformulated in the one-dimensional case.

Theorem 8.4.3 *Let $m \geq 1$ be an integer, and suppose that $F \in \mathbb{D}^{m,2^m}$. Then*

$$\kappa_m(F) = (m-1)! \, E[\Gamma_m(F)]. \tag{8.4.1}$$

Finally, Theorem 8.3.1 takes the following simpler form:

Theorem 8.4.4 *Let $q \geq 2$, and assume that $F = I_q(f)$, where $f \in \mathfrak{H}^{\odot q}$. Denote by $\kappa_m(F)$, $m \geq 1$, the cumulants of F. We have $\kappa_1(F) = 0$, $\kappa_2(F) = q! \| f \|^2_{\mathfrak{H}^{\otimes q}}$ and, for every $m \geq 3$,*

$$\kappa_m(F) = q!(m-1)! \sum c_q(r_2, \ldots, r_{m-1})$$
$$\times \langle (\ldots ((f \widetilde{\otimes}_{r_2} f) \widetilde{\otimes}_{r_3} f) \ldots \widetilde{\otimes}_{r_{m-2}} f) \widetilde{\otimes}_{r_{m-1}} f, f \rangle_{\mathfrak{H}^{\otimes q}}, \tag{8.4.2}$$

where the sum $\sum$ runs over all collections of integers $r_2, \ldots, r_{s-1}$ such that:

(i) $1 \leq r_2, \ldots, r_{m-1} \leq q$;

(ii) $r_2 + \ldots + r_{m-1} = \frac{(m-2)q}{2}$;

(iii) $r_2 < q$, $r_2 + r_3 < \frac{3q}{2}$, $\ldots$, $r_2 + \ldots + r_{m-2} < \frac{(m-2)q}{2}$;

(iv) $r_3 \leq 2q - 2r_2$, $\ldots$, $r_{m-1} \leq (m-2)q - 2r_2 - \ldots - 2r_{m-2}$;

and where the combinatorial constants $c_q(r_2, \ldots, r_{m-1})$ are recursively defined by the relations

$$c_q(r) = q(r-1)!\binom{q-1}{r-1}^2,$$

and, for $a \geq 2$,

$$c_q(r_2, \ldots, r_{a+1})$$
$$= q(r_{a+1} - 1)!\binom{aq - 2r_2 - \ldots - 2r_a - 1}{r_{a+1} - 1}\binom{q-1}{r_{a+1} - 1}c_q(r_1, \ldots, r_a).$$

Remark 8.4.5   1. If mq is odd, then $\kappa_m(F) = 0$; see condition (ii).

2. If $q = 2$ and $F = I_2(f)$, $f \in \mathfrak{H}^{\odot 2}$, then the only possible integers $r_1, \ldots, r_{s-2}$ satisfying (i)–(iv) in Theorem 8.4.4 are $r_1 = \ldots = r_{s-2} = 1$. On the other hand, it is easy to see that $c_2(1) = 2$, $c_2(1,1) = 4$, $c_2(1, 1, 1) = 8$, and so on. As a consequence,

$$\kappa_m(F) = 2^{m-1}(m-1)!\langle \underbrace{f \otimes_1 \ldots \otimes_1 f}_{m-1 \text{ copies of } f}, f\rangle_{\mathfrak{H}^{\otimes 2}}. \tag{8.4.3}$$

See Exercises 8.5.2 and 8.5.3.

3. If $q \geq 2$ and $F = I_q(f)$, $f \in \mathfrak{H}^{\odot q}$, then (8.4.2) for $s = 4$ reads

$$\kappa_4(I_q(f)) = 6q! \sum_{r=1}^{q-1} c_q(r, q-r)\langle (f\widetilde{\otimes}_r f)\widetilde{\otimes}_{q-r} f, f\rangle_{\mathfrak{H}^{\otimes q}}$$

$$= \frac{3}{q}\sum_{r=1}^{q-1} rr!^2\binom{q}{r}^4 (2q-2r)!\langle (f\widetilde{\otimes}_r f)\otimes_{q-r} f, f\rangle_{\mathfrak{H}^{\otimes q}}$$

$$= \frac{3}{q}\sum_{r=1}^{q-1} rr!^2\binom{q}{r}^4 (2q-2r)!\langle f\widetilde{\otimes}_r f, f\otimes_r f\rangle_{\mathfrak{H}^{\otimes(2q-2r)}}$$

$$= \frac{3}{q}\sum_{r=1}^{q-1} rr!^2\binom{q}{r}^4 (2q-2r)!\|f\widetilde{\otimes}_r f\|^2_{\mathfrak{H}^{\otimes(2q-2r)}},$$

and we recover expression (5.2.5) for $\kappa_4(F)$ by a different route.

8.5 Exercises

8.5.1 Complete the recursive argument at the beginning of the proof of Lemma 8.2.4 by showing that, if $\Gamma_{l_1,\dots,l_{k-1}}(F) \in \mathbb{D}^{j-k+2,2^{j-k+2}}$, $k \geq 2$, then $\Gamma_{l_1,\dots,l_k}(F) \in \mathbb{D}^{r,2} \cap L^{2^{j-k+1}}(\Omega)$, for every $r = 1, \dots, j - k + 1$.

8.5.2 Give a direct proof that formulae (8.4.3) and (2.7.17) are equivalent.

8.5.3 Generalize formula (8.4.3) and deduce an explicit expression for the joint cumulants of the vector $(I_2(f), I_2(g))$, where $f, g \in \mathfrak{H}^{\odot 2}$.

8.5.4 By proceeding as in Example 8.3.3, use Theorem 8.4.4 to deduce yet another proof of the implication (v)$\to$(i) of Theorem 5.2.7.

8.5.5 Let $m \geq 1$ be an integer, let $\varphi : \mathbb{R} \to \mathbb{R}$ belong to $\mathcal{C}^m$, and let $F \in \mathbb{D}^{m,2^m}$. Assume that $\varphi^{(s)}(F) \in L^1(\Omega)$ for all $s = 1, \dots, m-1$, and that $\varphi^{(m)}(F)\Gamma_{m+1}(F) \in L^1(\Omega)$. Prove, by induction on m, that $F\varphi(F) \in L^1(\Omega)$ and that

$$E[F\varphi(F)] = \sum_{s=0}^{m-1} \frac{1}{s!}\kappa_{s+1}(F)\, E[\varphi^{(s)}(F)] + E[\varphi^{(m)}(F)\Gamma_{m+1}(F)].$$

$$(8.5.1)$$

8.6 Bibliographic comments

The paper by Nourdin and Peccati [91] was the first to provide an explicit description of cumulants in terms of Malliavin operators: it contains, in particular, a proof of Theorem 8.4.4. The multidimensional Theorem 8.3.1 is taken from Noreddine and Nourdin [82]. In [91] one can also find a connection with the combinatorial diagram formulae for moments and cumulants, as discussed, for example, by Peccati and Taqqu in [110]. Formula (8.4.3) plays a crucial role in the classic paper by Fox and Taqqu [39]. A statement similar to (8.5.1) was proved by Barbour in [6]; see also Rotar [125].

9

Exact asymptotics and optimal rates

Throughout the following, we consider as given an isonormal Gaussian process $X = \{X(h) : h \in \mathfrak{H}\}$, and we adopt the Malliavin calculus notation introduced in Chapter 2. Let $\{F_n : n \geq 1\}$ be a sequence of random variables in $\mathbb{D}^{1,2}$ such that $E[F_n] = 0$, $\text{Var}(F_n) = 1$ and $F_n \overset{\text{law}}{\to} N \sim \mathcal{N}(0, 1)$ as $n \to \infty$. In this chapter, we shall develop techniques allowing us to compute an exact asymptotic expression for the (suitably normalized) sequence

$$P(F_n \leq z) - P(N \leq z), \quad n \geq 1,$$

when $z \in \mathbb{R}$ is fixed. As we will see, this yields lower bounds on the Kolmogorov distance between F_n and N, thus complementing, for example, the content of Theorem 5.1.3.

9.1 Some technical computations

Let $N \sim \mathcal{N}(0, 1)$. We start with some useful computations involving the class of Hermite polynomials $\{H_q : q \geq 0\}$ introduced in Definition 1.4.1. Rodrigues's formula (Proposition 1.4.2 (vii)) implies the induction relation

$$\frac{d}{dx}\left(H_q(x)e^{-x^2/2}\right) = -H_{q+1}(x)e^{-x^2/2}, \quad x \in \mathbb{R}, \tag{9.1.1}$$

yielding that the Hermite polynomials are related to the derivatives of $\Phi(x) := P(N \leq x)$, written as $\Phi^{(q)}(x)$ ($q = 1, 2, \ldots$), through the formula

$$\Phi^{(q)}(x) = (-1)^{q-1} H_{q-1}(x) \frac{e^{-x^2/2}}{\sqrt{2\pi}}. \tag{9.1.2}$$

For every fixed z, we denote by f_z the solution of the Stein's equation associated with the test function $\mathbf{1}_{(-\infty, z]}$, as given in (3.4.1). The following result, connecting f_z with the class of Hermite polynomials and the derivatives of Φ, will be used later.

Proposition 9.1.1 *For every $q \geq 1$ and every $z \in \mathbb{R}$,*

$$\int_{-\infty}^{+\infty} f_z'(x) H_q(x) \frac{e^{-x^2/2}}{\sqrt{2\pi}} dx = \frac{1}{q+2} H_{q+1}(z) \frac{e^{-z^2/2}}{\sqrt{2\pi}} = \frac{(-1)^{q+1}}{q+2} \Phi^{(q+2)}(z). \tag{9.1.3}$$

Proof   Integrating by parts using relation (9.1.1), we first obtain that

$$\int_{-\infty}^{+\infty} f_z'(x) H_q(x) \frac{e^{-x^2/2}}{\sqrt{2\pi}} dx = \int_{-\infty}^{+\infty} f_z(x) H_{q+1}(x) \frac{e^{-x^2/2}}{\sqrt{2\pi}} dx$$

$$= \frac{1}{\sqrt{2\pi}} \int_{-\infty}^{+\infty} H_{q+1}(x) \left(\int_{-\infty}^{x} [\mathbf{1}_{(-\infty,z]}(a) - P(N \leq z)] e^{-a^2/2} da \right) dx. \tag{9.1.4}$$

Integrating by parts once again, this time using the relation $H_{q+1}(x) = \frac{1}{q+2} H_{q+2}'(x)$ (see Proposition 1.4.2(i)), we deduce that

$$\int_{-\infty}^{+\infty} H_{q+1}(x) \left(\int_{-\infty}^{x} [\mathbf{1}_{(-\infty,z]}(a) - P(N \leq z)] e^{-a^2/2} da \right) dx$$

$$= -\frac{1}{q+2} \int_{-\infty}^{+\infty} H_{q+2}(x) [\mathbf{1}_{(-\infty,z]}(x) - P(N \leq z)] e^{-x^2/2} dx$$

$$= -\frac{1}{q+2} \left(\int_{-\infty}^{z} H_{q+2}(x) e^{-x^2/2} dx - P(N \leq z) \int_{-\infty}^{+\infty} H_{q+2}(x) e^{-x^2/2} dx \right)$$

$$= \frac{1}{q+2} H_{q+1}(z) e^{-z^2/2}, \quad \text{(by (9.1.1))}.$$

Using relation (9.1.2) concludes the proof of the proposition. $\qquad \square$

Remark 9.1.2   By specializing formula (9.1.3) to the case $q = 1$, we obtain:

$$E[f_z'(N) \times N] = \frac{1}{3}(z^2 - 1) \frac{e^{-z^2/2}}{\sqrt{2\pi}} = \frac{1}{3} \Phi^{(3)}(z). \tag{9.1.5}$$

9.2 A general result

Assume that $\{F_n : n \geq 1\}$ is a sequence of (sufficiently regular) centered random variables with unitary variance such that the sequence

$$\varphi(n) := \sqrt{\text{Var}[\langle DF_n, -DL^{-1}F_n \rangle_{\mathfrak{H}}]}, \quad n \geq 1, \tag{9.2.1}$$

converges to zero as $n \to \infty$. By Theorem 5.1.3 (in particular, formula (5.1.5)), therefore, as $n \to \infty$,

$$d_{\mathrm{Kol}}(F_n, N) \leq \varphi(n) \to 0, \qquad (9.2.2)$$

where $N \sim \mathscr{N}(0, 1)$. Theorem 9.2.2 provides useful criteria for dealing with the following tasks:

(i) computing an exact asymptotic expression (as $n \to \infty$) for the quantity

$$\frac{P(F_n \leq z) - P(N \leq z)}{\varphi(n)}, \quad n \geq 1;$$

(ii) determining whether the sequence $(\varphi(n))_{n \geq 1}$ provides an *optimal rate of convergence*, in the sense of Definition 9.2.1.

Definition 9.2.1   Assume that the sequence $\{\varphi(n) : n \geq 1\}$ appearing in (9.2.2) is strictly positive for every n large enough. We say that the rate of convergence associated with $\varphi(n)$ is **optimal** if there exists a constant $c \in (0, 1]$ such that, for n large enough,

$$c \leq \frac{d_{\mathrm{Kol}}(F_n, N)}{\varphi(n)} \leq 1. \qquad (9.2.3)$$

Theorem 9.2.2   *Let $\{F_n : n \geq 1\}$ be a sequence of random variables belonging to $\mathbb{D}^{1,2}$, and such that $E[F_n] = 0$, $\mathrm{Var}[F_n] = 1$. Suppose, moreover, that the following three conditions hold:*

(i) *(a) $\varphi(n)$ (as defined in (9.2.1)) is finite for every n; (b) as $n \to \infty$, $\varphi(n)$ converges to zero; and (c) there exists $m \geq 1$ such that $\varphi(n) > 0$ for $n \geq m$.*

(ii) *As $n \to \infty$, the two-dimensional vector $\left(F_n, \frac{1 - \langle DF_n, -DL^{-1}F_n \rangle_{\mathfrak{H}}}{\varphi(n)} \right)$ converges in distribution to a centered two-dimensional Gaussian vector (N_1, N_2), such that $E[N_1^2] = E[N_2^2] = 1$ and $E[N_1 N_2] = \rho$.*

(iii) *The law of F_n has a density with respect to Lebesgue measure for every n.*

Then, as $n \to \infty$, we have that $d_{\mathrm{Kol}}(F_n, N) \to 0$, where $N \sim \mathscr{N}(0, 1)$. Moreover, for every $z \in \mathbb{R}$,

$$\frac{P(F_n \leq z) - P(N \leq z)}{\varphi(n)} \longrightarrow \frac{\rho}{3}(z^2 - 1)\frac{e^{-z^2/2}}{\sqrt{2\pi}} = \frac{\rho}{3}\Phi^{(3)}(z) \quad as\ n \to \infty.$$

$$(9.2.4)$$

As a consequence, if $\rho \neq 0$ there exists a constant $c > 0$, as well as an integer $n_0 \geq 1$, such that (9.2.3) holds true for every $n \geq n_0$, that is, the rate of convergence associated with $\varphi(n)$ is optimal in the sense of Definition 9.2.1.

Proof For any integer n and any $\mathcal{C}^1$-function f with a bounded derivative, we know by Theorem 2.9.1 that $E[F_n f(F_n)] = E[f'(F_n)\langle DF_n, -DL^{-1}F_n\rangle_{\mathfrak{H}}]$. Fix $z \in \mathbb{R}$ and recall that the function f_z defined by (3.4.1) is not $\mathcal{C}^1$ due to the singularity in z. However, by using a regularization argument given assumption (iii) and taking into account that (3.4.2) holds, we get that the identity

$$E[F_n f_z(F_n)] = E[f'_z(F_n)\langle DF_n, -DL^{-1}F_n\rangle_{\mathfrak{H}}]$$

is true for any n. Therefore, since $P(F_n \leq z) - P(N \leq z) = E[f'_z(F_n)] - E[F_n f_z(F_n)]$, we get

$$\frac{P(F_n \leq z) - P(N \leq z)}{\varphi(n)} = E\left[f'_z(F_n) \times \frac{1 - \langle DF_n, -DL^{-1}F_n\rangle_{\mathfrak{H}}}{\varphi(n)}\right].$$

Since f'_z is bounded by 1 (see (3.4.2)) and $\varphi(n)^{-1}(1 - \langle DF_n, -DL^{-1}F_n\rangle_{\mathfrak{H}})$ has variance 1 by definition of $\varphi(n)$, the sequence

$$f'_z(F_n) \times \frac{1 - \langle DF_n, -DL^{-1}F_n\rangle_{\mathfrak{H}}}{\varphi(n)}, \quad n \geq 1,$$

is uniformly integrable. Identity (3.4.4) shows that $x \to f'_z(x)$ is continuous at every $x \neq z$. This yields that, as $n \to \infty$ and due to assumption (ii),

$$E\left[f'_z(F_n) \times \frac{1 - \langle DF_n, -DL^{-1}F_n\rangle_{\mathfrak{H}}}{\varphi(n)}\right] \longrightarrow E[f'_z(N_1)N_2] = \rho\, E[f'_z(N_1)N_1].$$

Consequently, relation (9.2.4) now follows from formula (9.1.5). If in addition $\rho \neq 0$, we can obtain the bound in (9.2.3) by using the elementary relation $|P(F_n \leq 0) - \Phi(0)| \leq d_{\mathrm{Kol}}(F_n, N)$. $\qquad\square$

9.3 Connections with Edgeworth expansions

In this section, we connect our results with Edgeworth expansions – see Section A.4 for an introduction to the subject. Let $N \sim \mathcal{N}(0, 1)$ and, as before, write Φ for the distribution function of N. Recall from Example A.4.4 that, if F is a random variable with finite third moment and such that $E(F) = 0$ and $E(F^2) = 1$, then the third-order Edgeworth expansion of the distribution function of F around Φ is given by

$$\mathcal{E}_3(F, N; z) = \Phi(z) - \frac{1}{3!}E(F_n^3)\Phi^{(3)}(z), \quad z \in \mathbb{R}.$$

The next proposition states that, under the assumptions of Theorem 9.2.2 plus some weak additional conditions, for every z the quantity $P(F_n \leq z)$

– $\mathcal{E}_3(F_n, N; z)$ converges to zero at a rate which is faster than that of the Kolmogorov distance $d_{\text{Kol}}(F_n, N)$.

Proposition 9.3.1 (One-term Edgeworth expansions) *Let $\{F_n : n \geq 1\}$, be a sequence of centered and square-integrable functionals of the isonormal Gaussian process $X = \{X(h) : h \in \mathfrak{H}\}$, such that $E(F_n^2) = 1$. Suppose that conditions (i)–(iii) of Theorem 9.2.2 are satisfied, and also that:*

(a) $E|F_n|^3 < \infty$, for every n;
(b) *there exists $\varepsilon > 0$ such that $\sup_{n \geq 1} E|F_n|^{2+\varepsilon} < \infty$.*

Then, as $n \to \infty$,

$$\frac{1}{2\varphi(n)} E(F_n^3) \longrightarrow -\rho, \tag{9.3.1}$$

and, for every $z \in \mathbb{R}$,

$$P(F_n \leq z) - \mathcal{E}_3(F, N; z) = P(F_n \leq z) - \Phi(z) + \frac{1}{3!} E(F_n^3) \Phi^{(3)}(z)$$

$$= o_z(\varphi(n)), \tag{9.3.2}$$

where $o_z(\varphi(n))$ denotes a numerical sequence (depending on z) such that $\varphi(n)^{-1} o_z(\varphi(n)) \to 0$, as $n \to \infty$.

Remark 9.3.2 Of course, relation (9.3.2) is interesting only when $\rho \neq 0$. Indeed, in this case we have that, thanks to Theorem 9.2.2, $P(F_n \leq z) - \Phi(z) \asymp \varphi(n)$ (the symbol $\asymp$ means asymptotic equivalence), so that, for a fixed z, replacing $P(F_n \leq z) - \Phi(z)$ with $P(F_n \leq z) - \mathcal{E}_3(F, N; z)$ actually increases the rate of convergence to zero.

Proof of Proposition 9.3.1 Since assumption (a) holds and $E(F_n) = 0$, we can integrate by parts and deduce that

$$E\left(F_n \times \frac{1 - \langle DF_n, -DL^{-1}F_n \rangle_{\mathfrak{H}}}{\varphi(n)}\right) = -\frac{1}{2\varphi(n)} E(F_n^3).$$

Assumption (b), combined with the fact that $\varphi(n)^{-1}(1 - \langle DF_n, -DL^{-1}F_n \rangle_{\mathfrak{H}})$ has variance 1, immediately yields that there exists $\delta > 0$ such that

$$\sup_{n \geq 1} E\left|F_n \times \varphi(n)^{-1}(1 - \langle DF_n, -DL^{-1}F_n \rangle_{\mathfrak{H}})\right|^{1+\delta} < \infty.$$

In particular, the sequence $\{F_n \times \varphi(n)^{-1}(1 - \langle DF_n, -DL^{-1}F_n \rangle_{\mathfrak{H}}) : n \geq 1\}$ is uniformly integrable. Therefore, since assumption (iii) in the statement of Theorem 9.2.2 holds, we deduce that, as $n \to \infty$,

$$\frac{1}{2\varphi(n)} E(F_n^3) \longrightarrow -E(N_1 N_2) = -\rho.$$

As a consequence,

$$\varphi(n)^{-1} \left| P(F_n \leq z) - \Phi(z) + \frac{1}{3!} E(F_n^3) \Phi^{(3)}(z) \right|$$

$$\leq \left| \frac{P(F_n \leq z) - \Phi(z)}{\varphi(n)} - \frac{\rho}{3} \Phi^{(3)}(z) \right| + \frac{|\Phi^{(3)}(z)|}{3} \left| \frac{1}{2\varphi(n)} E(F_n^3) + \rho \right|,$$

and the conclusion follows from Theorem 9.2.2. □

9.4 Double integrals

When applying Theorem 9.2.2 in concrete situations, the main issue is often to check that its condition (ii) holds true. For that purpose, Theorem 6.1.3 appears to be a perfect tool. In the particular case of sequences belonging to the second Wiener chaos, we can go further by using the material developed in Section 2.7.4.

Proposition 9.4.1 *Let $N \sim \mathcal{N}(0, 1)$ and let $F_n = I_2(f_n)$, $n \geq 1$, be such that $f_n \in \mathfrak{H}^{\odot 2}$. Write $\kappa_p(F_n)$, $p \geq 1$, for the sequence of the cumulants of F_n (as given in (2.7.15)). Assume that $\kappa_2(F_n) = E[F_n^2] = 1$ for all $n \geq 1$ and that $\kappa_4(F_n) \to 0$ as $n \to \infty$. If, in addition,*

$$\frac{\kappa_3(F_n)}{\sqrt{\kappa_4(F_n)}} \to \alpha \quad and \quad \frac{\kappa_8(F_n)}{(\kappa_4(F_n))^2} \to 0 \tag{9.4.1}$$

(recall that $\kappa_4(F_n) > 0$ for every n by Corollary 5.2.11), then

$$\frac{P(F_n \leq z) - P(N \leq z)}{\sqrt{\kappa_4(F_n)}} \longrightarrow \frac{\alpha}{6\sqrt{2\pi}} \left(1 - z^2 \right) e^{-\frac{z^2}{2}} \quad as \; n \to \infty. \tag{9.4.2}$$

As a consequence, if $\alpha \neq 0$ there exist a constant $c > 0$ and an integer $n_0 \geq 1$ such that $d_{\mathrm{Kol}}(F_n, N) \geq c\sqrt{\kappa_4(F_n)}$ for every $n \geq n_0$.

Remark 9.4.2 Due to (9.4.1), we see that (9.4.2) is equivalent to

$$\frac{P(F_n \leq z) - P(N \leq z)}{\kappa_3(F_n)} \to \frac{1}{6\sqrt{2\pi}} \left(1 - z^2 \right) e^{-\frac{z^2}{2}} \quad as \; n \to \infty.$$

Since each F_n is centered, we also have that $\kappa_3(F_n) = E[F_n^3]$.

Proof We shall apply Theorem 9.2.2. Thanks to (2.7.17) with $p = 4$, we get that $\frac{\kappa_4(F_n)}{6} = 8 \| f_n \otimes_1 f_n \|_{\mathfrak{H}^{\otimes 2}}^2$. By combining this identity with (5.2.3)

(it is worth observing here that $f_n \otimes_1 f_n$ is symmetric, so that the symmetrization $f_n \widetilde{\otimes}_1 f_n$ is immaterial), we see that the quantity $\varphi(n)$ appearing in (9.2.1)
is given by $\sqrt{\kappa_4(F_n)/6}$. In particular, condition (i) in Theorem 9.2.2 is met
(see also Corollary 5.2.11). On the other hand, since F_n is a non-zero double
integral, its law has a density with respect to Lebesgue measure, according
to Theorem 2.10.1. This means that condition (iii) in Theorem 9.2.2 is also
in order. Hence, it remains to check condition (ii). Assume that (9.4.1) holds.
Using (2.7.17) in the cases $p = 3$ and $p = 8$, we deduce that

$$\frac{\kappa_3(F_n)}{\sqrt{\kappa_4(F_n)}} = \frac{8 \langle f_n, f_n \otimes_1 f_n \rangle_{\mathfrak{H}^{\otimes 2}}}{\sqrt{6}\,\varphi(n)}$$

and

$$\frac{\kappa_8(F_n)}{(\kappa_4(F_n))^2} = \frac{17\,920 \| (f_n \otimes_1 f_n) \otimes_1 (f_n \otimes_1 f_n) \|_{\mathfrak{H}^{\otimes 2}}^2}{\varphi(n)^4}.$$

On the other hand, set

$$Y_n = \frac{\frac{1}{2} \| DF_n \|_{\mathfrak{H}}^2 - 1}{\varphi(n)}.$$

By (5.2.2), we have $\frac{1}{2} \| DY_n \|_{\mathfrak{H}}^2 - 1 = 2\, I_2(f_n \otimes_1 f_n)$. Therefore, by (5.2.3),

$$\mathrm{Var}\left(\frac{1}{2} \| DY_n \|_{\mathfrak{H}}^2 \right) = \frac{128}{\varphi(n)^4} \| (f_n \otimes_1 f_n) \otimes_1 (f_n \otimes_1 f_n) \|_{\mathfrak{H}^{\otimes 2}}$$

$$= \frac{\kappa_8(F_n)}{140\,(\kappa_4(F_n))^2} \longrightarrow 0, \quad \text{as } n \to \infty.$$

Hence, by Theorem 5.2.7, we deduce that $Y_n \overset{\text{Law}}{\longrightarrow} \mathcal{N}(0, 1)$. We also have

$$E[Y_n F_n] = \frac{4}{\varphi(n)} \langle f_n \otimes_1 f_n, f_n \rangle_{\mathfrak{H}^{\otimes 2}}$$

$$= \frac{\sqrt{6}\,\kappa_3(F_n)}{2\sqrt{\kappa_4(F_n)}} \longrightarrow \frac{\alpha\sqrt{6}}{2} =: \rho, \quad \text{as } n \to \infty.$$

Therefore, to conclude that condition (ii) in Theorem 9.2.2 holds, it suffices to
apply Theorem 6.2.3.   □

9.5 Further examples

We conclude this chapter by continuing the study initiated in Section 7.3.
Recall that $X = \{X_k : k \in \mathbb{Z}\}$ is a centered stationary Gaussian sequence
with unit variance, and that $\rho(v) = E[X_0 X_v]$, $v \in \mathbb{Z}$, so that in particular
$\rho(0) = E[X_0^2] = 1$ and $|\rho(v)| \leq 1$ for every v. Consider the sequence

$$V_n = \frac{1}{\sqrt{n}} \sum_{k=1}^{n} [X_k^2 - 1], \quad n \geq 1,$$

and define $v_n > 0$ by $E[V_n^2] = v_n^2$. As an application of Proposition 9.4.1, we can prove the following result.

Theorem 9.5.1 *Assume $\rho \in \ell^{\frac{4}{3}}(\mathbb{Z})$, and let $N \sim \mathcal{N}(0,1)$. Then $x \mapsto \sum_{k \in \mathbb{Z}} \rho(k)e^{ikx}$ is even, real-valued, positive, and belongs to $L^4([-\pi, \pi])$. Moreover, for any $z \in \mathbb{R}$, as $n \to \infty$,*

$$\sqrt{n}\left[P(V_n/v_n \leq z) - P(N \leq z)\right]$$

$$\longrightarrow \frac{\int_{-\pi}^{\pi} \left(\sum_{k \in \mathbb{Z}} \rho(k)e^{ikx}\right)^3 dx}{6\pi\sqrt{\pi} \left(\sum_{k \in \mathbb{Z}} \rho^2(k)\right)^{3/2}} \left(1 - z^2\right) e^{-\frac{z^2}{2}}. \tag{9.5.1}$$

Proof By virtue of Proposition 7.3.3, the first part of the theorem holds true and, moreover,

$$\sqrt{n}\, \kappa_3(V_n/v_n) \to \frac{\sqrt{2} \int_{-\pi}^{\pi} \left(\sum_{k \in \mathbb{Z}} \rho(k)e^{ikx}\right)^3 dx}{\pi \left(\sum_{k \in \mathbb{Z}} \rho^2(k)\right)^{\frac{3}{2}}} \quad \text{as } n \to \infty.$$

Hence, according to Proposition 9.4.1 (see also Remark 9.4.2), in order to show that (9.5.1) holds true we only need to check the two asymptotic relations in (9.4.1). The first follows directly from Proposition 7.3.3, so we shall concentrate on the second. By (2.7.17),

$$n^2 \kappa_8(V_n/v_n)$$

$$= \frac{1}{v_n^8 n^2} \sum_{i,j,k,l,i',j',k',l'=0}^{n-1} \rho(k-l)\rho(i-j)\rho(k-i)\rho(k'-l')\rho(i'-j')$$

$$\times \rho(k'-i')\rho(l-l')\rho(j-j')$$

$$\leq \frac{1}{v_n^8 n^2} \sum_{l,j'=0}^{n-1} \sum_{i,i',k,k',l',j \in \mathbb{Z}} |\rho(k-l)\rho(i-j)\rho(k-i)\rho(k'-l')$$

$$\times \rho(i'-j')\rho(k'-i')\rho(l-l')\rho(j-j')|$$

$$= \frac{1}{v_n^8 n^2} \sum_{l,j'=0}^{n-1} (|\rho| * |\rho| * |\rho| * |\rho|)(l-j')^2$$

$$\leq \frac{1}{v_n^8 n} \sum_{l=-n+1}^{n-1} (|\rho| * |\rho| * |\rho| * |\rho|)(l)^2. \tag{9.5.2}$$

On the other hand, for all $M \geq 1$,

$$
\begin{aligned}
(|\rho| * |\rho|)(n) &= \sum_{j \in \mathbb{Z}} |\rho(j)| |\rho(n-j)| \\
&\leq 2 \sum_{|j| \leq M} |\rho(j)| |\rho(n-j)| + \sum_{|j| > M, |j-n| > M} |\rho(j)| |\rho(n-j)| \\
&\leq (4M+2) \sup_{|j| \leq M} |\rho(n-j)| + \sqrt{\sum_{|j| > M} \rho(j)^2} \sqrt{\sum_{|j-n| > M} \rho(n-j)^2} \\
&= (4M+2) \sup_{|j| \leq M} |\rho(n-j)| + \sum_{|j| > M} \rho(j)^2,
\end{aligned}
$$

so that $\limsup_{n \to \infty} (|\rho| * |\rho|)(n) \leq \sum_{|j| > M} \rho(j)^2$, and then $\lim_{n \to \infty} (|\rho| * |\rho|)(n) = 0$ by letting $M \to \infty$. Similarly (just by replacing $|\rho|$ with $|\rho| * |\rho|$ in the previous computation), we can show that $\lim_{n \to \infty} (|\rho| * |\rho| * |\rho| * |\rho|)(n) = 0$. Hence, using Cesaro's theorem in (9.5.2) (recall also that $\lim_{n \to \infty} v_n^2 = 2 \|\rho\|_{\ell^2(\mathbb{Z})}^2 \geq 2$), we get that $n^2 \kappa_8(V_n/v_n) \to 0$ as $n \to \infty$, implying in turn, because of (7.3.3), that $\kappa_8(V_n/v_n)/\kappa_4(V_n/v_n)^2 \to 0$ as $n \to \infty$. The proof of Theorem 9.5.1 is complete. $\qquad\square$

9.6 Exercises

9.6.1 Let $\{W_t = t \in [0, 1]\}$ be a standard Brownian motion starting from zero. It is well known that, with probability one,

$$
\int_0^1 \frac{W_t^2}{t^2} dt = \infty
$$

(see, for example, [112] and the references therein). For every $\epsilon > 0$, we set

$$
B_\epsilon = \frac{1}{2\sqrt{\log(1/\epsilon)}} \left\{ \int_\epsilon^1 \frac{W_t^2}{t^2} dt - \log(1/\epsilon) \right\}.
$$

Use Proposition 9.4.1 to prove that there exist constants $0 < c < C < \infty$ such that, for ϵ small enough,

$$
\frac{c}{\sqrt{\log(1/\epsilon)}} \leq d_{\text{Kol}}(B_\epsilon, N) \leq \frac{C}{\sqrt{\log(1/\epsilon)}},
$$

where $N \sim \mathcal{N}(0, 1)$.

9.6.2 Fix $d \geq 2$. We recall that a Brownian sheet on $[0, 1]^d$ is a centered Gaussian process of the type $\{W(t_1, \ldots, t_d) : (t_1, \ldots, t_d) \in [0, 1]^d\}$ such that

$$E[W(t_1, \ldots, t_d)W(s_1, \ldots, s_d)] = \prod_{i=1}^{d} \min(t_i, s_i).$$

Prove a CLT (with optimal rate of convergence in the Kolmogorov distance, as $\epsilon \to 0$) for the quantity

$$\int_{\epsilon}^{1} \ldots \int_{\epsilon}^{1} \frac{W(t_1, \ldots, t_d)^2}{t_1^2 \ldots t_d^2} dt_1 \ldots dt_d.$$

9.7 Bibliographic comments

Theorem 9.2.2 and Proposition 9.3.1 are taken from Nourdin and Peccati [89]. In this reference one can also find a general version of Proposition 9.4.1, involving sequences of multiple integrals of arbitrary orders. Exercises 9.6.1 and 9.6.2, again taken from [89], build on previous findings by Peccati and Yor [112, 113]. The main results of this chapter are close in spirit to the 'reversed Berry–Esseen inequalities' proved by Barbour and Hall in [46]. Some early examples of Edgeworth-type techinques, as applied to non-linear functionals of Gaussian fields, can be found, for example, in Lieberman, Rousseau and Zucker [66] and Taniguchi [141].

10

Density estimates

In this chapter, we consider an isonormal Gaussian process $X = \{X(h) : h \in \mathfrak{H}\}$ (defined on $(\Omega, \mathscr{F}, P)$), and we make use of the Malliavin calculus terminology introduced in Chapter 2. We shall show how one can adapt the techniques developed in the previous chapters, in order to deduce explicit expressions and estimates for the densities of sufficiently regular functionals of X.

Throughout the following, we fix once and for all a random variable $F \in \mathbb{D}^{1,2}$, such that $E[F] = 0$. We associate with F the following function g_F, which will play a crucial role: for (almost) all x in the support of the law of F, we set

$$g_F(x) = E[\langle DF, -DL^{-1}F \rangle_{\mathfrak{H}} | F = x]. \qquad (10.0.1)$$

Recall that $g_F \geq 0$ – see Proposition 2.9.4.

10.1 General results

The main result of the chapter is the following statement, yielding an explicit expression for the density of F (when it exists).

Theorem 10.1.1 (Explicit expression for densities) *The law of F admits a density (with respect to Lebesgue measure), say ρ, if and only if the random variable $g_F(F)$ is P-almost surely strictly positive. In this case, the support of ρ, denoted by supp ρ, is a closed interval of $\mathbb{R}$ containing zero and, for (almost) all $x \in$ supp ρ,*

$$\rho(x) = \frac{E[|F|]}{2g_F(x)} \exp\left(-\int_0^x \frac{y \, dy}{g_F(y)}\right). \qquad (10.1.1)$$

170

Proof Assume that the random variable $g_F(F)$ is strictly positive almost surely. Combining Theorem 2.9.1 with an approximation argument (because $x \mapsto \int_{-\infty}^{x} \mathbf{1}_{B \cap [-n,n]}(y)dy$ is just Lipschitz, not C^1), we get, for any Borel set $B \in \mathscr{B}(\mathbb{R})$ and any $n \geq 1$, that

$$E\left[F \int_{-\infty}^{F} \mathbf{1}_{B \cap [-n,n]}(y)dy\right] = E\left[\mathbf{1}_{B \cap [-n,n]}(F)g_F(F)\right]. \tag{10.1.2}$$

Suppose that the Lebesgue measure of $B \in \mathcal{B}(\mathbb{R})$ is zero. Then $\int_{-\infty}^{F} \mathbf{1}_{B \cap [-n,n]}$ $(y)dy = 0$, so that $E\left[\mathbf{1}_{B \cap [-n,n]}(F)g_F(F)\right] = 0$ by (10.1.2). But, since $g_F(F) > 0$ almost surely, we get that $\mathbf{1}_{B \cap [-n,n]}(F) = 0$ P-a.s., that is, $P(F \in B \cap [-n, n]) = 0$. By letting $n \to \infty$, we get by monotone convergence that $P(F \in B) = 0$. Therefore, the Radon–Nikodym criterion is satisfied, hence implying that the law of F has a density.

Conversely, assume that the law of F has a density, say ρ. Let $\phi : \mathbb{R} \to \mathbb{R}$ be a continuous function with compact support, and let Φ denote any antiderivative of ϕ. Note that Φ is necessarily bounded. We can write:

$$E\left[\phi(F)g_F(F)\right] = E\left[\Phi(F)F\right] \quad \text{(by Theorem 2.9.1)}$$

$$= \int_{\mathbb{R}} \Phi(x)\, x\, \rho(x)dx \underset{(*)}{=} \int_{\mathbb{R}} \phi(x)\left(\int_{x}^{\infty} y\rho(y)dy\right)dx$$

$$= E\left[\phi(F)\frac{\int_{F}^{\infty} y\rho(y)dy}{\rho(F)}\right].$$

Equation $(*)$ was obtained by integrating by parts, after observing that

$$\int_{x}^{\infty} y\rho(y)dy \longrightarrow 0, \quad \text{as } |x| \to \infty$$

(for $x \to +\infty$, this is because $F \in L^1(\Omega)$; for $x \to -\infty$, it is because F has mean zero). Therefore, we have shown that, P-a.s.,

$$g_F(F) = \frac{\int_{F}^{\infty} y\rho(y)dy}{\rho(F)}. \tag{10.1.3}$$

(Notice that $P(\rho(F) > 0) = \int_{\mathbb{R}} \mathbf{1}_{\{\rho(x) > 0\}}\rho(x)dx = \int_{\mathbb{R}} \rho(x)dx = 1$, so that identity (10.1.3) always makes sense.) Since $F \in \mathbb{D}^{1,2}$, we have (see, for example, [98, Proposition 2.1.7]) that $\operatorname{supp} \rho = [\alpha, \beta]$ with $-\infty \leq \alpha < \beta \leq +\infty$. Since F has zero mean, note that $\alpha < 0$ and $\beta > 0$ necessarily. For every $x \in (\alpha, \beta)$, define

$$\varphi(x) = \int_{x}^{\infty} y\rho(y)\, dy. \tag{10.1.4}$$

The function φ is differentiable almost everywhere on (α, β), and its derivative is $-x\rho(x)$. In particular, since $\varphi(\alpha) = \varphi(\beta) = 0$ and φ is strictly increasing before 0 and strictly decreasing afterwards, we have $\varphi(x) > 0$ for all $x \in (\alpha, \beta)$. Hence, (10.1.3) implies that $g_F(F)$ is strictly positive almost surely.

Finally, let us prove (10.1.1). Let φ still be defined by (10.1.4). On the one hand, we have $\varphi'(x) = -x\rho(x)$ for almost all $x \in \operatorname{supp} \rho$. On the other hand, by (10.1.3), we have, for almost all $x \in \operatorname{supp} \rho$,

$$\varphi(x) = \rho(x)g_F(x). \tag{10.1.5}$$

By putting these two facts together, we get the following ordinary differential equation satisfied by φ:

$$\frac{\varphi'(x)}{\varphi(x)} = -\frac{x}{g_F(x)} \quad \text{for almost all } x \in \operatorname{supp} \rho.$$

Integrating this relation over the interval $[0, x]$ yields

$$\log \varphi(x) = \log \varphi(0) - \int_0^x \frac{y \, dy}{g_F(y)}.$$

Taking the exponential and using $0 = E(F) = E(F_+) - E(F_-)$ so that $E|F| = E(F_+) + E(F_-) = 2E(F_+) = 2\varphi(0)$, we get

$$\varphi(x) = \frac{1}{2} E|F| \exp\left(-\int_0^x \frac{y \, dy}{g_F(y)}\right).$$

Finally, the desired conclusion comes from (10.1.5). $\qquad\square$

A first consequence of Theorem 10.1.1 is the following statement, yielding sufficient conditions in order for the law of F to have a support equal to the real line.

Corollary 10.1.2 *Assume that there exists $\sigma_{\min} > 0$ such that*

$$g_F(F) \geq \sigma_{\min}^2, \quad \textit{P-a.s.} \tag{10.1.6}$$

Then the law of F, which has a density ρ by Theorem 10.1.1, has $\mathbb{R}$ for support and (10.1.1) holds almost everywhere in $\mathbb{R}$.

Proof Apart from $\operatorname{supp} \rho = \mathbb{R}$, this is an immediate consequence of Theorem 10.1.1. For the moment, we just know that $\operatorname{supp} \rho = [\alpha, \beta]$ with $-\infty \leq \alpha < 0 < \beta \leq +\infty$. Identity (10.1.3) yields

$$\int_x^\infty y\rho(y) \, dy \geq \sigma_{\min}^2 \rho(x) \quad \text{for almost all } x \in (\alpha, \beta). \tag{10.1.7}$$

Let φ be defined by (10.1.4), and recall that $\varphi(x) > 0$ for all $x \in (\alpha, \beta)$. When multiplied by $x \in [0, \beta)$, inequality (10.1.7) gives $\frac{\varphi'(x)}{\varphi(x)} \geq -\frac{x}{\sigma_{\min}^2}$. Integrating this relation over the interval $[0, x]$ yields $\log \varphi(x) - \log \varphi(0) \geq -\frac{x^2}{2\sigma_{\min}^2}$, i.e., since $\varphi(0) = \frac{1}{2}E|F|$,

$$\varphi(x) = \int_x^\infty y\rho(y)\,dy \geq \frac{1}{2}E|F|e^{-\frac{x^2}{2\sigma_{\min}^2}}. \tag{10.1.8}$$

Similarly, when multiplied by $x \in (\alpha, 0]$, inequality (10.1.7) gives $\frac{\varphi'(x)}{\varphi(x)} \leq -\frac{x}{\sigma_{\min}^2}$. Integrating this relation over the interval $[x, 0]$ yields $\log \varphi(0) - \log \varphi(x) \leq \frac{x^2}{2\sigma_{\min}^2}$, i.e. (10.1.8) still holds for $x \in (\alpha, 0]$. Now, let us prove that $\beta = +\infty$. If this were not the case, by definition, we would have $\varphi(\beta) = 0$; on the other hand, by letting x tend to β in the above inequality, because φ is continuous, we would have $\varphi(\beta) \geq \frac{1}{2}E|F|e^{-\frac{\beta^2}{2\sigma_{\min}^2}} > 0$, which contradicts $\beta < +\infty$. The proof of $\alpha = -\infty$ is similar. In conclusion, we have shown that $\operatorname{supp}\rho = \mathbb{R}$. $\square$

Using Corollary 10.1.2, we can deduce a neat criterion for normality. It should be compared with Theorem 5.1.3.

Corollary 10.1.3 (Characterization of normality) *Assume that F is not identically zero. Then F is Gaussian if and only if* $\operatorname{Var}(g_F(F)) = 0$.

Proof By (2.9.1) (choose $g(x) = x$, $G = F$ and recall that $E[F] = 0$),

$$E(\langle DF, -DL^{-1}F\rangle_{\mathfrak{H}}) = E(F^2) = \operatorname{Var}F. \tag{10.1.9}$$

Therefore, the condition $\operatorname{Var}(g_F(F)) = 0$ is equivalent to

$$g_F(F) = \operatorname{Var}F, \quad P\text{-a.s.}$$

Let $F \sim \mathcal{N}(0, \sigma^2)$ with $\sigma > 0$. Using (10.1.3), we immediately check that $g_F(F) = \sigma^2$, P-a.s. Conversely, if $g_F(F) = \sigma^2 > 0$ P-a.s., then Corollary 10.1.2 implies that the law of F has a density ρ, given by $\rho(x) = \frac{E|F|}{2\sigma^2}e^{-\frac{x^2}{2\sigma^2}}$ for almost all $x \in \mathbb{R}$, from which we immediately deduce that $F \sim \mathcal{N}(0, \sigma^2)$. $\square$

Observe that if $F \sim \mathcal{N}(0, \sigma^2)$ with $\sigma > 0$, then $E|F| = \sqrt{2/\pi}\,\sigma$, so that the formula (10.1.1) for ρ agrees, of course, with the usual one in this case.

When g_F can be bounded above and away from zero, we have the crude density estimates appearing in the following statement:

Corollary 10.1.4 *If there exist* $\sigma_{\min}, \sigma_{\max} > 0$ *such that*

$$\sigma_{\min}^2 \leq g_F(F) \leq \sigma_{\max}^2 \quad P\text{-a.s.},$$

then the law of F *has a density* ρ *satisfying, for almost all* $x \in \mathbb{R}$,

$$\frac{E|F|}{2\sigma_{\max}^2} \exp\left(-\frac{x^2}{2\sigma_{\min}^2}\right) \leq \rho(x) \leq \frac{E|F|}{2\sigma_{\min}^2} \exp\left(-\frac{x^2}{2\sigma_{\max}^2}\right).$$

Proof All that is necessary is to apply Corollary 10.1.2. $\square$

10.2 Explicit computations

We now show how to compute $g_F(F) = E(\langle DF, -DL^{-1}F \rangle_{\mathfrak{H}} | F)$ in practice. The next section contains an explicit example.

Proposition 10.2.1 *Write* $DF = \Phi_F(X)$, *where* $\Phi_F : \mathbb{R}^{\mathfrak{H}} \to \mathfrak{H}$ *is an a.s. uniquely defined measurable function. We have*

$$\langle DF, -DL^{-1}F \rangle_{\mathfrak{H}} = \int_0^\infty e^{-t} \langle \Phi_F(X), E'\big(\Phi_F(e^{-t}X + \sqrt{1 - e^{-2t}}X')\big) \rangle_{\mathfrak{H}} dt,$$

$$(10.2.1)$$

so that

$$g_F(F) = \int_0^\infty e^{-t} \mathbf{E}\big(\langle \Phi_F(X), \Phi_F(e^{-t}X + \sqrt{1 - e^{-2t}}X') \rangle_{\mathfrak{H}} | F\big) dt,$$

where X' *stands for an independent copy of* X, *and is such that* X *and* X' *are defined on the product probability space* $(\Omega \times \Omega', \mathcal{F} \otimes \mathcal{F}', P \times P')$. *Here,* $\mathbf{E}$ *denotes the mathematical expectation with respect to* $P \times P'$, *while* E' *is the mathematical expectation with respect to* P'.

Proof Recall from (2.9.2) that

$$-DL^{-1}F = \int_0^\infty e^{-t} P_t(DF) dt.$$

By Mehler's formula (2.8.1), and since $DF = \Phi_F(X)$ by assumption, we deduce that

$$-DL^{-1}F = \int_0^\infty e^{-t} E'\big(\Phi_F(e^{-t}X + \sqrt{1 - e^{-2t}}X')\big) dt,$$

and hence the formula for $\langle DF, -DL^{-1}F \rangle_{\mathfrak{H}}$ follows. Using $E(E'(\ldots)|F) = \mathbf{E}(\ldots|F)$, the formula for $g_F(F)$ holds. $\square$

By combining Theorem 10.1.1 with Proposition 10.2.1, we deduce the following statement:

Corollary 10.2.2 *Let the assumptions of Theorem 10.1.1 prevail, and assume that the law of F admits a density ρ. Let $\Phi_F : \mathbb{R}^{\mathfrak{H}} \to \mathfrak{H}$ be measurable and such that $DF = \Phi_F(X)$. Then, for (almost) all x in $\operatorname{supp}\rho$, the density ρ is given by*

$$\rho(x) = \frac{E|F|}{2\int_0^\infty e^{-t}\,\mathbf{E}\big(\langle \Phi_F(X), \Phi_F(e^{-t}X + \sqrt{1-e^{-2t}}X')\rangle_{\mathfrak{H}} | F = x\big)dt}$$

$$\times \exp\left(-\int_0^x \frac{y\,dy}{\int_0^\infty e^{-s}\,\mathbf{E}\big(\langle \Phi_F(X), \Phi_F(e^{-s}X+\sqrt{1-e^{-2s}}X')\rangle_{\mathfrak{H}}|F=y\big)ds}\right).$$

10.3 An example

Let $N \sim \mathcal{N}_m(0, C)$, with $C \in \mathcal{M}_m(\mathbb{R})$ positive definite. Consider an isonormal Gaussian process X over the Euclidean space $\mathfrak{H} = \mathbb{R}^m$, endowed with the inner product $\langle h_i, h_j\rangle_{\mathfrak{H}} = E[N_i N_j] = C_{i,j}$. Here, $\{h_i\}_{1\le i\le m}$ stands for the canonical basis of $\mathfrak{H} = \mathbb{R}^m$. Without loss of generality, we can identify N_i with $X(h_i)$ for any $i = 1, \ldots, m$. Since C is positive definite, note that $h_1, \ldots, h_m$ are necessarily pairwise different. Let

$$F = \max_{1\le i\le m} N_i - E\left[\max_{1\le i\le m} N_i\right],$$

and set, for any $t \ge 0$,

$$I_t = \operatorname{argmax}_{1\le i\le m}(e^{-t}X(h_i) + \sqrt{1-e^{-2t}}X'(h_i)),$$

where X' stands for an independent copy of X.

Lemma 10.3.1 *For any $t \ge 0$, I_t is a well-defined random element taking values in $\{1, \ldots, m\}$. Moreover, $F \in \mathbb{D}^{1,2}$ and $DF = \Phi_F(X) = h_{I_0}$.*

Proof Fix $t \ge 0$. Observe that $e^{-t}X + \sqrt{1-e^{-2t}}X'$ and X share the same law. Hence, for any $i \ne j$,

$$P\big(e^{-t}X(h_i) + \sqrt{1-e^{-2t}}X'(h_i) = e^{-t}X(h_j) + \sqrt{1-e^{-2t}}X'(h_j)\big)$$
$$= P\big(X(h_i) = X(h_j)\big) = 0,$$

so that the random variable I_t is well defined and takes values in $\{1, \ldots, m\}$. The rest of the proof actually corresponds to Example 2.3.9. $\qquad\square$

We deduce from Lemma 10.3.1 that

$$\langle \Phi_F(X), \Phi_F(e^{-t}X + \sqrt{1 - e^{-2t}}X') \rangle_{\mathfrak{H}} = C_{I_0, I_t}, \tag{10.3.1}$$

so that, by Corollary 10.2.2, the law of F has a density ρ which is given, for (almost) all x in supp ρ, by

$$\rho(x)$$
$$= \frac{E|F|}{2 \int_0^\infty e^{-t} \mathbf{E}(C_{I_0, I_t} | F = x) dt} \exp\left(-\int_0^x \frac{y dy}{\int_0^\infty e^{-s} \mathbf{E}(C_{I_0, I_s} | F = y) ds}\right).$$

As a by-product (see also Corollary 10.1.4), we obtain the density estimates of the next proposition, as well as a variance formula (which is immediately shown by combining (10.1.9), (10.2.1) and (10.3.1)).

Proposition 10.3.2 *Let $N \sim \mathcal{N}_m(0, C)$, with $C \in \mathcal{M}_m(\mathbb{R})$ positive definite.*

– *If there exist $\sigma_{\min}, \sigma_{\max} > 0$ such that $\sigma_{\min}^2 \leq C_{i,j} \leq \sigma_{\max}^2$ for any $i, j \in \{1, \ldots, m\}$, then the law of $F = \max_{1 \leq i \leq m} N_i - E\left[\max_{1 \leq i \leq m} N_i\right]$ has a density ρ satisfying*

$$\frac{E|F|}{2\sigma_{\max}^2} \exp\left(-\frac{x^2}{2\sigma_{\min}^2}\right) \leq \rho(x) \leq \frac{E|F|}{2\sigma_{\min}^2} \exp\left(-\frac{x^2}{2\sigma_{\max}^2}\right)$$

for almost all $x \in \mathbb{R}$.
– *With N' an independent copy of N and $I_t := \text{argmax}_{1 \leq i \leq m}\left(e^{-t}N_i + \sqrt{1 - e^{-2t}}\, N_i'\right)$, we have*

$$\text{Var}\left(\max_{1 \leq i \leq m} N_i\right) = \int_0^\infty e^{-t} \mathbf{E}(C_{I_0, I_t}) dt. \tag{10.3.2}$$

10.4 Exercises

10.4.1 Let $F \in \mathbb{D}^{1,2}$ be such that $\frac{DF}{\|DF\|_{\mathfrak{H}}^2} \in \text{Dom}\,\delta$. The goal is to show that F admits a density (with respect to Lebesgue measure) given by

$$\rho(x) = E\left[\mathbf{1}_{\{F > x\}}\delta\left(\frac{DF}{\|DF\|_{\mathfrak{H}}^2}\right)\right] \tag{10.4.1}$$

(compare with (10.1.1)).

1. Let ψ be a non-negative smooth function with compact support, and set $\varphi(y) = \int_{-\infty}^{y} \psi(z)dz$. Show that

$$E[\psi(F)] = E\left[\varphi(F)\,\delta\left(\frac{DF}{\|DF\|_{\mathfrak{H}}^2}\right)\right]. \tag{10.4.2}$$

2. Show that (10.4.2) continues to hold for $\psi = \mathbf{1}_{[a,b]}$, $a < b$.

3. Conclude by showing that F admits the density (with respect to Lebesgue measure) given by (10.4.1).

4. If $F \in \mathbb{D}^{2,4}$ is such that $E[\|DF\|_{\mathfrak{H}}^{-8}] < \infty$, show that $\frac{DF}{\|DF\|_{\mathfrak{H}}^2} \in$ Dom δ.

10.4.2 Let $N \sim \mathcal{N}_m(0, C)$, with $C \in \mathcal{M}_m(\mathbb{R})$ positive definite, and assume that $C_{i,j} \geq 0$ for any $i, j \in \{1, \ldots, m\}$. Let $\phi : \mathbb{R}^m \to \mathbb{R}$ be a $\mathcal{C}^1$ function, and assume that there exist $\alpha_i, \beta_i \geq 0$ such that $\alpha_i \leq \frac{\partial f}{\partial x_i}(x) \leq \beta_i$ for any $i \in \{1, \ldots, m\}$ and $x \in \mathbb{R}^m$. Suppose, moreover, that $\sum_{i,j=1}^{m} \alpha_i \alpha_j C_{i,j} > 0$. Show that $F = \phi(N) - E[\phi(N)]$ admits a density ρ satisfying, for (almost) $x \in \mathbb{R}$,

$$\frac{E|F|}{2\sum_{i,j=1}^{m} \beta_i \beta_j C_{i,j}} \exp\left(-\frac{x^2}{2\sum_{i,j=1}^{m} \alpha_i \alpha_j C_{i,j}}\right) \leq \rho(x)$$

$$\leq \frac{E|F|}{2\sum_{i,j=1}^{m} \alpha_i \alpha_j C_{i,j}} \exp\left(-\frac{x^2}{2\sum_{i,j=1}^{m} \beta_i \beta_j C_{i,j}}\right).$$

10.4.3 Let $F \in \mathbb{D}^{1,2}$ with $E[F] = 0$. Assume that $\alpha \geq 0$ and $\beta > 0$ are fixed real numbers such that $g_F(F) \leq \alpha F + \beta$ almost surely, with g_F defined by (10.0.1). The goal of this exercise is to show that

$$P(F \geq x) \leq \exp\left(-\frac{x^2}{2\alpha x + 2\beta}\right), \quad x \geq 0. \tag{10.4.3}$$

1. For any $A > 0$, define $m_A : [0, +\infty) \to \mathbb{R}$ by $m_A(\theta) = E[e^{\theta F}\mathbf{1}_{\{F \leq A\}}]$. Prove that

$$m_A'(\theta) = \int_{-\infty}^{A} dy\,\theta\,e^{\theta y} \int_{\mathbb{R}} dP_F(x)\,x\,\mathbf{1}_{\{x \geq y\}}$$

$$- \int_{-\infty}^{A} dy\,\theta\,e^{\theta y} \int_{\mathbb{R}} dP_F(x)\,x\,\mathbf{1}_{\{x > A\}}$$

$$\leq \int_{-\infty}^{A} dy\,\theta\,e^{\theta y} \int_{\mathbb{R}} dP_F(x)\,x\,\mathbf{1}_{\{x \geq y\}}.$$

2. By combining Theorem 2.9.1 with an approximation argument (because $x \mapsto \int_{-\infty}^{x} e^{\theta y} \mathbf{1}_{\{y \leq A\}} dy$ is just Lipschitz, not $\mathcal{C}^1$), show that

$$E\big[g_F(F) e^{\theta F} \mathbf{1}_{\{F \leq A\}}\big] = \int_{-\infty}^{A} dy\, e^{\theta y} \int_{\mathbb{R}} dP_F(x)\, x\, \mathbf{1}_{\{x \geq y\}}.$$

3. Deduce, for any $\theta \in (0, 1/\alpha)$, that

$$m'_A(\theta) \leq \frac{\theta \beta}{1 - \theta \alpha}\, m_A(\theta). \qquad (10.4.4)$$

4. Integrate (10.4.4) and use Fatou's lemma to get that

$$E\big[e^{\theta F}\big] \leq \exp\left(\frac{\beta \theta^2}{2(1 - \theta \alpha)}\right)$$

for all $\theta \in (0, 1/\alpha)$.

5. Conclude by showing that inequality (10.4.3) holds.

10.4.4 Let $N \sim \mathcal{N}_m(0, C)$ with $C \in \mathcal{M}_m(\mathbb{R})$ non-negative definite. By using (10.3.2), show that

$$\mathrm{Var}\left(\max_{1 \leq i \leq m} N_i\right) \leq \max_{1 \leq i \leq m} \mathrm{Var}\,(N_i).$$

10.5 Bibliographic comments

The main results of this chapter are taken from Nourdin and Viens [96]. Several generalizations in a multidimensional setting can be found in Airault, Malliavin and Viens [2]. Some applications to density estimates in connection with stochastic partial differential equations can be found in Nualart and Quer-Sardanyons [102,103]. There are a huge number of important explicit formulae and estimates for densities that are obtained by means of Malliavin calculus (see, for example, formula (10.4.1) in Exercise 10.4.1), and that should be compared with the formulae deduced in this chapter: for a detailed discussion of results and techniques, see Nualart [98, chapter 2]. For some recent developments that are related to the findings described in this chapter, see Nualart [104] and Malliavin and Nualart [71].

11

Homogeneous sums and universality

In this final chapter, we shall relate the results and techniques discussed so far to the so-called *universality phenomenon*, according to which the asymptotic behavior of large random systems does not depend on the distribution of its components. Distinguished examples of the universality phenomenon are the central limit theorem and its functional version (the so-called 'Donsker theorem'), as well as the semicircular and circular laws in random matrix theory. Other examples of universality will emerge as the chapter unfolds.

11.1 The Lindeberg method

The universality results discussed below concern random variables having the form of homogeneous sums, that is, of linear combinations of products of independent random variables (see Section 11.2 for precise definitions). Our approach relies quite heavily on a version (first developed in [79]) of the so-called 'Lindeberg method' for normal approximations. Roughly speaking, the Lindeberg method involves a discrete interpolation technique, allowing the distance between the laws of two sums of random variables to be assessed by means of a progressive replacement of the summands. The aim of this introductory section is to familiarize the reader with the Lindeberg method, in the basic setting of sums of independent elements. Since this section is merely intended for illustration purposes, we do not aim at generality (see the bibliographic remarks for pointers to more general statements).

Throughout this section, we shall denote by $Y = \{Y_i : i \geq 1\}$ a sequence of centered independent random variables, such that $E[Y_i^2] = 1$ for every i. We also fix an increasing sequence of integers $\{M_n : n \geq 1\}$. For every $n \geq 1$, $\{c_{n,i} : i = 1, \ldots, M_n\}$ is a collection of non-zero real numbers such that

$$\sum_{i=1}^{M_n} c_{n,i}^2 = 1, \tag{11.1.1}$$

and we also write $S_n = \sum_{i=1}^{M_n} c_{n,i} Y_i$ and $Z_{n,i} = c_{n,i} Y_i$. In view of (11.1.1), for every $i = 1, \ldots, M_n$, the quantity $c_{n,i}^2$ can be interpreted as the *influence* of the ith component of S_n, that is, as the measure of the impact of $Z_{n,i}$ on the overall fluctuations of S_n.

One natural problem is now to establish sufficient conditions in order to have that, as $n \to \infty$, S_n converges in distribution to a standard Gaussian random variable $G \sim \mathcal{N}(0, 1)$. The following statement, essentially due to Lindeberg, provides an exhaustive answer to this problem: it is a generalization of the usual CLT, as well as one of the most crucial results in modern probability. Note that condition (L-ii) below is the celebrated *Lindeberg condition*.

Theorem 11.1.1 (Lindeberg CLT)   *Under the above notation and assumptions, the following two conditions are equivalent as $n \to \infty$:*

(L-i) $S_n \xrightarrow{\text{Law}} G$ *and* $\max_{i=1,\ldots,M_n} |c_{n,i}| \to 0$;
(L-ii) *for every $\epsilon > 0$,*

$$\sum_{i=1}^{M_n} E[Z_{n,i}^2 \mathbf{1}_{\{|Z_{n,i}|>\epsilon\}}] = \sum_{i=1}^{M_n} c_{n,i}^2 E[Y_i^2 \mathbf{1}_{\{|Y_i|>\epsilon/|c_{n,i}|\}}] \to 0. \tag{11.1.2}$$

A complete proof of Theorem 11.1.1 is omitted, and can be found, for example, in Kallenberg [60, chapter 5]. Here, we shall merely focus on the implication (L-ii)$\Rightarrow$(L-i) under some special moment assumptions.

Remark 11.1.2   In what follows, we shall actually prove that, under the additional assumption that $\max_{i \geq 1} E|Y_i|^3 < \infty$, the following chain of implications holds:

$$(\text{L-ii}) \quad \Longrightarrow \quad \max_{i=1,\ldots,M_n} |c_{n,i}| \to 0 \quad \Longrightarrow \quad S_n \xrightarrow{\text{Law}} G.$$

For every n, write j_n for the index such that $|c_{n,j_n}| = \max_{i=1,\ldots,M_n} |c_{n,i}|$: since for every $\epsilon > 0$,

$$c_{n,j_n}^2 \leq \epsilon^2 + E[Z_{n,j_n}^2 \mathbf{1}_{\{|Z_{n,j_n}|>\epsilon\}}] \leq \epsilon^2 + \sum_{i=1}^{M_n} E[Z_{n,i}^2 \mathbf{1}_{\{|Z_{n,i}|>\epsilon\}}], \tag{11.1.3}$$

we obtain immediately (since ϵ is arbitrary) that condition (L-ii) implies that

$$\max_{i=1,\ldots,M_n} |c_{n,i}| \to 0,$$

that is, the influences of the variables composing S_n converge uniformly to zero. The next statement, whose proof is based on a particular instance of the general Lindeberg method adopted in this chapter, provides a complete proof of the implication (L-ii)⇒(L-i) in the case of random variables having uniformly bounded third moments.

Proposition 11.1.3 *In addition to the previous assumptions, suppose that*

$$A := \max_{i \geq 1} E|Y_i|^3 < \infty.$$

Then, for every function $h : \mathbb{R} \to \mathbb{R}$ with bounded third derivative, we have

$$|E[h(S_n)] - E[h(G)]| \leq \left(\frac{A}{6} + \frac{1}{3}\sqrt{\frac{2}{\pi}} \right) \|h'''\|_\infty \max_{i=1,\ldots,M_n} |c_{n,i}|. \quad (11.1.4)$$

In particular, if condition (11.1.2) is satisfied, then S_n converges in distribution to G.

Proof Let $\mathbf{G} = \{G_i : i \geq 1\}$ be a collection of i.i.d copies of G. Without loss of generality, we may assume that the random variables in $\mathbf{G}$ are defined on the same probability space as $Y = \{Y_i : i \geq 1\}$, and that Y and $\mathbf{G}$ are independent. In view of (11.1.1), the random variables G and $U_n := \sum_{i=1}^{M_n} c_{n,i} G_i$ have the same distribution for every n. For every $j = 0, 1, \ldots, M_n$ set

$$S_n(j) := \sum_{i=1}^{j} c_{n,i} G_i + \sum_{i=j+1}^{M_n} c_{n,i} Y_i,$$

where we adopt the usual convention that $\sum_{i=1}^{0} = \sum_{i=M_n+1}^{M_n} \equiv 0$, and also, for $j = 1, \ldots, M_n$,

$$\widehat{S}_n(j) := S_n(j) - c_{n,j} G_j.$$

With this notation, we have that: (a) $S_n(0) = S_n$ and $S_n(M_n) = U_n$; (b) $\widehat{S}_n(j)$ is independent of Y_j and G_j for every j; and (c) for every $j = 1, \ldots, M_n$, $S_n(j) = \widehat{S}_n(j) + c_{n,j} G_j$ and $S_n(j-1) = \widehat{S}_n(j) + c_{n,j} Y_j$. Using a Taylor expansion, for every $j = 1, \ldots, M_n$, we have that

$$E[h(S_n(j))] = E[h(\widehat{S}_n(j))] + E[h'(\widehat{S}_n(j))c_{n,j} G_j]$$
$$+ \frac{1}{2} E[h''(\widehat{S}_n(j))c_{n,j}^2 G_j^2] + R_1(j),$$

$$E[h(S_n(j-1))] = E[h(\widehat{S}_n(j))] + E[h'(\widehat{S}_n(j))c_{n,j} Y_j]$$
$$+ \frac{1}{2} E[h''(\widehat{S}_n(j))c_{n,j}^2 Y_j^2] + R_2(j),$$

where $|R_1(j)| \leq \frac{1}{6}\|h'''\|_\infty |c_{n,j}|^3 E|G|^3 = \frac{1}{3}\sqrt{\frac{2}{\pi}}\|h'''\|_\infty |c_{n,j}|^3$ and $|R_2(j)| \leq \frac{1}{6}\|h'''\|_\infty |c_{n,j}|^3 A$. Also, using (b) above as well as the fact that $E[G_j] = E[Y_j] = 0$ and $E[G_j^2] = E[Y_j^2] = 1$, we infer that

$$E[h(S_n(j))] - E[h(S_n(j-1))] = R_1(j) - R_2(j).$$

Using a telescopic sum, this last relation entails that

$$|E[h(S_n)] - E[h(G)]| \le \sum_{j=1}^{M_n} |E[h(S_n(j))] - E[h(S_n(j-1))]|$$

$$\le \sum_{j=1}^{M_n} [|R_1(j)| + |R_2(j)|],$$

so that the desired conclusion follows from (11.1.1), since

$$\sum_{i=1}^{M_n} |c_{n,i}|^3 \le \max_{i=1,\dots,M_n.} |c_{n,i}| \times \sum_{i=1}^{M_n} c_{n,i}^2 = \max_{i=1,\dots,M_n.} |c_{n,i}|.$$

The last assertion in the statement is a consequence of (11.1.3). □

Remark 11.1.4   The Lindeberg-type proof of Proposition 11.1.3 suggests the following heuristic interpretation of the inequality (11.1.3): *if n is large and the influences of the random variables Y_i are uniformly small, then replacing each Y_i with its 'Gaussian counterpart' G_i does not change significantly the distribution of S_n.*

In the following sections, we will prove several analogs of Proposition 11.1.3, involving homogeneous sums of arbitrary orders.

11.2 Homogeneous sums and influence functions

For the sake of simplicity, from now on we write $[M] = \{1, \dots, M\}$.

Definition 11.2.1 (Multilinear polynomials)   Let $1 \le d \le M$ be integers. A symmetric polynomial $Q \in \mathbb{R}[x_1, \dots, x_M]$ in M variables is said to be a **multilinear polynomial of degree** d if it has the form

$$Q(x_1, \dots, x_M) = Q(f; x_1, \dots, x_M) \tag{11.2.1}$$

$$= \sum_{1 \le i_1, \dots, i_d \le M} f(i_1, \dots, i_d) x_{i_1} x_{i_2} \dots x_{i_d}$$

$$= d! \sum_{\{i_1, \dots, i_d\} \subset [M]^d} f(i_1, \dots, i_d) x_{i_1} x_{i_2} \dots x_{i_d}$$

$$= d! \sum_{1 \le i_1 < i_2 < \dots < i_d \le M} f(i_1, \dots, i_d) x_{i_1} x_{i_2} \dots x_{i_d},$$

where the mapping $f : [M]^d \to \mathbb{R}$ satisfies the following properties:

(i) f is symmetric, that is, $f(i_1, \ldots, i_d) = f(i_{\sigma(1)}, \ldots, i_{\sigma(d)})$, for every permutation σ of $\{1, \ldots, d\}$;

(ii) f vanishes on diagonals, that is, $f(i_1, \ldots, i_d) = 0$ whenever there exist $k \neq l$ such that $i_k = i_l$.

Plainly, a multilinear polynomial of degree $d = 1$ is just a linear combination of the variables $x_1, \ldots, x_M$. By convention, a multilinear polynomial of degree $d = 0$ is a real constant.

Example 11.2.2 For every $d \geq 2$, the polynomial

$$Q(x_1, \ldots, x_M) = \sum_{1 \leq i_1 < i_2 < \ldots < i_d \leq M} x_{i_1} x_{i_2} \ldots x_{i_d}$$

is a homogeneous polynomial of degree d, corresponding to the kernel

$$f(i_1, \ldots, i_d) = \frac{1}{d!} \mathbf{1}_{\{1 \leq i_k \neq i_l \leq M, \ \forall 1 \leq k \neq l \leq d\}}. \tag{11.2.2}$$

Definition 11.2.3 (Homogeneous sums) Let $(Y_1, \ldots, Y_M)$ be a collection of random variables, and let Q be a homogeneous polynomial of degree $d \leq M$ as in (11.2.1). The random variable

$$Q(Y_1, \ldots, Y_M) = Q(f; Y_1, \ldots, Y_M) = \sum_{1 \leq i_1, \ldots, i_d \leq M} f(i_1, \ldots, i_d) Y_{i_1} Y_{i_2} \ldots Y_{i_d} \tag{11.2.3}$$

is called the **homogeneous sum** of order d based on Q and $(Y_1, \ldots, Y_M)$. If the random variables Y_i are independent, centered and with unit variance, then

$$E[Q(Y_1, \ldots, Y_M)] = 0 \quad \text{and} \quad E[Q(Y_1, \ldots, Y_M)^2] = d! \|f\|_d^2,$$

where we have used the shorthand notation $\|f\|_d^2 = \sum_{1 \leq i_1, \ldots, i_d \leq M} f(i_1, \ldots, i_d)^2$.

Example 11.2.4 Homogeneous sums are recurring objects in modern probability theory. The following three examples are relevant for the theory developed in this book.

(i) According to Exercise 2.7.6, the elements of the dth Wiener chaos associated with a Gaussian field can be approximated in square norm by random variables of type (11.2.3), where the Y_i are independent and centered Gaussian random variables.

(ii) Analogously, the elements of the so-called *Poisson homogeneous chaos* (see, for example, [110, chapter 5]) associated with a general Poisson measure can be approximated in square norm by homogeneous sums such as (11.2.3), where the Y_i are independent and centered Poisson random variables.

(iii) When $(Y_1, \ldots, Y_M)$ is a *Rademacher vector* (that is, the Y_i are independent and such that $P(Y_i = 1) = \frac{1}{2} = P(Y_i = -1)$), the random variable in (11.2.3) is an element of the so-called dth *Walsh chaos*. The Walsh chaos plays a role analogous to that of Wiener chaos in the analysis of functionals of infinite Rademacher sequences.

Remark 11.2.5 Let $G = \{G_i : i \geq 1\}$ be an i.i.d. $\mathcal{N}(0, 1)$ sequence. There is no loss of generality in assuming that G is such that, for every i, $G_i = X(e_i)$, where $X = \{X(h) : h \in \mathfrak{H}\}$ is an isonormal Gaussian process over some separable Hilbert space $\mathfrak{H}$, and $\{e_i : i \geq 1\}$ is an orthonormal basis of $\mathfrak{H}$. Since $X(e_i) = I_1(e_i)$, Theorem 2.7.10 implies that homogeneous sums of order $d \geq 2$ can always be written as multiple Wiener–Itô integrals of order d (and therefore as the elements of some Gaussian Wiener chaos of order d), namely:

$$Q(f; G_1, \ldots, G_M) = \sum_{1 \leq i_1, \ldots, i_d \leq M} f(i_1, \ldots, i_d) X(e_{i_1}) \ldots X(e_{i_d}) = I_d(g),$$

where

$$g = \sum_{1 \leq i_1, \ldots, i_d \leq M} f(i_1, \ldots, i_d) e_{i_1} \otimes e_{i_2} \otimes \ldots \otimes e_{i_d} \in \mathfrak{H}^{\odot d},$$

with $\otimes = \otimes_0$ indicating a standard tensor product.

Definition 11.2.6 (Influences) Let $Q(\cdot) = Q(f; \cdot)$ be the d-homogeneous polynomial in M variables appearing in (11.2.1). For every $i = 1, \ldots, M$, the **influence** of the ith variable of $Q(f; \cdot)$ is given by the quantity

$$\text{Inf}_i(f) = \sum_{\{i_2, \ldots, i_d\} \subset [M]^{d-1}} f^2(i, i_2, \ldots, i_d) \tag{11.2.4}$$

$$= \frac{1}{(d-1)!} \sum_{1 \leq i_2, \ldots, i_d \leq M} f^2(i, i_2, \ldots, i_d).$$

Let $(Y_1, \ldots, Y_M)$ be a vector of independent random variables, with zero mean and unit variance. We have the following probabilistic interpretation of the quantity $\text{Inf}_i(f)$:

$$d!^2 \text{Inf}_i(f) = E\{\text{Var}[Q(Y_1, \ldots, Y_M) \mid Y_k, k \neq i]\}, \tag{11.2.5}$$

where we have used the notation

$$\mathrm{Var}[Q(Y_1,\ldots,Y_M) \mid Y_k, \; k \neq i]$$
$$= E\big[(Q(Y_1,\ldots,Y_M) - E[Q(Y_1,\ldots,Y_M) \mid Y_k, \; k \neq i])^2 \mid Y_k, \; k \neq i\big].$$
$$(11.2.6)$$

From formula (11.2.5) we see that $\mathrm{Inf}_i(f)$ roughly measures the impact of the variable Y_i on the overall fluctuations of the homogenous sum $Q(f; W_1,\ldots,W_M)$.

Example 11.2.7 Let us compute $\mathrm{Inf}_i(f)$ when f is the kernel appearing in (11.2.2). In this case, for every $i = 1,\ldots,M$, $\mathrm{Inf}_i(f)$ is equal to the factor $d!^{-2}(d-1)!^{-1}$ multiplied by the number of vectors $(i_1,\ldots,i_{d-1}) \in \{1,\ldots,M\}^{d-1}$ such that $i_k \neq i_l$ for every $1 \leq k \neq l \leq d-1$ and $i \neq i_k$ for every $k = 1,\ldots,d-1$. It follows that

$$\mathrm{Inf}_i(f) = \frac{(M-1)(M-2)\ldots(M-d+1)}{d!^2(d-1)!}.$$

11.3 The universality result

The main achievement of this chapter is the following universality result for homogeneous sums.

Theorem 11.3.1 (Universality of the Gaussian Wiener chaos) *Let $d_m,\ldots,d_1 \geq 2$ be fixed integers, and let $\{M_n : n \geq 1\}$ be an increasing integer sequence. For each $k = 1,\ldots,m$, let $f_k^{(n)} : [M_n]^{d_k} \to \mathbb{R}$, $n \geq 1$, be a symmetric function vanishing on diagonals (see Definition 11.2.1), and define, for $x_1,\ldots,x_n \in \mathbb{R}$,*

$$Q_k(n, x_1,\ldots,x_{M_n}) := Q\big(f_k^{(n)}; x_1,\ldots,x_{M_n}\big)$$
$$= \sum_{1 \leq i_1,\ldots,i_{d_k} \leq M_n} f_k^{(n)}(i_1,\ldots,i_k) x_{i_1} \ldots x_{i_{d_k}}.$$

Assume that, for every $k = 1,\ldots,m$,

$$\sup_{n \geq 1} \sum_{1 \leq i_1,\ldots,i_{d_k} \leq M_n} f_k^{(n)}(i_1,\ldots,i_{d_k})^2 < \infty. \qquad (11.3.1)$$

Let C be an $m \times m$ non-negative symmetric matrix. Then, as $n \to \infty$, the following two conditions are equivalent:

(i) *The vector* $\{Q_k(n, G_1, \ldots, G_{M_n})\}_{1 \le k \le m}$ *converges in law to* $\mathcal{N}_m(0, C)$
     (that is, to an m-dimensional centered Gaussian vector with covariance
     C), where $G_1, G_2, \ldots$ *are independent* $\mathcal{N}(0, 1)$ *random variables.*

(ii) *The vector* $\{Q_k(n, Y_1, \ldots, Y_{M_n})\}_{1 \le k \le m}$ *converges in law to* $\mathcal{N}_m(0, C)$,
     for every sequence $Y_1, Y_2, \ldots$ *of independent random variables such that*
     $E[Y_i] = 0$, $E[Y_i^2] = 1$ *and*

$$\max_{i \ge 1} E[Y_i^4] < \infty.$$

Remark 11.3.2 In what follows, we shall provide a self-contained proof of
Theorem 11.3.1. In [94] it is shown that, by using the results of [79, 124], one
can indeed relax the assumptions underlying this result. For instance, the statement of Theorem 11.3.1 continues to hold if the the random variables $\{Y_i\}$ have
uniformly bounded third moments. Also, in the one-dimensional case (that is,
when $m = 1$), one only needs that the moments of order $2 + \delta$ are uniformly
bounded, for some $\delta > 0$. Finally, again in the one-dimensional case, if the
$\{Y_i\}$ are also assumed to be identically distributed, then the statement holds
under the minimal assumption that $E[Y_i] = 0$ and $E[Y_i^2] = 1$. Note that
the statements of Theorem 11.3.1 and of its refinements *do not* involve any
Lindeberg-type condition.

Exercise 11.3.3 Prove by means of a counterexample that the statement of
Theorem 11.3.1 is false whenever $d_k = 1$ for some $k = 1, \ldots, m$.

Exercise 11.3.4 Assume that the sequence $\{Y_i : i \ge 1\}$ is composed
of independent random variables such that $E[Y_i] = 0$, $E[Y_i^2] = 1$ and
$\max_{i \ge 1} E[Y_i^4] < \infty$. Consider, moreover, an array of real numbers $\{c_{n,i} :
i = 1, \ldots, M_n\}$ satisfying both (11.1.1) and the Lindeberg condition (11.1.2).
Write $Q_0(n, Y_1, \ldots, Y_{M_n}) = \sum_{i=1}^{M_n} c_{n,i} Y_i$, $n \ge 1$, and assume that condition
(i) in the statement of Theorem 11.3.1 is verified. Prove that, as $n \to \infty$, the
vector $\{Q_k(n, Y_1, \ldots, Y_{M_n})\}_{0 \le k \le m}$ converges in law to an $(m+1)$-dimensional
Gaussian vector $(G_0, \ldots, G_m)$ such that $G_0 \sim \mathcal{N}(0, 1)$, G_0 is independent of
$(G_1, \ldots, G_m)$ and $(G_1, \ldots, G_m) \sim \mathcal{N}_m(0, C)$.

Remark 11.3.5 Consider again a sequence $\{Y_i : i \ge 1\}$ composed
of independent random variables such that $E[Y_i] = 0$, $E[Y_i^2] = 1$ and
$\max_{i \ge 1} E[Y_i^4] < \infty$, fix an integer $q \ge 2$, and select a sequence of random
variables of the type $A_n = P_n(Y_i : i = 1, \ldots, M_n)$, $n \ge 1$, where $M_n \to \infty$
and, for every n, P_n is a polynomial in M_n variables of degree q (we stress that
q is independent of n). Exercise 11.3.4 suggests the following strategy in order
to prove that $A_n - E[A_n]$ obeys a CLT:

(a) Write $A_n = E[A_n] + \sum_{k=1}^{q} A_n(k; Y_1, \ldots, Y_{M_n})$, where $A_n(k; Y_1, \ldots, Y_{M_n})$ is a homogeneous sum of degree k, in the variables $(Y_1, \ldots, Y_{M_n})$.

(b) Replace $(Y_1, \ldots, Y_{M_n})$ with an i.i.d. $\mathcal{N}(0, 1)$ Gaussian vector $(G_1, \ldots, G_{M_n})$ in each $A_n(k; \cdot)$, $k = 2, \ldots, q$.

(c) Prove that $\{A_n(k; G_1, \ldots, G_{M_n}) : k = 2, \ldots, m\}$ converges in distribution to a Gaussian vector with dimension m. To do so, one usually exploits the fact that $A_n(k; G_1, \ldots, G_{M_n})$ can be written as an element of the kth chaos of some isonormal Gaussian process (see Remark 11.2.5)

(d) Check that the sequence $A_n(1; Y_1, \ldots, Y_{M_n})$, $n \geq 1$, satisfies a Lindeberg-type condition.

We now illustrate the method described in Remark 11.3.5 by means of a simple example.

Example 11.3.6 (Two-runs) Let $\{\xi_i : i \geq 1\}$ be a sequence of independent Bernoulli trials (that is, $P(\xi_i = 1) = \frac{1}{2} = P(\xi_i = 0)$) and consider the random variables

$$Z_n = \sum_{i=1}^{n} \xi_i \xi_{i+1}, \quad n \geq 1,$$

counting, for every n, the number of pairs of consecutive 1s in the first $n + 1$ trials. It is immediately seen that $E[Z_n] = n/4$. We shall prove that, as $n \to \infty$, the sequence

$$\widehat{Z_n} = \frac{Z_n - \frac{n}{4}}{\sqrt{n}}, \quad n \geq 1, \tag{11.3.2}$$

converges in distribution to a centered Gaussian random variable with variance $5/16$. To do so, we simply exploit the fact that $\xi_i = \frac{1}{2} Y_i + 1$, where $\{Y_i\}$ is a i.i.d. Rademacher sequence, so that elementary computations show that the limit of $\widehat{Z_n}$ coincides with the limit of

$$\frac{1}{4\sqrt{n}} \sum_{i=1}^{n} Y_i Y_{i+1} + \frac{1}{2\sqrt{n}} \sum_{i=1}^{n} Y_i, \quad n \geq 1.$$

Since the sequence $\frac{1}{\sqrt{n}} \sum_{i=1}^{n} Y_i$ clearly satisfies a Lindeberg condition, by Theorem 11.3.1 and Exercise 11.3.4 our claim is proved if we manage to show that the homogeneous sum

$$J_n = \frac{1}{\sqrt{n}} \sum_{i=1}^{n} G_i G_{i+1}, \quad n \geq 1,$$

where $\{G_i\}$ is an i.i.d. standard Gaussian sequence, converges in distribution to $N \sim \mathcal{N}(0, 1)$. Since each J_n has unit variance and is an element of the second

Wiener chaos of some Gaussian field (see Remark 11.2.5), Theorem 5.2.7 implies that it is sufficient to check that $E[J_n^4] \to 3$. This asymptotic relation can easily be verified by, for example, using the Wick formula (A.2.13).

11.4 Some technical estimates

This section contains some technical estimates that are needed in order to prove Theorem 11.3.1.

Lemma 11.4.1 (Hypercontractivity) *Let $Q \in \mathbb{R}[x_1, \ldots, x_M]$ be a multilinear polynomial of degree $d \leq M$. Let $Y_1, \ldots, Y_M$ be independent real random variables such that $E[Y_i] = 0$, $E[Y_i^2] = 1$ and $E[Y_i^4] < \infty$ for $i = 1, 2, \ldots, M$. Then*

$$E\left[Q(Y_1, \ldots, Y_M)^4\right] \leq \left(3 + 2\max_{1 \leq i \leq M} E[Y_i^4]\right)^{2d} E\left[Q(Y_1, \ldots, Y_M)^2\right]^2.$$
(11.4.1)

Proof The proof is by induction on the number of variables M. The case $M = 0$ is trivial, as Q is just a constant. We therefore assume $M \geq 1$. We can write

$$Q(x_1, \ldots, x_M) = R(x_1, \ldots, x_{M-1}) + x_M S(x_1, \ldots, x_{M-1}),$$

where $R, S \in \mathbb{R}[x_1, \ldots, x_{M-1}]$ are multilinear polynomials in at most $M - 1$ variables, $\deg(R) \leq d$ and $\deg(S) \leq d - 1$. Now write $\mathbf{Q} = Q(Y_1, \ldots, Y_M)$, $\mathbf{R} = R(Y_1, \ldots, Y_{M-1})$, $\mathbf{S} = S(Y_1, \ldots, Y_{M-1})$ and $\alpha = \max_{1 \leq i \leq M} E[Y_i^4]$. Clearly, $\mathbf{R}$ and $\mathbf{S}$ are independent of Y_M. We have, using $E[Y_M] = 0$ and $E[Y_M^2] = 1$,

$$E[\mathbf{Q}^2] = E[(\mathbf{R} + \mathbf{S}Y_M)^2] = E[\mathbf{R}^2] + E[\mathbf{S}^2],$$
$$E[\mathbf{Q}^4] = E[(\mathbf{R} + \mathbf{S}Y_M)^4] = E[\mathbf{R}^4] + 6E[\mathbf{R}^2\mathbf{S}^2] + 4E[Y_M^3]E[\mathbf{R}\mathbf{S}^3]$$
$$+ E[Y_M^4]E[\mathbf{S}^4].$$

Observe that $E[\mathbf{R}^2\mathbf{S}^2] \leq \sqrt{E[\mathbf{R}^4]}\sqrt{E[\mathbf{S}^4]}$ and

$$E[Y_M^3]E[\mathbf{R}\mathbf{S}^3] \leq \alpha^{\frac{3}{4}}\left(E[\mathbf{R}^4]\right)^{\frac{1}{4}}\left(E[\mathbf{S}^4]\right)^{\frac{3}{4}} \leq \alpha\sqrt{E[\mathbf{R}^4]}\sqrt{E[\mathbf{S}^4]} + \alpha E[\mathbf{S}^4],$$

where the last inequality uses both $x^{\frac{1}{4}}y^{\frac{3}{4}} \leq \sqrt{xy} + y$ (by considering $x < y$ and $x > y$) and $\alpha^{\frac{3}{4}} \leq \alpha$ (because $\alpha \geq E[Y_M^4] \geq E[Y_M^2]^2 = 1$). Hence

$$E[\mathbf{Q}^4] \leq E[\mathbf{R}^4] + 2(3 + 2\alpha)\sqrt{E[\mathbf{R}^4]}\sqrt{E[\mathbf{S}^4]} + 5\alpha E[\mathbf{S}^4]$$
$$\leq E[\mathbf{R}^4] + 2(3 + 2\alpha)\sqrt{E[\mathbf{R}^4]}\sqrt{E[\mathbf{S}^4]} + (3 + 2\alpha)^2 E[\mathbf{S}^4]$$
$$= \left(\sqrt{E[\mathbf{R}^4]} + (3 + 2\alpha)\sqrt{E[\mathbf{S}^4]}\right)^2.$$

By induction, we have $\sqrt{E[\mathbf{R}^4]} \leq (3 + 2\alpha)^d E[\mathbf{R}^2]$ and $\sqrt{E[\mathbf{S}^4]} \leq (3 + 2\alpha)^{d-1} E[\mathbf{S}^2]$. Therefore

$$E[\mathbf{Q}^4] \leq (3 + 2\alpha)^{2d} \left(E[\mathbf{R}^2] + E[\mathbf{S}^2] \right)^2 = (3 + 2\alpha)^{2d} E[\mathbf{Q}^2]^2,$$

and the lemma is proved. $\qquad\square$

Before stating and proving the next proposition, we need to recall the standard multi-index notation. A multi-index is a vector $\alpha \in \mathbb{N}^m$. We write

$$|\alpha| = \sum_{j=1}^m \alpha_j, \quad \alpha! = \prod_{j=1}^m \alpha_j!, \quad \partial_j = \frac{\partial}{\partial x_j}, \quad \partial^\alpha = \partial_1^{\alpha_1} \ldots \partial_d^{\alpha_d}, \quad x^\alpha = \prod_{j=1}^m x_j^{\alpha_j}.$$

Note that, by convention, $0^0 = 1$. Also note that $|x^\alpha| = y^\alpha$, where $y_j = |x_j|$ for all j. Finally, whenever $\varphi : \mathbb{R}^m \to \mathbb{R}$ is sufficiently regular and $k \geq 1$, we put

$$\|\varphi^{(k)}\|_\infty = \max_{|\alpha|=k} \sup_{x \in \mathbb{R}^m} |\partial^\alpha \varphi(x)|.$$

By a slight abuse of notation, we shall sometimes write $\|\varphi'\|_\infty = \|\varphi^{(1)}\|_\infty$, $\|\varphi''\|_\infty = \|\varphi^{(2)}\|_\infty$, $\|\varphi'''\|_\infty = \|\varphi^{(3)}\|_\infty$, and so on.

Proposition 11.4.2 *Fix integers $M, m \geq 1$, as well as $2 \leq d_m, \ldots, d_1 \leq M$. Write $[M] = \{1, \ldots, M\}$. For each k, let $f_k : [M]^{d_k} \to \mathbb{R}$ be a symmetric function vanishing on diagonals, and let $Q_k \in \mathbb{R}[x_1, \ldots, x_M]$ be the multilinear polynomial of degree d_k built from f_k, given by*

$$Q_k(x_1, \ldots, x_M) = \sum_{1 \leq i_1, \ldots, i_{d_k} \leq M} f_k(i_1, \ldots, i_{d_k}) x_{i_1} \ldots x_{i_{d_k}}. \qquad (11.4.2)$$

Write $\mathbf{Q}(\mathbf{x}) = (Q_1(x_1, \ldots, x_n), \ldots, Q_k(x_1, \ldots, x_n))$. Let $Y_1, \ldots, Y_M$ be independent real random variables with $E[Y_i] = 0$, $E[Y_i^2] = 1$ and $E[Y_i^4] < \infty$ for $i = 1, 2, \ldots, M$. Let $G_1, \ldots, G_M$ be independent $\mathcal{N}(0, 1)$ distributed random variables. Write $\mathbf{Y} = (Y_1, \ldots, Y_M)$ and $\mathbf{G} = (G_1, \ldots, G_M)$. Let $h : \mathbb{R}^m \to \mathbb{R}$ belong to C^3 with $\|h'''\|_\infty < \infty$. Then

$$|E[h(\mathbf{Q}(\mathbf{Y}))] - E[h(\mathbf{Q}(\mathbf{G}))]|$$

$$\leq 2m^3 \gamma (3 + 2\gamma)^{\frac{3}{2}(\max\{d_i\}-1)} \frac{\max\{d_i\}^{\frac{3}{2}}}{\sqrt{\min\{d_i d_i!\}}} \|h'''\|_\infty \max_{1 \leq k \leq m} \|f_k\|_{d_k}^2$$

$$\times \max_{1 \leq k \leq m} \max_{1 \leq j \leq M} \sqrt{\mathrm{Inf}_j(f_k)} \qquad (11.4.3)$$

$$\leq 2m^3 \gamma (3 + 2\gamma)^{\frac{3}{2}(\max\{d_i\}-1)} \frac{\max\{d_i\}^{\frac{3}{2}}}{\sqrt{\min\{d_i d_i!\}}} \|h'''\|_\infty \max_{1 \leq k \leq m} \|f_k\|_{d_k}^2$$

$$\times \max_{1 \leq k \leq m} \left\{ E[Q_k(\mathbf{G})^4] - 3E[Q_k(\mathbf{G})^2]^2 \right\}^{\frac{1}{4}}, \qquad (11.4.4)$$

where we use the notation (11.2.4) for $\mathrm{Inf}_j(f_k)$, *and where* $\gamma = \max\left(3, \max_{1 \leq i \leq M} E[Y_i^4]\right)$ *and*

$$\|f_k\|_{d_k}^2 = \sum_{1 \leq i_1, \dots, i_{d_k} \leq M} f_k(i_1, \dots, i_{d_k})^2.$$

Proof Without loss of generality, we assume that $d_m \geq \dots \geq d_1$ and that $\mathbf{Y}$ and $\mathbf{G}$ are stochastically independent. For $i = 0, \dots, M$, let $\mathbf{W}^{(i)} = (Y_1, \dots, Y_i, G_{i+1}, \dots, G_M)$. Fix a particular $i \in \{1, \dots, M\}$, and write, for $k = 1, \dots, m$,

$$U_{k,i} = \sum_{\substack{1 \leq i_1, \dots, i_{d_k} \leq M \\ i_1 \neq i, \dots, i_{d_k} \neq i}} f_k(i_1, \dots, i_{d_k}) W_{i_1}^{(i)} \dots W_{i_{d_k}}^{(i)},$$

$$V_{k,i} = \sum_{\substack{1 \leq i_1, \dots, i_{d_k} \leq M \\ \exists j : i_j = i}} f_k(i_1, \dots, i_{d_k}) W_{i_1}^{(i)} \dots \widehat{W_i^{(i)}} \dots W_{i_{d_k}}^{(i)},$$

where $\widehat{W_i^{(i)}}$ means that this particular term is dropped (observe that this notation is unambiguous: indeed, since f_k vanishes on diagonals, each string $i_1, \dots, i_{d_k}$ contributing to the definition of $V_{k,i}$ contains the symbol i exactly once). For each i, note that $\mathbf{U}_i = (U_{k,i})_{1 \leq k \leq m}$ and $\mathbf{V}_i = (V_{k,i})_{1 \leq k \leq m}$ are independent of the variables Y_i and G_i, and that $\mathbf{Q}(\mathbf{W}^{(i-1)}) = \mathbf{U}_i + G_i \mathbf{V}_i$ and $\mathbf{Q}(\mathbf{W}^{(i)}) = \mathbf{U}_i + Y_i \mathbf{V}_i$. By Taylor's theorem, using the independence of Y_i from $\mathbf{U}_i$ and $\mathbf{V}_i$, we have

$$\left| E\big[h(\mathbf{U}_i + Y_i \mathbf{V}_i)\big] - \sum_{|\alpha| \leq 2} \frac{1}{\alpha!} E\big[\partial^\alpha h(\mathbf{U}_i) \mathbf{V}_i^\alpha\big] E[Y_i^{|\alpha|}] \right|$$

$$\leq \sum_{|\alpha|=3} \frac{1}{\alpha!} \sup_{x \in \mathbb{R}^m} \left|\partial^\alpha h(x)\right| E[|Y_i|^3] E[|\mathbf{V}_i^\alpha|] \leq \|h'''\|_\infty E[|Y_i|^3] \sum_{|\alpha|=3} E[|\mathbf{V}_i^\alpha|].$$

Similarly,

$$\left| E\big[h(\mathbf{U}_i + G_i \mathbf{V}_i)\big] - \sum_{|\alpha| \leq 2} \frac{1}{\alpha!} E\big[\partial^\alpha h(\mathbf{U}_i) \mathbf{V}_i^\alpha\big] E[G_i^{|\alpha|}] \right|$$

$$\leq \|h'''\|_\infty E[|G_i|^3] \sum_{|\alpha|=3} E[|\mathbf{V}_i^\alpha|].$$

Due to the matching moments up to second order on the one hand, and using the fact that $E[|Y_i|^3] \leq E[Y_i^4]^{\frac{3}{4}} \leq \gamma^{\frac{3}{4}} \leq \gamma$ and $E[|G_i|^3] \leq \gamma$ on the other hand, we obtain that

$$
\left| E\big[h(\mathbf{Q}(\mathbf{W}^{(i-1)}))\big] - E\big[h(\mathbf{Q}(\mathbf{W}^{(i)}))\big] \right|
$$
$$
= \left| E\big[h(\mathbf{U}_i + Y_i\mathbf{V}_i)\big] - E\big[h(\mathbf{U}_i + G_i\mathbf{V}_i)\big] \right| \leq 2\gamma \|h'''\|_\infty \sum_{|\alpha|=3} E[|\mathbf{V}_i^\alpha|].
$$

By Lemma 11.4.2, we have $E[|V_{k,i}|^3] \leq E[V_{k,i}^4]^{\frac{3}{4}} \leq (3+2\gamma)^{\frac{3}{2}(d_k-1)} E[V_{k,i}^2]^{\frac{3}{2}}$. Thus

$$
\sum_{|\alpha|=3} E\big[|\mathbf{V}_i^\alpha|\big] = \sum_{j,k,l=1}^m E[|V_{j,i} V_{k,i} V_{l,i}|]
$$
$$
\leq \sum_{j,k,l=1}^m E[|V_{j,i}|^3]^{\frac{1}{3}} E[|V_{k,i}|^3]^{\frac{1}{3}} E[|V_{l,i}|^3]^{\frac{1}{3}}
$$
$$
= \left[\sum_{k=1}^m E[|V_{k,i}|^3]^{\frac{1}{3}} \right]^3 \leq (3+2\gamma)^{\frac{3}{2}(d_m-1)} \left[\sum_{k=1}^m E[V_{k,i}^2]^{\frac{1}{2}} \right]^3
$$
$$
\leq m^2 (3+2\gamma)^{\frac{3}{2}(d_m-1)} \sum_{k=1}^m E[V_{k,i}^2]^{\frac{3}{2}}.
$$

Moreover, using the independence between $\mathbf{Y}$ and $\mathbf{G}$, the properties of f_k (which is symmetric and vanishes on diagonals) as well as $E[Y_i] = E[G_i] = 0$ and $E[Y_i^2] = E[G_i^2] = 1$, we get

$$
E[V_{k,i}^2]^{3/2} = \left(\sum_{\substack{1 \leq i_1,\dots,i_{d_k} \leq n \\ \exists j : i_j = i}} f_k(i_1,\dots,i_{d_k})^2 \right)^{3/2}
$$
$$
= d_k^{3/2} \left(\sum_{1 \leq i_2,\dots,i_{d_k} \leq M} f_k(i, i_2,\dots,i_{d_k})^2 \right)^{3/2}
$$
$$
\leq d_k^{3/2} \sqrt{ \max_{1 \leq j \leq M} \sum_{1 \leq j_2,\dots,j_{d_k} \leq M} f_k(j, j_2,\dots,j_{d_k})^2 }
$$
$$
\times \sum_{1 \leq i_2,\dots,i_{d_k} \leq M} f_k(i, i_2,\dots,i_{d_k})^2,
$$

implying in turn that

$$\sum_{i=1}^{M} E[V_{k,i}^2]^{3/2} \leq d_k^{3/2} \|f_k\|_{d_k}^2 \max_{1 \leq j \leq M} \sqrt{\sum_{1 \leq j_2,\ldots,j_{d_k} \leq M} f_k(j,j_2,\ldots,j_{d_k})^2}$$

$$= d_k^{3/2} \|f_k\|_{d_k}^2 \max_{1 \leq j \leq M} \sqrt{\mathrm{Inf}_j(f_k)}.$$

By collecting the previous bounds, we get

$$|E[h(\mathbf{Q}(\mathbf{Y}))] - E[h(\mathbf{Q}(\mathbf{G}))]|$$

$$\leq \sum_{i=1}^{M} \left| E[h(\mathbf{Q}(\mathbf{W}^{(i-1)}))] - E[h(\mathbf{Q}(\mathbf{W}^{(i)}))] \right|$$

$$\leq 2m^2 \gamma (3 + 2\gamma)^{\frac{3}{2}(d_m-1)} \|h'''\|_\infty \sum_{k=1}^{m} \sum_{i=1}^{M} E[V_{k,i}^2]^{\frac{3}{2}}$$

$$\leq 2m^3 \gamma (3 + 2\gamma)^{\frac{3}{2}(d_m-1)} d_m^{\frac{3}{2}} \|h'''\|_\infty \max_{1 \leq k \leq m} \|f_k\|_{d_k,M}^2 \max_{1 \leq k \leq m} \max_{1 \leq j \leq M} \sqrt{\mathrm{Inf}_j(f_k)}.$$

Hence, all that remains is to show the following bound: for all $1 \leq k \leq m$,

$$\max_{1 \leq j \leq n} \sqrt{\mathrm{Inf}_j(f_k)} \leq \frac{1}{\sqrt{d_k d_k!}} \left\{ E[Q_k(\mathbf{G})^4] - 3E[Q_k(\mathbf{G})^2]^2 \right\}^{\frac{1}{4}}. \qquad (11.4.5)$$

Without loss of generality, we assume from now on that $G_i = X(e_i)$, with $X = \{X(h), h \in \mathfrak{H}\}$ an isonormal Gaussian process and $\{e_1, \ldots, e_n\}$ an orthonormal family of $\mathfrak{H}$. We use the notation introduced in Chapter 2 (in particular, I_{d_k} stands for d_kth multiple Wiener integral with respect to X). As observed before, because f_k vanishes on diagonals, it is readily verified that (see Remark 11.2.5)

$$Q_k(\mathbf{G}) = I_{d_k}(h_k), \quad \text{with } h_k = \sum_{1 \leq i_1,\ldots,i_{d_k} \leq M} f_k(i_1,\ldots,i_{d_k}) e_{i_1} \otimes \ldots \otimes e_{i_{d_k}}.$$

$$(11.4.6)$$

Thanks to (5.2.6), we then have that

$$Q_k(\mathbf{G})^4 - 3E[Q_k(\mathbf{G})^2]^2 \geq d_k^2 d_k!^2 \|h_k \otimes_{d_k-1} h_k\|_{\mathfrak{H}^{\otimes 2}}^2. \qquad (11.4.7)$$

Let us compute $\|h_k \otimes_{d_k-1} h_k\|_{\mathfrak{H}^{\otimes 2}}^2$ in terms of f_k. We have that

$$h_k \otimes_{d_k-1} h_k = \sum_{1 \leq i_1,\ldots,i_{d_k} \leq M} \sum_{1 \leq j_1,\ldots,j_{d_k} \leq M} f_k(i_1,\ldots,i_{d_k}) f_k(j_1,\ldots,j_{d_k})$$

$$(e_{i_1} \otimes \ldots \otimes e_{i_{d_k}}) \otimes_{d_k-1} (e_{j_1} \otimes \ldots \otimes e_{j_{d_k}})$$

$$= \sum_{1 \leq a,b \leq M} \sum_{1 \leq i_1,\ldots,i_{d_k-1} \leq M} f_k(i_1,\ldots,i_{d_k-1},a)$$

$$f(i_1,\ldots,i_{d_k-1},b) e_a \otimes e_b,$$

so that

$$\|h_k \otimes_{d_k-1} h_k\|^2_{\mathfrak{H}^{\otimes 2}}$$

$$= \sum_{1 \leq a,b \leq M} \left(\sum_{1 \leq i_1, \ldots, i_{d_k-1} \leq M} f_k(i_1, \ldots, i_{d_k-1}, a) f(i_1, \ldots, i_{d_k-1}, b) \right)^2.$$

By first removing the off-diagonal elements and then using the fact that a sum of non-negative number is always bigger than the maximum of those numbers, we deduce that

$$\|h_k \otimes_{d_k-1} h_k\|^2_{\mathfrak{H}^{\otimes 2}} \geq \sum_{j=1}^{M} \left(\sum_{1 \leq i_1, \ldots, i_{d_k-1} \leq M} f_k(i_1, \ldots, i_{d_k-1}, j)^2 \right)^2$$

$$\geq \left(\max_{1 \leq j \leq M} \sum_{1 \leq i_1, \ldots, i_{d_k-1} \leq M} f_k(i_1, \ldots, i_{d_k-1}, j)^2 \right)^2$$

$$= \max_{1 \leq j \leq M} [\mathrm{Inf}_j(f_k)]^2. \tag{11.4.8}$$

Using the symmetry of f_k, the desired conclusion (11.4.5) follows by combining the two inequalities (11.4.7) and (11.4.8). The proof of the proposition is done. $\qquad\square$

Theorem 11.4.3 *Let the assumptions and notation of Proposition 11.4.2 prevail. Assume, moreover, that $\|h''\|_\infty < \infty$ (in addition to $\|h'''\|_\infty < \infty$), and let $N \sim \mathcal{N}_m(0, C)$, with the covariance matrix $C \in \mathcal{M}_m(\mathbb{R})$ given by $C_{k,l} = E[Q_k(\mathbf{G})Q_l(\mathbf{G})]$. Then*

$$\left| E[h(\mathbf{Q}(\mathbf{Y}))] - E[h(N)] \right| \leq 2m^3 \gamma (3 + 2\gamma)^{\frac{3}{2}(\max\{d_i\}-1)} \frac{\max\{d_i\}^{\frac{3}{2}}}{\sqrt{\min\{d_i d_i!\}}} \|h'''\|_\infty$$

$$\times \max_{1 \leq k \leq m} \|f_k\|^2_{d_k} \max_{1 \leq k \leq m} \left\{ E[Q_k(\mathbf{G})^4] - 3E[Q_k(\mathbf{G})^2]^2 \right\}^{\frac{1}{4}}$$

$$+ \frac{1}{2}\|h''\|_\infty \sum_{k,l=1}^{m} \mathbf{1}_{\{d_k=d_l\}} \sqrt{2 \sum_{r=1}^{d_k-1} \binom{2r}{r} \sqrt{E[Q_k(\mathbf{G})^4] - 3E[Q_k(\mathbf{G})^2]^2}}$$

$$+ \frac{1}{2}\|h''\|_\infty \sum_{k,l=1}^{m} \mathbf{1}_{\{d_k \neq d_l\}} \Bigg[\sqrt{2d_l!} \|f_l\|_{d_l} \left\{ E[Q_k(\mathbf{G})^4] - 3E[Q_k(\mathbf{G})^2]^2 \right\}^{\frac{1}{4}}$$

$$+ \sum_{r=1}^{d_k \wedge d_l-1} \sqrt{2(d_k + d_l - 2r)!} \binom{d_l}{r} \sqrt{E[Q_k(\mathbf{G})^4] - 3E[Q_k(\mathbf{G})^2]^2} \Bigg].$$

Proof Without loss of generality, we assume that $d_m \geq \ldots \geq d_1$ and that $G_i = X(e_i)$, where $X = \{X(h), h \in \mathfrak{H}\}$ is a given isonormal Gaussian process and $\{e_1, \ldots, e_n\}$ is an orthonormal family of $\mathfrak{H}$. We again adopt the representation (11.4.6). Hence, Theorem 6.2.2 applies, and we deduce that

$$
\left| E[h(\mathbf{Q}(\mathbf{G}))] - E[h(N)] \right| \leq \frac{1}{2} \|h''\|_\infty
$$

$$
\times \sum_{k,l=1}^{m} \mathbf{1}_{\{d_k = d_l\}} \sqrt{2 \sum_{r=1}^{d_k-1} \binom{2r}{r} \sqrt{E[Q_k(\mathbf{G})^4] - 3E[Q_k(\mathbf{G})^2]^2}}
$$

$$
+ \frac{1}{2} \|h''\|_\infty \sum_{k,l=1}^{m} \mathbf{1}_{\{d_k \neq d_l\}} \left[\sqrt{2 d_l!} \| f_l \|_{d_l} \left\{ E[Q_k(\mathbf{G})^4] - 3E[Q_k(\mathbf{G})^2]^2 \right\}^{\frac{1}{4}} \right.
$$

$$
\left. + \sum_{r=1}^{d_k \wedge d_l - 1} \sqrt{2(d_k + d_l - 2r)!} \binom{d_l}{r} \sqrt{E[Q_k(\mathbf{G})^4] - 3E[Q_k(\mathbf{G})^2]^2} \right].
$$

$$
(11.4.9)
$$

On the other hand, the triangle inequality yields

$$
\left| E[h(\mathbf{Q}(\mathbf{Y}))] - E[h(N)] \right| \leq \left| E[h(\mathbf{Q}(\mathbf{Y}))] - E[h(\mathbf{Q}(\mathbf{G}))] \right|
$$
$$
+ \left| E[h(\mathbf{Q}(\mathbf{G}))] - E[h(N)] \right|.
$$

Therefore, the desired conclusion follows by combining (11.4.9) and (11.4.4). $\qquad\square$

11.5 Proof of Theorem 11.3.1

Given random variables Y, Z with values in $\mathbb{R}^m$, we denote by $d(Y, Z)$ the quantity

$$
d(Y, Z) = \sup_h |E[h(Y)] - E[h(Z)]|,
$$

where the supremum runs over all thrice-differentiable functions $h : \mathbb{R}^m \to \mathbb{R}$ such that $\|h^{(j)}\|_\infty \leq 1$, $j = 1, 2, 3$. It is easy to check that $d(\cdot, \cdot)$ defines a distance on the class of probability measures on $\mathbb{R}^m$, and that the topology induced by $d(\cdot, \cdot)$ is stronger than the topology of convergence in distribution. Of course, in the statement of Theorem 11.3.1, we have that (ii) implies (i), so we shall only prove the converse implication.

Assume that (i) holds, and observe that

$$
E[Q_k(n, G_1, \ldots, G_{M_n}) Q_l(n, G_1, \ldots, G_{M_n})] = E[Q_k(n, Y_1, \ldots, Y_{M_n})
$$
$$
\times Q_l(n, Y_1, \ldots, Y_{M_n})]
$$
$$
= \mathbf{1}_{\{d_k = d_l\}} d_k! \sum_{1 \leq i_1, \ldots, i_{d_k} \leq M_n} f_k(i_1, \ldots, i_{d_k}) f_l(i_1, \ldots, i_{d_k}).
$$

Adopting the previous notation, we need to show that $d\left(\{Q_k(n, Y_1, \ldots, Y_{M_n})\}_{1 \le k \le m}, N_C\right) \to 0$ as $n \to \infty$, where $N_C \sim \mathcal{N}_m(0, C)$. For each n, let $C_n \in \mathcal{M}_m(\mathbb{R})$ be the matrix given by $C_n(k, l) = E[Q_k(n, G_1, \ldots, G_{M_n}) Q_l(n, G_1, \ldots, G_{M_n})]$. Because of (11.3.1), the convergence in law (i) implies, by bounded convergence, that, for all $k, l = 1, \ldots, m$, $C_n(k, l) \to C(k, l)$ as $n \to \infty$. In particular, if $N_{C_n} \sim \mathcal{N}(0, C_n)$, then $d\left(N_{C_n}, N_C\right) \to 0$ (use estimate (6.4.1), for example). On the other hand, because of (11.3.1) and (11.4.1), we also have by bounded convergence that, for all $k = 1, \ldots, m$,

$$E[Q_k(n, G_1, \ldots, G_{M_n})^4] - 3E[Q_k(n, G_1, \ldots, G_{M_n})^2]^2 \to 0, \quad \text{as } n \to \infty.$$

As a consequence, Theorem 11.4.3 yields that $d\left(\{Q_k(n, Y_1, \ldots, Y_{M_n})\}_{1 \le k \le m}, N_{C_n}\right) \to 0$ as $n \to \infty$ (note that, without loss of generality, we may assume that $d_m \ge \ldots \ge d_1$). Using the inequality

$$d\left(\{Q_k(n, Y_1, \ldots, Y_{M_n})\}_{1 \le k \le m}, N_C\right)$$
$$\le d\left(\{Q_k(n, Y_1, \ldots, Y_{M_n})\}_{1 \le k \le m}, N_{C_n}\right), +d\left(N_{C_n}, N_C\right),$$

we deduce the desired conclusion.

11.6 Exercises

11.6.1 (Walsh chaos is not universal) Let $\varepsilon = \{\varepsilon_i : i \ge 1\}$ be an i.i.d. Rademacher sequence (that is, $P(\varepsilon_1 = 1) = P(\varepsilon_1 = -1) = \frac{1}{2}$). Prove that, for every $d \ge 1$ and as $n \to \infty$, the sequence

$$\varepsilon_1 \times \ldots \times \varepsilon_d \times \sum_{i=d+1}^{n} \frac{\varepsilon_i}{\sqrt{n-d}}, \quad n \ge 1,$$

converges in distribution to a standard Gaussian random variable $G \sim \mathcal{N}(0, 1)$. Now let $\{G_i : i \ge 1\}$ be a sequence of i.i.d. copies of G. Prove that, as $n \to \infty$, the sequence

$$G_1 \times \ldots \times G_d \times \sum_{i=d+1}^{n} \frac{G_i}{\sqrt{n-d}}, \quad n \ge 1,$$

converges in distribution to a non-Gaussian random variable. Deduce that the Walsh chaos associated with ε is not universal – in other words, the statement of Theorem 11.3.1 becomes false whenever one replaces **G** with ε.

11.6.2 Let $\{\xi_i : i \ge 1\}$ be a sequence of independent Bernoulli trials (that is, $P(\xi_i = 1) = P(\xi_i = 0) = \frac{1}{2}$), and fix $d \ge 2$. Generalize the computations contained in Example 11.3.6 in order to prove a CLT for the sequence of d-runs,

$$V_n = \sum_{i=1}^{n} \xi_i \xi_{i+1} \cdots \xi_{i+d}, \quad n \geq 1,$$

counting the number of strings of d consecutive 1s in the first $n + d$ trials.

11.7 Bibliographic comments

A comprehensive reference for universality results in random matrix theory is the monograph by Anderson, Guionnet and Zeitouni [3]. A classical reference for the Donsker theorem and related functional invariance principles is the book by Billingsley [14]. The 'Lindeberg method' was first developed by Lindeberg in [67], where a first version of Theorem 11.1.1 was proved, and later popularized by Trotter in his classical 1959 paper [145]. Another detailed discussion of Lindeberg's approach can be found in Feller [38, chapter VIII]. Standard discussions of the Lindeberg condition (11.1.2) and of many associated convergence results are contained, for example, in Dudley [32, section 9.6] and Kallenberg [60, ch. 5]. The version of the Lindeberg method used in this chapter is largely inspired by the paper by Mossel, O'Donnel and Oleszkiewicz [79], to which the reader is referred for further information about influence functions and their use in modern probability. A pioneering use of influence functions in order to assess the proximity of homogeneous sums appears in the 1979 paper by Rotar [124]. An introduction to the homogeneous Poisson chaos can be found, for example, in Peccati and Taqqu [110] and Kwapień and Woyczyński [63]. A fascinating discussion of the properties of the Walsh chaos, as well as of its connections with harmonic analysis, is contained in the monograph by Blei [15]. Theorem 11.3.1 was first proved (in a more general form) by Nourdin, Peccati and Reinert [94]. An application of Theorem 11.3.1 to the study of the Gaussian fluctuations of non-Hermitian matrix ensembles can be found in [90]. One should note that homogeneous sums based on centered independent random variables are also *degenerate U-statistics*: it follows that Theorem 11.3.1 should be compared with some well-known criteria for the normal approximations of U-statistics, such as those established in de Jong [58] and Jammalamadaka and Janson [56].

Appendix A

Gaussian elements, cumulants and Edgeworth expansions

Throughout this appendix, we consider as given a probability space $(\Omega, \mathscr{F}, P)$.

A.1 Gaussian random variables

We assume that the reader has a basic knowledge of the main properties of Gaussian random variables and random vectors. Here, we shall only recall some elementary facts. Many crucial properties of Gaussian random variables are proved and explored throughout the book.

Definition A.1.1 Let $\mu, \sigma \in \mathbb{R}$. A real-valued random variable X is said to have a **Gaussian distribution** with mean μ and variance σ^2, written $X \sim \mathscr{N}(\mu, \sigma^2)$ or $X \sim \mathscr{N}_1(\mu, \sigma^2)$, if

$$E[\exp(itX)] = \exp\{it\mu - t^2\sigma^2/2\}, \quad \forall t \in \mathbb{R}.$$

Alternatively, we say that X is a **Gaussian random variable**. When $X \sim \mathscr{N}(0, 1)$, we say that X is a **standard Gaussian** random variable.

Note that, if $X \sim \mathscr{N}(\mu, 0)$, then necessarily $X = \mu$, a.s.-P. If $\sigma \neq 0$, then $X \sim \mathscr{N}(\mu, \sigma^2)$ has a density f with support equal to $\mathbb{R}$, namely

$$f(x) = \frac{1}{\sqrt{2\pi\sigma^2}} e^{-\frac{1}{2}\left(\frac{x-\mu}{\sigma}\right)^2}, \quad x \in \mathbb{R}.$$

Definition A.1.2 Let $m \geq 2$. A random vector $(X_1, \ldots, X_m)$ is said to have an m-**dimensional Gaussian distribution** if, for every $(t_1, \ldots, t_m) \in \mathbb{R}^m$, the random variable $\sum_{i=1}^{m} t_i X_i$ has a one-dimensional Gaussian distribution. If $(X_1, \ldots, X_m)$ has an m-dimensional Gaussian distribution, then we say that the random variables $X_1, \ldots, X_m$ are **jointly Gaussian** or, alternatively, that $(X_1, \ldots, X_m)$ is a **Gaussian vector**.

The previous definition implies that the distribution of an m-dimensional Gaussian vector $(X_1, \ldots, X_m)$ is uniquely determined by the vector of the means $\mu = (\mu_1, \ldots, \mu_m)$, where

$$\mu_i = E(X_i), \quad i = 1, \ldots, m,$$

and by the covariance matrix $C = \{C(i, j) : i, j = 1, \ldots, m\}$, given by

$$C(i, j) = \mathrm{Cov}(X_i, X_j), \quad i, j = 1, \ldots, m.$$

In particular, we have that

$$E\left[\exp\left(i \sum_{j=1}^{m} t_j X_j\right)\right] = \exp\left\{i \sum_{j=1}^{m} t_j \mu_j - \frac{1}{2} \sum_{j,k=1}^{m} t_j t_k C(j, k)\right\},$$

$$\forall (t_1, \ldots, t_m) \in \mathbb{R}^m. \tag{A.1.1}$$

When $(X_1, \ldots, X_m)$ satisfies (A.1.1), we write $(X_1, \ldots, X_m) \sim \mathcal{N}_m(\mu, C)$. Finally, if C is positive definite and $(X_1, \ldots, X_m) \sim \mathcal{N}_m(\mu, C)$, then $(X_1, \ldots, X_m)$ admits a density with support equal to $\mathbb{R}^m$, given by

$$f(x_1, \ldots, x_m) = \frac{1}{(2\pi)^{m/2} |C|^{1/2}} \exp\left\{-\frac{1}{2}(x - \mu)' C^{-1}(x - \mu)\right\},$$

$$(x_1, \ldots, x_m) \in \mathbb{R}^m,$$

where $x = (x_1, \ldots, x_m)$, and $|C|$ stands for the determinant of C.

Definition A.1.3 Let I be an arbitrary set. A **Gaussian family** indexed by I is a collection of random variables $X = \{X_i : i \in I\}$ such that, for every $m \geq 1$ and every $(i_1, \ldots, i_m) \in I^m$, the vector $(X_{i_1}, \ldots, X_{i_m})$ has an m-dimensional Gaussian distribution. When I is equal to either $\mathbb{Z}$ or $\mathbb{N}$, we say that X is a **Gaussian sequence**. A Gaussian sequence X is said to be **stationary** if the mapping $k \mapsto E[X_k], k \in I$, is constant and, for every $j, k \in I$, the covariance $\mathrm{Cov}(X_j, X_k)$ depends only on the quantity $|j - k|$.

A.2 Cumulants

In this section, we recall some classical results concerning cumulants of random variables. For complete proofs and further results, see [110, chapter 3]. The formalism developed below is the basis of the results discussed in Chapter 8, for example.

A.2.1 Basic definitions

For $n \geq 1$, we consider a vector of real-valued random variables $\mathbf{X}_{[n]} = (X_1, \ldots, X_n)$ such that $E\left|X_j\right|^n < \infty$, $\forall j = 1, \ldots, n$. For every $b = \{j_1, \ldots, j_k\} \subset [n] = \{1, \ldots, n\}$, we write

$$\mathbf{X}_b = \left(X_{j_1}, \ldots, X_{j_k}\right) \quad \text{and} \quad \mathbf{X}^b = X_{j_1} \ldots X_{j_k}. \tag{A.2.1}$$

For every $b = \{j_1, \ldots, j_k\} \subset [n]$ and $(t_1, \ldots, t_k) \in \mathbb{R}^k$, we write

$$\phi_{\mathbf{X}_b}(t_1, \ldots, t_k) = E\left[\exp\left(i \sum_{\ell=1}^{k} t_\ell X_{j_\ell}\right)\right],$$

to denote the characteristic function of $\mathbf{X}_b$. The *cumulant* of the components of the vector $\mathbf{X}_b$ is defined as

$$\kappa(\mathbf{X}_b) = (-i)^k \frac{\partial^k}{\partial t_1 \ldots \partial t_k} \log \phi_{\mathbf{X}_b}(t_1, \ldots, t_k)\,|_{t_1 = \ldots = t_k = 0}. \tag{A.2.2}$$

The following facts are easily checked.

1. The application $\mathbf{X}_b \mapsto \kappa(\mathbf{X}_b)$ is *homogeneous*, meaning that, for every $(h_1, \ldots, h_k) \in \mathbb{R}^k$,

$$\kappa\left(h_1 X_{j_1}, \ldots, h_k X_{j_k}\right) = h_1 \ldots h_k \, \kappa(\mathbf{X}_b).$$

2. The application $\mathbf{X}_b \mapsto \kappa(\mathbf{X}_b)$ is invariant with respect to the permutations of b.
3. If the vector $\mathbf{X}_b$ has the form $\mathbf{X}_b = \mathbf{X}_{b'} \cup \mathbf{X}_{b''}$, with $b', b'' \neq \varnothing$, $b' \cap b'' = \varnothing$ and $\mathbf{X}_{b'}$ and $\mathbf{X}_{b''}$ independent, then $\kappa(\mathbf{X}_b) = 0$.
4. If $\mathbf{Y} = \{Y_i : i \in I\}$ is a Gaussian family and if $\mathbf{X}$ is a vector obtained by juxtaposing $n \geq 3$ elements of $\mathbf{Y}$ (with possible repetitions), then $\kappa(\mathbf{X}) = 0$ (just use (A.1.1)).

A.2.2 Cumulants of a random variable

When $\mathbf{X}_{[n]} = (X_1, \ldots, X_n)$ is such that $X_j = X$, $\forall j = 1, \ldots, n$, where X is a random variable in $L^n(\Omega)$, one uses the notation κ_n and writes

$$\kappa\left(\mathbf{X}_{[n]}\right) = \kappa(\underbrace{X, \ldots, X}_{n \text{ times}}) = \kappa_n(X) \tag{A.2.3}$$

and we say that $\kappa_n(X)$ is the nth cumulant (or the *cumulant of order n*) of X. We have that

$$\kappa_n(X) = (-i)^n \frac{\partial^n}{\partial t^n} \ln \phi_X(t)\bigg|_{t=0},$$

where $\phi_X(t) = E\left[\exp(itX)\right]$. Note that if $X, Y \in L^n(\Omega)$, $n \geq 1$, are independent random variables, then (A.2.2) implies that

$$\kappa_n(X+Y) = \kappa_n(X) + \kappa_n(Y).$$

In particular, by setting Y equal to a constant $c \in \mathbb{R}$, we obtain that

$$\kappa_1(X+c) = E(X+c) = \kappa(X) + c = \kappa_1(X) + c, \qquad \text{(A.2.4)}$$

$$\kappa_n(X+c) = \kappa_n(X), \quad n \geq 2, \qquad \text{(A.2.5)}$$

because $\kappa_1(X) = E(X)$, and $\kappa_n(c) = 0$ for every $n \geq 2$.

A.2.3 Joint cumulants of a random vector

For the rest of this section, we fix an integer $d \geq 2$, as well as a random vector $X = (X_1, \ldots, X_d)$.

We need some standard multi-index notation. A multi-index (of length d) is a vector of non-negative integers of the type $m = (m_1, \ldots, m_d) \in \mathbb{N}^d$. We write

$$|m| = \sum_{j=1}^d m_j, \quad \partial_j = \frac{\partial}{\partial t_j}, \quad \partial^m = \partial_1^{m_1} \ldots \partial_d^{m_d}, \quad x^m = \prod_{j=1}^d x_j^{m_j}. \quad \text{(A.2.6)}$$

By convention, we have $0^0 = 1$. Also, note that $|x^m| = y^m$, where $y_i = |x_i|$ for all i. If $s \in \mathbb{N}^d$, we say that $s \leq m$ if and only if $s_i \leq m_i$ for all i. For any $i = 1, \ldots, d$, we let $e_i \in \mathbb{N}^d$ be the multi-index defined by $(e_i)_j = \delta_{ij}$, with δ_{ij} the Kronecker symbol.

Definition A.2.1   Assume that $E|X|^m < \infty$ for some multi-index $m \in \mathbb{N}^d \setminus \{0\}$, and let $\phi_X(t) = E[e^{i\langle t, F \rangle_{\mathbb{R}^d}}]$, $t \in \mathbb{R}^d$, stand for the characteristic function of F. The **joint cumulant of order** m of X is (well) defined as

$$\kappa_m(X_1, \ldots, X_d) = \kappa_m(X) = (-i)^{|m|} \partial^m \log \phi_X(t)|_{t=0}.$$

In other words, with the notation of Section A.2.1, we have that

$$\kappa_m(X) = \kappa(\mathbf{X}_{[|m|]}), \qquad \text{(A.2.7)}$$

where $\mathbf{X}_{[|m|]}$ is the $|m|$-dimensional vector obtained by juxtaposing m_i copies of each X_i, that is,

$$\mathbf{X}_{[|m|]} = (\underbrace{X_1, \ldots, X_1}_{m_1 \text{ times}}, \underbrace{X_2, \ldots, X_2}_{m_2 \text{ times}}, \ldots, \underbrace{X_d, \ldots, X_d}_{m_d \text{ times}}).$$

Example A.2.2   If X_i, X_j belong to $L^2(\Omega)$, then

$$\kappa_{e_i}(X) = E[X_i] \quad \text{and} \quad \kappa_{e_i + e_j}(X) = \text{Cov}[X_i, X_j].$$

A.2.4 Relations linking cumulants and moments

Given a finite set b, we denote by $\mathcal{P}(b)$ the class of all *partitions* of b. By definition, an element of $\mathcal{P}(b)$ is an object of the form $\pi = \{b_1, \ldots, b_k\}$, where the b_i are non-empty disjoint subsets of b such that $b_1 \cup \ldots \cup b_k = b$.

The next result, whose content is known as the *Leonov and Shiryaev formulae*, provides an exact relation between moments and cumulants. In particular, it shows that *the joint moments of a given random vector are completely determined by the joint cumulants, and vice versa.*

Proposition A.2.3 (Leonov and Shiryaev formulae) *Let* $\mathbf{X}_{[n]} = (X_1, \ldots, X_n)$ *be such that* $E\left|X_j\right|^n < \infty$, $\forall j = 1, \ldots, n$. *For every* $b \subset [n]$,

$$E[\mathbf{X}^b] = \sum_{\pi = \{b_1, \ldots, b_k\} \in \mathcal{P}(b)} \kappa\left(\mathbf{X}_{b_1}\right) \ldots \kappa\left(\mathbf{X}_{b_k}\right); \tag{A.2.8}$$

$$\kappa\left(\mathbf{X}_b\right) = \sum_{\sigma = \{a_1, \ldots, a_r\} \in \mathcal{P}(b)} (-1)^{r-1} (r-1)! E\left(\mathbf{X}^{a_1}\right) \ldots E\left(\mathbf{X}^{a_r}\right); \tag{A.2.9}$$

For a single random variable X, we deduce from (A.2.3) and Proposition A.2.3 the following statement:

Corollary A.2.4 *Let* X *be a random variable such that* $E\,|X|^n < \infty$. *Then*

$$E\left[X^n\right] = \sum_{\pi = \{b_1, \ldots, b_k\} \in \mathcal{P}([n])} \kappa_{|b_1|}(X) \ldots \kappa_{|b_k|}(X) \tag{A.2.10}$$

$$= \sum_{s=0}^{n-1} \binom{n-1}{s} \kappa_{s+1}(X) \times E[X^{n-s-1}], \tag{A.2.11}$$

$$\kappa_n(X) = \sum_{\sigma = \{a_1, \ldots, a_r\} \in \mathcal{P}([n])} (-1)^{r-1} (r-1)! E[X^{|a_1|}] \ldots E[X^{|a_r|}]. \tag{A.2.12}$$

Example A.2.5 Corollary A.2.4 implies that the first four cumulants of a random variable X are the following:

$\kappa_1(X) = E[X];$

$\kappa_2(X) = E[X^2] - E[X]^2 = \text{Var}(X);$

$\kappa_3(X) = E[X^3] - 3E[X^2]E[X] + 2E[X]^3;$

$\kappa_4(X) = E[X^4] - 4E[X]E[X^3] - 3E[X^2]^2 + 12E[X]^2 E[X^2] - 6E[X]^4.$

In particular, when $E[X] = 0$, we have that $\kappa_3(X) = E[X^3]$ and $\kappa_4(X) = E[X^4] - 3E[X^2]^2$. We also have $\kappa(X, Y) = \text{Cov}(X, Y)$.

Example A.2.6 Let $\mathbf{G} = (G_1, \ldots, G_n)$, $n \geq 3$, be jointly Gaussian and such that $E(G_i) = 0$, $i = 1, \ldots, n$. Then, for every $b \subset [n]$ such that $|b| \geq 3$, we know that

$$\kappa(\mathbf{G}_b) = \kappa(G_i : i \in b) = 0.$$

By applying this relation and formula (A.2.8) to $\mathbf{G}$, we therefore obtain the well-known *Wick formula*:

$$E[G_1 G_2 \ldots G_n] \tag{A.2.13}$$
$$= \begin{cases} \sum_{\pi = \{\{i_1, j_1\}, \ldots, \{i_k, j_k\}\} \in \mathcal{P}([n])} E(G_{i_1} G_{j_1}) \ldots E(G_{i_k} G_{j_k}), & n \text{ even} \\ 0, & n \text{ odd.} \end{cases}$$

Observe that, on the right-hand side of (A.2.13), the sum is taken over all partitions π of $[n]$ such that each block of π contains exactly two elements.

A.3 The method of moments and cumulants

Given a random vector $Z = (Z_1, \ldots, Z_d)$ whose components have finite moments of every order, we say that the law of Z is *determined by its moments* (or, equivalently, by its cumulants) if any other vector $Y = (Y_1, \ldots, Y_d)$, such that

$$E[Z_1^{m_1} Z_2^{m_2} \ldots Z_d^{m_d}] = E[Y_1^{m_1} Y_2^{m_2} \ldots Y_d^{m_d}],$$
$$\text{for every multi-index } m = (m_1, \ldots, m_d),$$

(or, equivalently,

$$\kappa_m(Z) = \kappa_m(Y) \text{ for every multi-index } m = (m_1, \ldots, m_d))$$

is such that Z and Y have the same law.

The following general statement provides a version of the so-called 'method of moments and cumulants', which is an effective tool for establishing probabilistic limit theorems.

Theorem A.3.1 (Method of moments and cumulants) *Fix $d \geq 1$, assume that the law of $Z = (Z_1, \ldots, Z_d)$ is determined by its moments, and consider a sequence $Y_n = (Y_{n,1}, \ldots, Y_{n,d})$, $n \geq 1$ such that, for every multi-index $m = (m_1, \ldots, m_d)$,*

1. *the sequence of monomials $Y_{n,1}^{m_1} Y_{n,2}^{m_2} \ldots Y_{n,d}^{m_d}$, $n \geq 1$, is uniformly integrable;*

2. *as* $n \to \infty$,

$$E[Y_{n,1}^{m_1} Y_{n,2}^{m_2} \cdots Y_{n,d}^{m_d}] \longrightarrow E[Z_1^{m_1} Z_2^{m_2} \cdots Z_d^{m_d}]$$

or, equivalently, $\kappa_m(Y_n) \longrightarrow \kappa_m(Z)$.

Then Y_n *converges in distribution to* Z *as* $n \to \infty$.

Proof Condition 2 in the statement implies that the sequence of the laws of the random variables Y_n is tight, and therefore relatively compact with respect to the topology of the convergence in distribution. Condition 1 then ensures that the limit of every weakly converging subsequence $\{Y_{n_k}\}$ must necessarily have the same moments as Z, so that the conclusion follows from the fact that the distribution of Z is determined by its moments. $\square$

Remark A.3.2 By virtue of Lemma 3.1.1 and Exercise 4.1.4, any Gaussian distribution is determined by its moments.

A.4 Edgeworth expansions in dimension one

Let X_1, X_2 be two real-valued random variables, with finite moments up to some order $M \in \{1, 2, \ldots\} \cup \{+\infty\}$. For $i = 1, 2$, write $\{\kappa_n(X_i) : n = 1, \ldots, M\}$ for the associated sequence of cumulants. Define the *formal cumulants* associated with (X_1, X_2) as

$$\kappa_n^f(X_1, X_2) = \kappa_n(X_1) - \kappa_n(X_2), \quad n = 1, 2, \ldots, M.$$

The *formal moments* associated with the sequence $\{\kappa_n(X_1, X_2)\}$, denoted by

$$m_n^f(X_1, X_2), \quad n = 0, 1, \ldots, M,$$

are (recursively) obtained by setting $m_0^f(X_1, X_2) = 1$, and then by replacing every $E[X^{n-\alpha}]$ with $m_{n-\alpha}^f(X_1, X_2)$ and every $\kappa_{s+1}(X)$ with $\kappa_{s+1}^f(X_1, X_2)$ in (A.2.11):

$$m_n^f(X_1, X_2) = \sum_{s=0}^{n-1} \binom{n-1}{s} \kappa_{s+1}^f(X_1, X_2) \times m_{n-s-1}^f(X_1, X_2). \quad \text{(A.4.1)}$$

Example A.4.1 We have that

$$m_1^f(X_1, X_2) = \kappa_1^f(X_1, X_2), \quad \text{(A.4.2)}$$

$$m_2^f(X_1, X_2) = \kappa_1^f(X_1, X_2)^2 + \kappa_2^f(X_1, X_2), \quad \text{(A.4.3)}$$

$$m_3^f(X_1, X_2) = \kappa_1^f(X_1, X_2)^3 + 3\kappa_1^f(X_1, X_2)\kappa_2^f(X_1, X_2) + \kappa_3^f(X_1, X_2). \quad \text{(A.4.4)}$$

Remark A.4.2   In general, the sequence $\{\kappa_n^f(X, Y)\}$ cannot be represented as the sequence of cumulants associated with some real-valued random variable. For instance, if $E[X_1] = E[X_2]$, $E[X_1^2] = E[X_2^2]$ and $E[X_1^3] \neq E[X_2^3]$, then $\kappa_1^f(X_1, X_2) = \kappa_2^f(X_1, X_2) = 0$, and $\kappa_3^f(X_1, X_2) = E[X_1^3] - E[X_2^3] \neq 0$. It follows that, if $\kappa_1^f(X_1, X_2)$, $\kappa_2^f(X_1, X_2)$ and $\kappa_3^f(X_1, X_2)$ were the first three cumulants of some real random variable Z, then we would have that $E[Z] = E[Z^2] = 0$ and $E[Z^3] \neq 0$, which is absurd.

Definition A.4.3   For $i = 1, 2$, write $F_{X_i}(z) = P[X_i \leq z]$, $z \in \mathbb{R}$. The Mth-order **Edgeworth expansion** of F_{X_1} around F_{X_2} is given by:

$$\mathcal{E}_M(X_1, X_2; z) = F_{X_2}(z) + \sum_{i=1}^{M} \frac{(-1)^i}{i!} m_i^f(X_1, X_2) \frac{d^i F_{X_2}}{dz^i}(z), \quad z \in \mathbb{R}.$$

(A.4.5)

Note that the series appearing on the right-hand side of (A.4.5) is formal for $M = +\infty$. The mapping $z \mapsto \mathcal{E}_M(X_1, X_2; z)$ has to be interpreted as a pointwise approximation of F_{X_1}, becoming more and more precise as M grows to infinity.

Example A.4.4   Assume that $X_2 = N \sim \mathcal{N}(0, 1)$, and that $X_1 = X$ has the same mean and variance as N. Write $F_N = \Phi$. Then, using (A.4.2)–(A.4.4), as well as the relation $\frac{d^3\Phi}{dz^3}(z) = \frac{e^{-z^2/2}}{\sqrt{2\pi}}(z^2 - 1)$, we obtain that the third-order Edgeworth expansion of F_X around Φ is given by

$$\mathcal{E}_3(X, N; z) = \Phi(z) - \frac{E[X^3]}{3!} \frac{e^{-z^2/2}}{\sqrt{2\pi}}(z^2 - 1), \quad z \in \mathbb{R}.$$

Observe that $H_2(z) = z^2 - 1$ is the second Hermite polynomial. Using Rodrigues's formula (see, for example, Proposition 1.4.2(vii)), we can always represent Edgeworth expansions around the normal distribution in terms of Hermite polynomials with increasing orders.

A.5 Bibliographic comments

An elementary presentation of the (possibly multidimensional) Gaussian distribution is contained, for example, in Dudley [32, Section 9.5]. An exhaustive discussion of the properties of cumulants, with a specific emphasis on combinatorial aspects, can be found in the monograph by Peccati and Taqqu [110]. A very clear presentation of formal Edgeworth expansions can be found in Chapter 5 of McCullagh's monograph [77]. A reference for the use of Edgeworth expansions in mathematical statistics is Hall [45].

Appendix B

Hilbert space notation

B.1 General notation

Given a real, separable, infinite-dimensional Hilbert space $\mathfrak{H}$, we denote by $\langle \cdot, \cdot \rangle_\mathfrak{H}$ and $\| \cdot \|_\mathfrak{H} = \langle \cdot, \cdot \rangle_\mathfrak{H}^{1/2}$, respectively, the associated inner product and norm. Given an integer $q \geq 2$, we denote by $\mathfrak{H}^{\otimes q}$ and $\mathfrak{H}^{\odot q}$, respectively, the qth tensor product and the qth *symmetric* tensor product of $\mathfrak{H}$. Given an orthonormal basis $\{e_j\}_{j \geq 1}$ of $\mathfrak{H}$, the set $\{e_{j_1} \otimes \ldots \otimes e_{j_q} : j_1, \ldots, j_q \geq 1\}$ is an orthonormal basis of $\mathfrak{H}^{\otimes q}$. Throughout the book, we also adopt the notation $\mathfrak{H}^{\otimes 1} = \mathfrak{H}^{\odot 1} = \mathfrak{H}$ and $\mathfrak{H}^{\otimes 0} = \mathfrak{H}^{\odot 0} = \mathbb{R}$.

B.2 L^2 spaces

We recall that, if $\mathfrak{H} = L^2(A, \mathscr{A}, \mu)$ where $(A, \mathscr{A}, \mu)$ is a non-atomic measure space, then: (i) $\mathfrak{H}^{\otimes q}$ can be identified with $L^2(A^q, \mathscr{A}^q, \mu^q)$, that is, with the space of functions on A^q that are square-integrable with respect to μ^q; and (ii) $\mathfrak{H}^{\odot q}$ can be identified with $L_s^2(A^q, \mathscr{A}^q, \mu^q)$, that is, with the space of functions f on A^q that are square-integrable with respect to μ^q and a.e. symmetric, meaning that $f = \tilde{f}$ a.e.-μ^q, where $\tilde{f}$ is the *canonical symmetrization* of f, given by

$$\tilde{f}(a_1, \ldots, a_q) = \frac{1}{q!} \sum_\sigma f(a_{\sigma(1)}, \ldots, a_{\sigma(q)}), \qquad \text{(B.2.1)}$$

and the sum runs over all permutations σ of $\{1, \ldots, q\}$.

B.3 More on symmetrization

The notion of 'symmetrization' can be extended to the case of general Hilbert spaces. Let $\mathfrak{H}$ be a real separable Hilbert space, let $\{e_j\}_{j \geq 1}$ be an orthonormal basis of $\mathfrak{H}$, and fix $q \geq 2$. Let the expansion of $f \in \mathfrak{H}^{\otimes q}$ be given by $f = \sum_{j_1,\ldots,j_q} a(j_1,\ldots,j_q) e_{j_1} \otimes \ldots \otimes e_{j_q}$. Then the canonical symmetrization of f, denoted by $\tilde{f}$, is the element of $\mathfrak{H}^{\odot q}$ given by

$$\tilde{f} = \frac{1}{q!} \sum_{\sigma} \sum_{j_1,\ldots,j_q=1}^{\infty} a(j_1,\ldots,j_q) e_{j_{\sigma(1)}} \otimes \ldots \otimes e_{j_{\sigma(q)}}, \tag{B.3.1}$$

where the first sum runs over all permutations σ of $\{1,\ldots,q\}$. It is immediate that (B.2.1) and (B.3.1) coincide whenever $\mathfrak{H} = L^2(A, \mathscr{A}, \mu)$.

B.4 Contractions

Let $1 \leq p \leq q$ be integers and let $\{e_j\}_{j \geq 1}$ be an orthonormal basis of the real separable Hilbert space $\mathfrak{H}$. Given integers $j_1,\ldots,j_p$, $k_1,\ldots,k_q$ and $r = 0,\ldots,p$, we define the rth *contraction* of the two tensor products $e_{j_1} \otimes \ldots \otimes e_{j_p}$ and $e_{k_1} \otimes \ldots \otimes e_{k_q}$, denoted by

$$(e_{j_1} \otimes \ldots \otimes e_{j_p}) \otimes_r (e_{k_1} \otimes \ldots \otimes e_{k_q})$$

as the element of $\mathfrak{H}^{\otimes(p+q-2r)}$ given by

$$(e_{j_1} \otimes \ldots \otimes e_{j_p}) \otimes_0 (e_{k_1} \otimes \ldots \otimes e_{k_q}) = e_{j_1} \otimes \ldots \otimes e_{j_p} \otimes e_{k_1} \otimes \ldots \otimes e_{k_q},$$

and, for $r = 1,\ldots,p$,

$$(e_{j_1} \otimes \ldots \otimes e_{j_p}) \otimes_r (e_{k_1} \otimes \ldots \otimes e_{k_q})$$
$$= \left[\prod_{l=1}^{r} \langle e_{j_l}, e_{k_l} \rangle_{\mathfrak{H}} \right] e_{j_{r+1}} \otimes \ldots \otimes e_{j_p} \otimes e_{k_{r+1}} \otimes \ldots \otimes e_{k_q}.$$

If $r = q = p$, then

$$(e_{j_1} \otimes \ldots \otimes e_{j_p}) \otimes_p (e_{k_1} \otimes \ldots \otimes e_{k_p})$$
$$= \prod_{l=1}^{p} \langle e_{j_l}, e_{k_l} \rangle_{\mathfrak{H}} = \langle e_{j_1} \otimes \ldots \otimes e_{j_p}, e_{k_1} \otimes \ldots \otimes e_{k_p} \rangle_{\mathfrak{H}^{\otimes p}}.$$

We now wish to define the contraction of two general elements $f \in \mathfrak{H}^{\otimes p}$ and $g \in \mathfrak{H}^{\otimes q}$. First of all, observe that there exist (unique) arrays of real numbers $\{a(j_1,\ldots,j_p) : j_1,\ldots,j_p \geq 1\}$ and $\{b(k_1,\ldots,k_q) : k_1,\ldots,k_q \geq 1\}$ such that $\sum_{j_1,\ldots,j_p} a(j_1,\ldots,j_p)^2 < \infty$, $\sum_{k_1,\ldots,k_q} b(k_1,\ldots,k_q)^2 < \infty$, and

$$f = \sum_{j_1,\ldots,j_p=1}^{\infty} a(j_1,\ldots,j_p)e_{j_1} \otimes \ldots \otimes e_{j_p} \quad \text{and}$$

$$g = \sum_{k_1,\ldots,k_q=1}^{\infty} b(k_1,\ldots,k_q)e_{k_1} \otimes \ldots \otimes e_{k_q}. \tag{B.4.1}$$

For $r = 0, 1, \ldots, p$, we define the kernel $a \star_r b(z_1,\ldots,z_{p+q-2r})$, $z_1,\ldots,z_{p+q-2r} \geq 1$, as

$$a \star_0 b(z_1,\ldots,z_{p+q}) = a(z_1,\ldots,z_p)b(z_{p+1},\ldots,z_{p+q}) \tag{B.4.2}$$

and, for $r = 1, \ldots, p$,

$$a \star_r b(z_1,\ldots,z_{p+q-2r})$$
$$= \sum_{l_1,\ldots,l_r=1}^{\infty} a(l_1,\ldots,l_r,z_1,\ldots,z_{p-r})b(l_1,\ldots,l_r,z_{p-r+1},\ldots,z_{p+q-2r}).$$

$$\tag{B.4.3}$$

A simple application of the Cauchy–Schwarz inequality shows that

$$\sum_{z_1,\ldots,z_{p+q-2r}=1}^{\infty} a \star_r b(z_1,\ldots,z_{p+q-2r})^2 < \infty.$$

Finally, for $r = 0, \ldots, p$, the *contraction* of the two elements f, g in (B.4.1) is given by

$$f \otimes_r g = \sum_{j_1,\ldots,j_p=1}^{\infty} \sum_{k_1,\ldots,k_q=1}^{\infty} a(j_1,\ldots,j_p)b(k_1,\ldots,k_q)(e_{j_1} \otimes \ldots \otimes e_{j_p})$$
$$\otimes_r (e_{k_1} \otimes \ldots \otimes e_{k_q})$$
$$= \sum_{z_1,\ldots,z_{p+q-2r}=1}^{\infty} a \star_r b(z_1,\ldots,z_{p+q-2r})e_{z_1} \otimes \ldots \otimes e_{z_{q+p-2r}}.$$

$$\tag{B.4.4}$$

When $f \in \mathfrak{H}^{\odot p}$ and $g \in \mathfrak{H}^{\odot q}$ (that is, when f, g are symmetric), one usually adopts the shorter (and more suggestive) notation

$$f \otimes_r g = \sum_{l_1,\ldots,l_r=1}^{\infty} \langle f, e_{l_1} \otimes \ldots \otimes e_{l_r} \rangle_{\mathfrak{H}^{\otimes r}} \otimes \langle g, e_{l_1} \otimes \ldots \otimes e_{l_r} \rangle_{\mathfrak{H}^{\otimes r}} \tag{B.4.5}$$

Loosely speaking, in (B.4.5) the symmetry of f and g allows the expressions $\langle f, e_{l_1} \otimes \ldots \otimes e_{l_r} \rangle_{\mathfrak{H}^{\otimes r}}$ and $\langle g, e_{l_1} \otimes \ldots \otimes e_{l_r} \rangle_{\mathfrak{H}^{\otimes r}}$, to be written without specifying which r-segments of the tensors composing f and g are involved in the inner products. Note that, if $p = q = r$, then $f \otimes_p g = \langle f, g \rangle_{\mathfrak{H}^{\otimes p}}$; if $r = 0$, then $f \otimes_0 g = f \otimes g$. Also, even if f and g are symmetric, the contraction $f \otimes_r g$ is not necessarily symmetric. We therefore denote by $f \tilde{\otimes}_r g$ its symmetrization, which (according to (B.3.1)) is given by

$$f \tilde{\otimes}_r g = \frac{1}{(p + q - 2r)!} \sum_{\sigma} \sum_{z_1, \ldots, z_{p+q-2r} = 1}^{\infty} a \star_r b(z_1, \ldots, z_{p+q-2r})$$
$$\times e_{z_{\sigma}(1)} \otimes \ldots \otimes e_{z_{\sigma(q+p-2r)}},$$

(B.4.6)

where the first sum runs over all permutations σ of $\{1, \ldots, p + q - 2r\}$. In the special case where $\mathfrak{H} = L^2(A, \mathscr{A}, \mu)$, with μ non-atomic, we have that, for every $1 \leq p \leq q$, every $f \in L^2(A^p, \mathscr{A}^p, \mu^p)$, every $g \in L^2(A^q, \mathscr{A}^q, \mu^q)$ and every $r = 1, \ldots, p$,

$$f \otimes_r g(a_1, \ldots, a_{p+q-2r}) = \int_{A^r} f(x_1, \ldots, x_r, a_1, \ldots, a_{p-r}) \qquad (B.4.7)$$
$$\times g(x_1, \ldots, x_r, a_{p-r+1}, \ldots, a_{p+q-2r}) d\mu(x_1) \ldots d\mu(x_r).$$

In this case, the symmetrization $f \tilde{\otimes}_r g$ is obtained by applying an appropriate version of formula (B.2.1) to the function $f \otimes_r g$.

B.5 Random elements

Given a probability space $(\Omega, \mathscr{F}, P)$ and a generic real separable Hilbert space $\mathfrak{H}$, we denote by $L^q(\Omega, \mathfrak{H}) = L^q(\Omega, \mathscr{F}, P; \mathfrak{H})$ $(q \geq 1)$ the class of those $\mathfrak{H}$-valued random elements Y that are $\mathscr{F}$-measurable and such that $E\|Y\|_{\mathfrak{H}}^q < \infty$. Observe that $L^2(\Omega, \mathfrak{H})$ is itself a Hilbert space, with respect to the inner product $\langle Y, Z \rangle_{L^2(\Omega, \mathfrak{H})} = E \langle Y, Z \rangle_{\mathfrak{H}}$.

B.6 Bibliographic comments

Two excellent references for basic Hilbert space theory are the books by Dudley [32] and Rudin [126]. A concise and probability-oriented discussion of tensor products is contained in Appendix E of Janson's book [57].

Appendix C

Distances between probability measures

Every random element in this appendix is defined on an adequate probability space $(\Omega, \mathscr{F}, P)$.

C.1 General definitions

Fix an integer $d \geq 1$. We first need the notion of a *separating* class of functions.

Definition C.1.1 Let $\mathscr{H}$ be a collection of Borel-measurable complex-valued functions on $\mathbb{R}^d$. We say that the class $\mathscr{H}$ is **separating** if the following property holds: any two $\mathbb{R}^d$-valued random variables F, G satisfying $h(F)$, $h(G) \in L^1(\Omega)$ and $E[h(F)] = E[h(G)]$, for every $h \in \mathscr{H}$, are necessarily such that F and G have the same law.

Instances of separating classes include: $\mathscr{H} = \{h : h \text{ is Borel-measurable and } bounded\}$, $\mathscr{H} = \{\mathbf{1}_B : B \in \mathscr{B}(\mathbb{R}^d)\}$, $\mathscr{H} = \{e^{i\langle v, \cdot \rangle} : v \in \mathbb{R}^d\}$ (where $\langle \cdot, \cdot \rangle$ is the inner product in $\mathbb{R}^d$); see Section C.2 for more examples.

Definition C.1.2 Let $\mathscr{H}$ be separating, and let F, G be random variables with values in $\mathbb{R}^d$ such that $h(F)$, $h(G) \in L^1(\Omega)$ for every $h \in \mathscr{H}$. The **distance** between the laws of F and G, induced by $\mathscr{H}$, is given by the quantity

$$d_{\mathscr{H}}(F, G) = \sup\{|E[h(F)] - E[h(G)]| : h \in \mathscr{H}\}. \tag{C.1.1}$$

It is easily checked that, for every separating class $\mathscr{H}$, the mapping $d_{\mathscr{H}}$ is an actual distance (or *metric*) on a subset of the class of probability measures on $\mathbb{R}^d$, that is: for all $\mathbb{R}^d$-valued random variables F, G, J such that $h(F), h(G), h(J) \in L^1(\Omega)$ for every $h \in \mathscr{H}$,

– $d_{\mathscr{H}}(F, G) = d_{\mathscr{H}}(G, F)$ (symmetry),
– $d_{\mathscr{H}}(F, G) = 0$ if and only if F and G have the same law,
– $d_{\mathscr{H}}(F, G) \leq d_{\mathscr{H}}(F, J) + d_{\mathscr{H}}(J, G)$ (triangle inequality).

C.2 Some special distances

Throughout this book, we adopt special symbols for the distances associated with some remarkable classes $\mathscr{H}$. Our main conventions are brought together in the next definition.

Definition C.2.1 Fix $d \geq 1$. All random variables considered below have values in $\mathbb{R}^d$.

1. The **Kolmogorov distance** between the law of two random variables F, G, denoted by $d_{\mathrm{Kol}}(F, G)$, is obtained from (C.1.1) by taking $\mathscr{H}$ to be the class of all functions $h : \mathbb{R}^d \to \mathbb{R}$ of the type $h(x_1, \ldots, x_d) = \mathbf{1}_{(-\infty, z_1]}(x_1) \ldots \mathbf{1}_{(-\infty, z_d]}(x_d)$, where $z_1, \ldots, z_d \in \mathbb{R}$. That is,

$$d_{\mathrm{Kol}}(F, G) = \sup_{z_1, \ldots, z_d \in \mathbb{R}} \left| P\left(F \in (-\infty, z_1] \times \ldots \times (-\infty, z_d]\right) \right. \quad \text{(C.2.1)}$$
$$\left. - P\left(G \in (-\infty, z_1] \times \ldots \times (-\infty, z_d]\right) \right|.$$

2. The **total variation distance** between the laws of F and G, denoted by $d_{\mathrm{TV}}(F, G)$, is obtained from (C.1.1) by taking $\mathscr{H}$ to be the collection of all functions $h : \mathbb{R}^d \to \mathbb{R}$ of the type $h(x_1, \ldots, x_d) = \mathbf{1}_B(x_1, \ldots, x_d)$, where $B \in \mathscr{B}(\mathbb{R}^d)$. That is,

$$d_{\mathrm{TV}}(F, G) = \sup_{B \in \mathscr{B}(\mathbb{R}^d)} \left| P[F \in B] - P[G \in B] \right|. \quad \text{(C.2.2)}$$

3. The **Wasserstein distance** between the laws of F and G, denoted by $d_{\mathrm{W}}(F, G)$, is obtained from (C.1.1) by taking $\mathscr{H}$ to be the set of all functions $h : \mathbb{R}^d \to \mathbb{R}$ such that $\|h\|_{\mathrm{Lip}} \leq 1$, where

$$\|h\|_{\mathrm{Lip}} = \sup_{\substack{\mathbf{x}_d \neq \mathbf{y}_d \\ \mathbf{x}_d, \mathbf{y}_d \in \mathbb{R}^d}} \frac{|h(\mathbf{x}_d) - h(\mathbf{y}_d)|}{\|\mathbf{x}_d - \mathbf{y}_d\|_{\mathbb{R}^d}} \quad \text{(C.2.3)}$$

(with $\| \cdot \|_{\mathbb{R}^d}$ the usual Euclidean norm on $\mathbb{R}^d$).

4. The **Fortet–Mourier distance** (or **bounded Wasserstein distance**) between the laws of F and G, denoted by $d_{\mathrm{FM}}(F, G)$, is obtained from

(C.1.1) by taking $\mathscr{H}$ to be the class of all functions $h : \mathbb{R}^d \to \mathbb{R}$ such that $\|h\|_{\text{Lip}} + \|h\|_\infty \le 1$, where

$$\|h\|_\infty = \sup_{\mathbf{x}_d \in \mathbb{R}^d} |h(\mathbf{x}_d)|. \qquad (C.2.4)$$

We can immediately see that the four classes $\mathscr{H}$ taken into account in Definition C.2.1 are all separating.

Remark C.2.2 Note that $d_{\text{W}}(\cdot, \cdot) \ge d_{\text{FM}}(\cdot, \cdot)$ and $d_{\text{TV}}(\cdot, \cdot) \ge d_{\text{Kol}}(\cdot, \cdot)$. Moreover, if $d = 1$, then

$$d_{\text{Kol}}(F, G) = \sup_{z \in \mathbb{R}} \left| P(F \le z) - P(G \le z) \right|. \qquad (C.2.5)$$

One can also prove the following relation: if F is any real-valued random variable and $N \sim \mathcal{N}(0, 1)$ is standard Gaussian, then

$$d_{\text{Kol}}(F, N) \le 2\sqrt{d_{\text{W}}(F, N)}. \qquad (C.2.6)$$

(See, for example, [22, Theorem 3.3] for a proof.)

C.3 Some further results

In this section, we gather together some useful information about the four distances d_{Kol}, d_{TV}, d_{W} and d_{FM}, as defined in the previous section.

Proposition C.3.1 *For every $d \ge 1$, the topologies induced by the three distances d_{Kol}, d_{TV} and d_{W}, on the set of probability measures on $\mathbb{R}^d$, are strictly stronger than the topology of the convergence in distribution.*

Proof (First part: d_{Kol} and d_{TV}) Due to the usual characterization of the convergence in distribution, it is clear that, if either $d_{\text{Kol}}(F_n, F)$ or $d_{\text{TV}}(F_n, F)$ converges to zero as $n \to \infty$, then F_n converges in distribution to F. To prove the desired claim it is now sufficient to consider a sequence F_n, $n \ge 1$, such that F_n equals $(\frac{1}{n}, \ldots, \frac{1}{n})$ with probability one. Indeed, in this case we have that the F_n converges almost surely to a random variable F such that $P[F = (0, \ldots, 0)] = 1$, but $d_{\text{TV}}(F_n, F) \ge d_{\text{Kol}}(F_n, F) = 1$.
(Second part: d_{W}) Since, for every $\lambda = (\lambda_1, \ldots, \lambda_d) \ne 0$, the functions $f_\lambda(x_1, \ldots, x_d) = (\sqrt{d}\|\lambda\|_{\mathbb{R}^d})^{-1} \sin(\lambda_1 x_1 + \ldots + \lambda_d x_d)$ and $g_\lambda(x_1, \ldots, x_d) = (\sqrt{d}\|\lambda\|_{\mathbb{R}^d})^{-1} \cos(\lambda_1 x_1 + \ldots + \lambda_d x_d)$ are such that $\|f_\lambda\|_{\text{Lip}} \le 1$ and $\|g_\lambda\|_{\text{Lip}} \le 1$, it is clear (through a characteristic function argument) that, if $d_{\text{W}}(F_n, F)$ converges to zero as $n \to \infty$, then F_n converges in distribution to F. To

prove that the topology induced by d_W is stronger than the topology of convergence in distribution, consider a random variable U uniformly distributed on the hypercube $[0, 1]^d$, and define U_n to be equal to $(n, \ldots, n)$ if U belongs to $[0, \frac{1}{n}]^d$, and zero otherwise. Then, as $n \to \infty$, U_n converges almost surely, hence in law, to zero. However, if we consider the function $f(x_1, \ldots, x_d) = x_1$, we have that $\|f\|_{\mathrm{Lip}} \leq 1$, and $E[f(U_n)] = 1$ for every n, albeit that $f(0) = 0$. $\qquad\square$

The next statement shows that the Kolmogorov distance metrizes the convergence in distribution towards real-valued random variables whose distribution function is *continuous*.

Proposition C.3.2 *Let $\{F, F_n : n \geq 1\}$ be a collection of real-valued random variables such that the mapping $z \mapsto P(F \leq z)$ is continuous for every $z \in \mathbb{R}$. Then, F_n converges in distribution to F, as $n \to \infty$, if and only if $d_{\mathrm{Kol}}(F_n, F) \to 0$.*

Proof Thanks to Proposition C.3.1, we only have to show that, if F_n converges in distribution to F, then $d_{\mathrm{Kol}}(F_n, F) \to 0$. So, let us assume that F_n converges in distribution to F, and write $\Phi_F(z) = P(F \leq z)$ and $\Phi_{F_n}(z) = P(F_n \leq z)$, $z \in \mathbb{R}$. We define increasing functions f, f_n on $[-\pi/2, \pi/2]$, $n \geq 1$, as follows: $f(-\pi/2) = f_n(-\pi/2) = 0$, $f(\pi/2) = f_n(\pi/2) = 1$ and, for every $x \in (-\pi/2, \pi/2)$,

$$f(x) = \Phi_F(\tan(x)), \quad f_n(x) = \Phi_{F_n}(\tan(x)).$$

It is easy to see that the mapping f is continuous on the compact interval $[-\pi/2, \pi/2]$. Now assume that F_n converges in distribution to F. Since the distribution function of F is continuous, we have that $\Phi_{F_n}(z) \to \Phi_F(z)$ for every real z, and consequently $f_n(x) \to f(x)$ for every $x \in [-\pi/2, \pi/2]$. Dini's second theorem therefore yields that

$$\sup_{x \in [-\pi/2, \pi/2]} |f_n(x) - f(x)| \to 0,$$

and the proof is concluded by observing that

$$\sup_{x \in [-\pi/2, \pi/2]} |f_n(x) - f(x)| = d_{\mathrm{Kol}}(F_n, F). \qquad\square$$

Remark C.3.3 It is important to note that the content of Proposition C.3.2 does not extend to the total variation and Wasserstein distances. To see this, consider, for instance, a sequence $\{\epsilon_i : i \geq 1\}$ of i.i.d. random variables such that $P[\epsilon_1 = 1] = 1/2 = P[\epsilon_1 = -1]$, and define $F_n = n^{-1/2} \sum_{i=1}^n \epsilon_i$. Then, by the central limit theorem, as $n \to \infty$, F_n converges in distribution to $N \sim \mathcal{N}(0, 1)$, although $d_{\mathrm{TV}}(F_n, N) = 1$ for every n. On the other hand, let $N \sim \mathcal{N}(0, 1)$ and consider a random variable U that is uniformly distributed on $[0, 1]$ (we can take N and U to be independent, but the joint law of the vector (N, U) is immaterial). For every $n \geq 1$, define the random variable U_n to be equal to n if $U \in [0, \frac{1}{n}]$, and zero otherwise. Then $F_n := N + U_n$ converges almost surely (and therefore in law) to N, as $n \to \infty$, although $E[F_n] = 1 \neq E[N] = 0$ for every n, and therefore the convergence does not take place in the sense of the Wasserstein distance.

Proposition C.3.4 *For every $d \geq 1$, the Fortet–Mourier distance d_{FM} metrizes the convergence in distribution, that is, a sequence $\{F_n : n \geq 1\}$ converges in distribution to F, as $n \to \infty$, if and only if $d_{\mathrm{FM}}(F_n, F) \to 0$.*

Proof See [32, Theorem 11.3.3]. □

Proposition C.3.5 *For every $d \geq 1$, we have the representation*

$$d_{\mathrm{TV}}(F, G)$$
$$= \frac{1}{2} \sup\{|E[h(F)] - E[h(G)]| : h \text{ is Borel-measurable and } \|h\|_\infty \leq 1\}.$$
$$(C.3.1)$$

Proof Denote by **D** the right-hand side of (C.3.1); for $B \in \mathscr{B}(\mathbb{R}^d)$, we write B^c for the complement of B. We have that

$$P(F \in B) - P(G \in B) = \frac{1}{2}\{E[(\mathbf{1}_B - \mathbf{1}_{B^c})(F)] - E[(\mathbf{1}_B - \mathbf{1}_{B^c})(G)]\},$$

from which we infer that $d_{\mathrm{TV}}(F, G) \leq \mathbf{D}$. To prove that $d_{\mathrm{TV}}(F, G) \geq \mathbf{D}$, fix h such that $\|h\|_\infty \leq 1$, and observe that the set function $B \mapsto \alpha(B) := P(F \in B) - P(G \in B)$ is a signed measure. We therefore have that

$$|E[h(F)] - E[h(G)]| \leq \int_{\mathbb{R}^d} |h(x)| d|\alpha|(x) \leq |\alpha|(C),$$

where C is the support of h and $|\alpha|$ is the total variation of α. In particular (see, for example, [32, Theorem 5.6.1]),

$$|\alpha|(C) = \sup\{\alpha(B) : B \subset C\} - \inf\{\alpha(B) : B \subset C\} \leq 2d_{\mathrm{TV}}(F, G),$$

where the inequality follows from the relations

$$|\sup\{\alpha(B) : B \subset C\}| \leq \sup_{B \subset C} |P(F \in B) - P(G \in B)| \leq d_{\mathrm{TV}}(F, G)$$

and

$$| \inf\{\alpha(B) : B \subset C\}| \leq \sup_{B \subset C} |P(F \in B) - P(G \in B)| \leq d_{\mathrm{TV}}(F, G),$$

thus concluding the proof.   □

C.4 Bibliographic comments

Our main references on the topics of Appendix C are the books by Billingsley [13] and Dudley [32, chapter 11].

Appendix D

Fractional Brownian motion

In this appendix, we assume that all random objects are defined on an adequate probability space $(\Omega, \mathscr{F}, P)$.

D.1 Definition and immediate properties

Definition D.1.1 Let $H \in (0, 1]$. A fractional Brownian motion (fBm) of Hurst parameter H is a centered Gaussian process $B^H = (B_t^H)_{t \in \mathbb{R}}$ with covariance function

$$E[B_t^H B_s^H] = \frac{1}{2}\big(|t|^{2H} + |s|^{2H} - |t - s|^{2H}\big), \tag{D.1.1}$$

and such that the paths of B^H are continuous with probability one.

Remark D.1.2 When $H > \frac{1}{2}$, it is easy to check that, for any $s, t \geq 0$,

$$H(2H - 1)\int_0^t du \int_0^s dv\, |v - u|^{2H-2} = \frac{1}{2}\big(t^{2H} + s^{2H} - |t - s|^{2H}\big). \tag{D.1.2}$$

Observe, however, that (D.1.2) is not valid when $H \leq \frac{1}{2}$ since, in this case, the kernel $|v - u|^{2H-2}$ is not integrable.

It is not evident *a priori* that the fBm exists. Let us prove this point.

Proposition D.1.3   (i) *For any $H \in (0, 1]$, the fBm of Hurst parameter H exists (and is unique in law). Moreover, it admits a modification that is P-a.s. α-Hölder continuous, for every $\alpha < H$.*

  (ii) *If $H = \frac{1}{2}$, then the fBm is nothing more than a standard (two-sided) Brownian motion.*

 (iii) *If $H = 1$ then $B_t^H = t\, B_1^H$ almost surely for all $t \in \mathbb{R}$.*

Since the case $H = 1$ is trivial, we will henceforth assume that $0 < H < 1$.

Proof of Proposition D.1.3 (i) We must first show that the function

$$r_H : (t, s) \mapsto \frac{1}{2}\left(|t|^{2H} + |s|^{2H} - |t - s|^{2H}\right)$$

is non-negative definite. When $H = 1$, we observe that $r_H(t, s) = st$, so that the conclusion is easily shown. Assume now that $H < 1$. Using the change of variable $v = u|t|$ (whenever $t \neq 0$), we immediately deduce that

$$|t|^{2H} = \frac{1}{c_H} \int_0^\infty \frac{1 - e^{-u^2 t^2}}{u^{1+2H}} du,$$

with $c_H = \int_0^\infty (1 - e^{-u^2}) u^{-1-2H} du = \frac{\Gamma(1-H)}{2H} < \infty$. Therefore,

$$
\begin{aligned}
r_H(t, s) &= \frac{1}{2c_H} \int_0^\infty \frac{(1 - e^{-u^2 t^2})(1 - e^{-u^2 s^2})}{u^{1+2H}} du \\
&\quad + \frac{1}{2c_H} \int_0^\infty \frac{e^{-u^2 t^2}(e^{2u^2 ts} - 1)e^{-u^2 s^2}}{u^{1+2H}} du \\
&= \frac{1}{2c_H} \int_0^\infty \frac{(1 - e^{-u^2 t^2})(1 - e^{-u^2 s^2})}{u^{1+2H}} du \\
&\quad + \frac{1}{2c_H} \sum_{n=1}^\infty \frac{2^n}{n!} \int_0^\infty \frac{t^n e^{-u^2 t^2} s^n e^{-u^2 s^2}}{u^{1-2n+2H}} du,
\end{aligned}
$$

so that, for all $s_1, \ldots, s_m, v_1, \ldots, v_m \in \mathbb{R}$,

$$
\begin{aligned}
\sum_{i,j=1}^m r_H(s_i, s_j) v_i v_j &= \frac{1}{2c_H} \int_0^\infty \frac{\left(\sum_{i=1}^m (1 - e^{-u^2 s_i^2}) v_i\right)^2}{u^{1+2H}} du \\
&\quad + \frac{1}{2c_H} \sum_{n=1}^\infty \frac{2^n}{n!} \int_0^\infty \frac{\left(\sum_{i=1}^m s_i^n e^{-u^2 s_i^2} v_i\right)^2}{u^{1-2n+2H}} du \geq 0.
\end{aligned}
$$

That is, r_H is non-negative definite. To prove the remaining part of the statement, assume that B^H is a centered Gaussian process with covariance function given by (D.1.1). Elementary computations show that, for every $q > 0$,

$$E[|B_t^H - B_s^H|^q] = E[|B_1^H|^q] \times |t - s|^{qH}.$$

Selecting $q > 1/H$, we may therefore apply the Kolmogorov–Chentsov criterion (see, for example, [61, Theorem 2.8]) to deduce that B^H admits an

α-Hölder continuous modification for every $\alpha < (qH - 1)/q$. The desired conclusion is obtained by letting $q \to \infty$.

(ii) When $H = \frac{1}{2}$, we immediately see that the covariance of B^H reduces to that of a standard (two-sided) Brownian motion.

(iii) Assume $H = 1$. Then, for all $t \in \mathbb{R}$,

$$E[(B_t^H - t\, B_1^H)^2] = E[(B_t^H)^2] - 2t\, E[B_t^H B_1^H] + t^2 E[(B_1^H)^2]$$
$$= t^2 - t(t^2 + 1 - (1 - t)^2) + t^2 = 0,$$

that is, $B_t^H = t\, B_1^H$ almost surely. $\qquad\square$

Proposition D.1.4 *Let B^H be a fractional Brownian motion of Hurst parameter $H \in (0, 1)$. Then:*

(i) *[Self-similarity] for all $a \in \mathbb{R} \setminus \{0\}$, $(|a|^{-H} B_{at}^H)_{t \in \mathbb{R}} \overset{\text{law}}{=} (B_t^H)_{t \in \mathbb{R}}$.*

(ii) *[Stationarity of increments] for all $h > 0$, $(B_{t+h}^H - B_h^H)_{t \in \mathbb{R}} \overset{\text{law}}{=} (B_t^H)_{t \in \mathbb{R}}$.*

Conversely, any continuous Gaussian process $B^H = (B_t^H)_{t \in \mathbb{R}}$ with $B_0^H = 0$, $\mathrm{Var}(B_1^H) = 1$, and such that (i) and (ii) hold, is a fractional Brownian motion of index H.

Proof Let B^H be an fBm of index H. Both (i) and (ii) are proved by showing that the process on the left-hand side is centered, Gaussian and has a covariance given by (D.1.1). Conversely, let $B^H = (B_t^H)_{t \in \mathbb{R}}$ be a Gaussian process with $B_0^H = 0$ and $\mathrm{Var}(B_1^H) = 1$, and satisfying (i) and (ii). We need to show that B^H is centered and has (D.1.1) for covariance. From (ii) with $t = h > 0$, we get that $E[B_{2t}^H] = 2E[B_t^H]$, whereas from (i) we infer that $E[B_{2t}^H] = 2^H E[B_t^H]$. Combining these two equalities gives $E[B_t^H] = 0$ for all $t > 0$. Using (i) with $a = -1$, we get that $E[B_t^H] = 0$ for $t < 0$ as well. That is, B^H is centered. Now, let $s, t \in \mathbb{R}$. We have

$$E[B_s^H B_t^H] = \frac{1}{2}\Big(E[(B_t^H)^2] + E[(B_s^H)^2] - E[(B_t^H - B_s^H)^2]\Big)$$
$$= \frac{1}{2}\Big(E[(B_t^H)^2] + E[(B_s^H)^2] - E[(B_{|t-s|}^H)^2]\Big) \quad \text{(because of (ii))}$$
$$= \frac{1}{2}E[(B_1^H)^2]\big(|t|^{2H} + |s|^{2H} - |t - s|^{2H}\big) \quad \text{(because of (i))}$$
$$= \frac{1}{2}\big(|t|^{2H} + |s|^{2H} - |t - s|^{2H}\big).$$

The proof of the proposition is done $\qquad\square$

D.2 Hurst phenomenon and invariance principle

Fractional Brownian motion has been successfully used in order to model a variety of natural phenomena. Following [127], we shall now describe how it was introduced historically, and why its parameter is called the 'Hurst exponent'.

In the 1950s, Harold Hurst studied the flow of water in the Nile, and empirically highlighted a somewhat curious phenomenom. Let us denote by $X_1, X_2, \ldots$ the set of data observed by Hurst. The statistic he looked at is the so-called R/S-statistic (for 'rescaled range of the observations'), defined as

$$\frac{R}{S}(X_1, \ldots, X_n) := \frac{\max_{1 \le i \le n} \left(S_i - \frac{i}{n} S_n \right) - \min_{1 \le i \le n} \left(S_i - \frac{i}{n} S_n \right)}{\sqrt{\frac{1}{n} \sum_{i=1}^{n} \left(X_i - \frac{1}{n} S_n \right)^2}},$$

where $S_n = X_1 + \ldots + X_n$. This quantity measures the ratio between the highest and lowest positions of the partial sums with respect to the straight line of uniform growth and the sample standard deviation.

As a first approximation, let us assume that the X_i are i.i.d., with common mean $\mu \in \mathbb{R}$ and common variance $\sigma^2 > 0$. Because $t \mapsto \frac{1}{\sqrt{n}}(S_{[nt]} - [nt]\mu)$ is constant on each interval $(i/n, (i+1)/n)$ whereas $t \mapsto \frac{t}{\sqrt{n}}(S_n - n\mu)$ is monotone, it is easy to see that the maximum of $t \mapsto \frac{1}{\sqrt{n}}\big(S_{[nt]} - [nt]\mu - t(S_n - n\mu)\big)$ on $[0, 1]$ is necessarily attained at a point t of the type $t = \frac{i}{n}$, $i = 0, \ldots, n$. Therefore,

$$\sup_{t \in [0,1]} \frac{1}{\sqrt{n}}\big(S_{[nt]} - [nt]\mu - t(S_n - n\mu)\big) = \max_{0 \le i \le n} \left(S_i - \frac{i}{n} S_n \right).$$

Similarly,

$$\inf_{t \in [0,1]} \frac{1}{\sqrt{n}}\big(S_{[nt]} - [nt]\mu - t(S_n - n\mu)\big) = \min_{0 \le i \le n} \left(S_i - \frac{i}{n} S_n \right).$$

Hence,

$$\frac{1}{\sigma\sqrt{n}} \left\{ \max_{1 \le i \le n} \left(S_i - \frac{i}{n} S_n \right) - \min_{1 \le i \le n} \left(S_i - \frac{i}{n} S_n \right) \right\}$$
$$= \phi \left(t \mapsto \frac{1}{\sigma\sqrt{n}}\big(S_{[nt]} - [nt]\mu\big) \right),$$

where $\phi(f) = \sup_{0 \le t \le 1}\{f(t) - tf(1)\} - \inf_{0 \le t \le 1}\{f(t) - tf(1)\}$. Therefore by applying a functional version of the classical central limit theorem (Donsker's theorem), we deduce that

$$\frac{1}{\sigma\sqrt{n}} \left\{ \max_{1 \le i \le n} \left(S_i - \frac{i}{n} S_n \right) - \min_{1 \le i \le n} \left(S_i - \frac{i}{n} S_n \right) \right\}$$

converges in distribution to $\phi(W) = \sup_{0 \le t \le 1}\{W_t - tW_1\} - \inf_{0 \le t \le 1}\{W_t - tW_1\}$, where W stands for a standard Brownian motion on $[0, 1]$. Finally, because $\sqrt{\frac{1}{n}\sum_{i=1}^{n}\left(X_i - \frac{1}{n}S_n\right)^2} \to \sigma$ almost surely (strong law of large numbers), we infer that

$$\frac{1}{\sqrt{n}} \times \frac{R}{S}(X_1, \ldots, X_n) \overset{\text{law}}{\to} \sup_{0 \le t \le 1}\{W_t - tW_1\} - \inf_{0 \le t \le 1}\{W_t - tW_1\} \quad \text{as } n \to \infty.$$

That is, in the case of i.i.d. observations the R/S-statistic grows as $\sqrt{n}$, where n denotes the sample size. However, this is not what Hurst observed when he calculated the R/S-statistic on the Nile river data (between 622 and 1469). Instead, he found a growth of order $n^{0.74}$.

Is it possible to find a stochastic model explaining this fact? The answer is yes. Indeed, it turns out that fBm allows us to do so. More precisely, suppose now that $X_1, X_2, \ldots$ take the form

$$X_i = \mu + \sigma(B_i^H - B_{i-1}^H), \tag{D.2.1}$$

with B a fractional Brownian motion of index $H = 0.74$. That is, the X_i are again distributed according to the Gaussian law with mean μ and variance σ^2, but without being independent. Due to the specific form of (D.2.1), it is readily checked that

$$\frac{R}{S}(X_1, \ldots, X_n) = \frac{\max_{1 \le i \le n}\left(B_i^H - \frac{i}{n}B_n^H\right) - \min_{1 \le i \le n}\left(B_i^H - \frac{i}{n}B_n^H\right)}{\sqrt{\frac{1}{n}\sum_{i=1}^{n}\left(B_i^H - B_{i-1}^H - \frac{1}{n}B_n^H\right)^2}}.$$

Using the self-similarity property of B (Proposition D.1.4), we get that

$$\frac{1}{n^H}\left\{\max_{1 \le i \le n}\left(B_i^H - \frac{i}{n}B_n^H\right) - \min_{1 \le i \le n}\left(B_i^H - \frac{i}{n}B_n^H\right)\right\}$$
$$\overset{\text{law}}{=} \max_{1 \le i \le n}\left(B_{i/n}^H - \frac{i}{n}B_1^H\right) - \min_{1 \le i \le n}\left(B_{i/n}^H - \frac{i}{n}B_1^H\right)$$
$$\overset{\text{a.s.}}{\to} \sup_{0 \le t \le 1}\{B_t^H - tB_1^H\} - \inf_{0 \le t \le 1}\{B_t^H - tB_1^H\} \quad \text{as } n \to \infty.$$

On the other hand, as $n \to \infty$ we have that $n^{-1}B_n^H \overset{\text{law}}{=} n^{H-1}B_1^H \overset{\text{a.s.}}{\to} 0$ whereas, by Theorem D.3.1 (with mesh $\frac{1}{n}$ instead of 2^{-n}), $\frac{1}{n}\sum_{i=1}^{n}(B_i^H - B_{i-1}^H)^2 \overset{L^2}{\to} 1$. By putting all these facts together, we get that

$$\frac{1}{n^H} \times \frac{R}{S}(X_1, \ldots, X_n) \overset{\text{law}}{\to} \sup_{0 \le t \le 1}\{B_t^H - tB_1^H\} - \inf_{0 \le t \le 1}\{B_t^H - tB_1^H\} \quad \text{as } n \to \infty;$$

hence model (D.2.1) represents a plausible explanation of the phenomenon observed by Hurst in [53].

We conclude this section with a result showing that fBm may appear naturally in various situations.

Theorem D.2.1 *Let $X_1, X_2, \ldots$ be a stationary Gaussian sequence with mean 0 and covariance $\rho(i - j) = E[X_i X_j]$ satisfying*

$$\sum_{i,j=1}^{n} \rho(i - j) \sim K n^{2H}, \quad \text{as } n \to \infty, \tag{D.2.2}$$

for some $0 < H < 1$ and $K > 0$. Then $\left(n^{-H} \sum_{i=1}^{[nt]} X_i \right)_{t \geq 0}$ converges in the sense of the finite-dimensional distributions, as $n \to \infty$, to $\left(\sqrt{K} B_t^H \right)_{t \geq 0}$, where B^H denotes an fBm of Hurst parameter H.

Proof Fix $t_p > \ldots > t_1 \geq 0$ and set $Z_n(t) = \frac{1}{n^H} \sum_{i=1}^{[nt]} X_i$. For $k \geq l$, as $n \to \infty$,

$$E[Z_n(t_k) Z_n(t_l)]$$

$$= \frac{1}{2} E[Z_n(t_k)^2] + \frac{1}{2} E[Z_n(t_l)^2] - \frac{1}{2} E[(Z_n(t_k) - Z_n(t_l))^2]$$

$$= \frac{K}{2n^{2H}} \sum_{i,j=1}^{[nt_l]} \rho(i - j) + \frac{K}{2n^{2H}} \sum_{i,j=1}^{[nt_k]} \rho(i - j) - \frac{K}{2n^{2H}} \sum_{i,j=[nt_l]+1}^{[nt_k]} \rho(i - j)$$

$$= \frac{K}{2n^{2H}} \sum_{i,j=1}^{[nt_l]} \rho(i - j) + \frac{K}{2n^{2H}} \sum_{i,j=1}^{[nt_k]} \rho(i - j) - \frac{K}{2n^{2H}} \sum_{i,j=1}^{[nt_k]-[nt_l]} \rho(i - j)$$

$$\to \frac{K}{2} \left(t_k^{2H} + t_l^{2H} - (t_k - t_l)^{2H} \right) = K \, E[B_{t_k}^H B_{t_l}^H].$$

Since $X_1, X_2, \ldots$ are centered and jointly Gaussian, so are $Z_n(t_1), \ldots, Z_n(t_p)$. Therefore, the previous convergence suffices to deduce that

$$\left(Z_n(t_1), \ldots, Z_n(t_p) \right) \overset{\text{law}}{\to} \left(\sqrt{K} B_{t_1}^H, \ldots, \sqrt{K} B_{t_p}^H \right), \quad \text{as } n \to \infty,$$

which is the desired conclusion. □

Remark D.2.2 1. When the stationary sequence $X_1, X_2, \ldots$ satisfies (D.2.2) without being Gaussian, it may happen that, once adequately renormalized, $\sum_{i=1}^{[nt]} X_i$ converges in distribution to a process that is not an fBm. This is, for instance, the case with $X_i = H_q(B_i^\alpha - B_{i-1}^\alpha)$, when H_q stands for the qth Hermite polynomial and B^α is an fBm of Hurst parameter $\alpha > 1 - \frac{1-H}{q}$.

2. When $\sum_{k \in \mathbb{Z}} |\rho(k)| < \infty$ then, by dominated convergence,

$$\frac{1}{n} \sum_{i,j=1}^{n} \rho(i-j) = \sum_{k \in \mathbb{Z}} \rho(k) \left(1 - \frac{|k|}{n} \right) \mathbf{1}_{\{|k|<n\}} \to \sum_{k \in \mathbb{Z}} \rho(k), \quad \text{as } n \to \infty,$$

that is, (D.2.2) holds with $H = \frac{1}{2}$ and $K = \sum_{r \in \mathbb{Z}} \rho(k)$. When $\rho(k) \sim C|k|^{-\alpha}$ with $\alpha \in (0, 1)$ and $C > 0$, then, using Riemann sums,

$$\sum_{i,j=1}^{n} \rho(i-j)$$

$$= n \sum_{k=-(n-1)}^{n-1} \rho(k) \left(1 - \frac{|k|}{n} \right) \sim 2Cn \sum_{k=1}^{n} k^{-\alpha} \left(1 - \frac{k}{n} \right)$$

$$\sim n^{2-\alpha} 2C \int_{0}^{1} x^{-\alpha}(1-x)dx = \frac{2C}{(1-\alpha)(2-\alpha)} n^{2-\alpha}, \quad \text{as } n \to \infty.$$

This time, (D.2.2) holds with $H = 1 - \frac{\alpha}{2}$ and $K = \frac{2C}{(1-\alpha)(2-\alpha)}$.

D.3 Fractional Brownian motion is not a semimartingale

We shall now show that fBm is never a semimartingale, except of course when it is the standard Brownian motion ($H = \frac{1}{2}$). Using Hermite polynomials, we start with a general preliminary result.

Theorem D.3.1 *Let $f : \mathbb{R} \to \mathbb{R}$ be a Borel function such that $E[f^2(N)] < \infty$ with $N \sim \mathcal{N}(0, 1)$. Let B^H be an fBm of Hurst index $H \in (0, 1)$. Then*

$$2^{-n} \sum_{k=1}^{2^n} f(B_k^H - B_{k-1}^H) \to E[f(N)] \quad \text{in } L^2(\Omega) \text{ as } n \to \infty. \qquad \text{(D.3.1)}$$

Remark D.3.2 Thanks to the self-similarity property of B (Proposition D.1.4), we also deduce that, if the assumptions of Theorem D.3.1 hold, then

$$2^{-n} \sum_{k=1}^{2^n} f\left(2^{nH}(B_{k2^{-n}}^H - B_{(k-1)2^{-n}}^H)\right) \to E[f(N)] \quad \text{in } L^2(\Omega) \text{ as } n \to \infty.$$

$$\text{(D.3.2)}$$

Proof of Theorem D.3.1 When $H = \frac{1}{2}$, the desired convergence (D.3.1) is easily obtained thanks to the independence of the increments of B^H. So, for the rest of the proof, we assume without loss of generality that $H \neq \frac{1}{2}$.

Because $E[f^2(N)] < \infty$, we can expand f in terms of Hermite polynomials (Proposition 1.4.2(iv)) and write:

$$f(x) = \sum_{l=0}^{\infty} \frac{c_l}{\sqrt{l!}} H_l(x), \quad x \in \mathbb{R}. \tag{D.3.3}$$

Due to the orthogonality property of Hermite polynomials of different orders (see, for example, Proposition 1.4.2(iii)), we observe that $\sum_{l=0}^{\infty} c_l^2 = E[f^2(N)] < \infty$. Also, choosing $x = N$ and taking the expectation in (D.3.3) leads to $c_0 = E[f(N)]$. Hence,

$$-E[f(N)] + 2^{-n} \sum_{k=1}^{2^n} f(B_k^H - B_{k-1}^H) = 2^{-n} \sum_{k=1}^{2^n} \left(f(B_k^H - B_{k-1}^H) - E[f(N)] \right)$$

$$= 2^{-n} \sum_{l=1}^{\infty} \frac{c_l}{\sqrt{l!}} \sum_{k=1}^{2^n} H_l(B_k^H - B_{k-1}^H),$$

so that, using Proposition 2.2.1 to go from the second line to the third,

$$E\left[\left(-E[f(N)] + 2^{-n} \sum_{k=1}^{2^n} f(B_k^H - B_{k-1}^H) \right)^2 \right]$$

$$= 2^{-2n} \sum_{l=1}^{\infty} \frac{c_l^2}{l!} \sum_{k,k'=1}^{2^n} E[H_l(B_k^H - B_{k-1}^H) H_l(B_{k'}^H - B_{k'-1}^H)]$$

$$= 2^{-2n} \sum_{l=1}^{\infty} c_l^2 \sum_{k,k'=1}^{2^n} E[(B_k^H - B_{k-1}^H)(B_{k'}^H - B_{k'-1}^H)]^l$$

$$= 2^{-2n} \sum_{l=1}^{\infty} c_l^2 \sum_{k,k'=1}^{2^n} \rho(k - k')^l,$$

with

$$\rho(x) = \rho(|x|) = \frac{1}{2} \left(|x + 1|^{2H} + |x - 1|^{2H} - 2|x|^{2H} \right), \quad x \in \mathbb{Z}.$$

Because $\rho(x) = E[(B_1^H (B_{|x|+1}^H - B_{|x|}^H)]$, we have, by the Cauchy–Schwarz inequality, that

$$|\rho(x)| \le \sqrt{E[(B_1^H)^2]} \sqrt{E[(B_{|x|+1}^H - B_{|x|}^H)^2]} = 1.$$

This leads to

$$
E\left[\left(-E[f(N)] + 2^{-n}\sum_{k=1}^{2^n} f(B_k^H - B_{k-1}^H)\right)^2\right]
$$

$$
\leq 2^{-2n}\sum_{l=1}^{\infty} c_l^2 \sum_{k,k'=1}^{2^n} |\rho(k-k')| = \mathrm{Var}(f(N))2^{-2n}\sum_{k,k'=1}^{2^n} |\rho(k-k')|
$$

$$
= \mathrm{Var}(f(N))2^{-2n}\sum_{k'=1}^{2^n}\sum_{k=1-k'}^{2^n-k'} |\rho(k)| \leq 2\,\mathrm{Var}(f(N))2^{-n}\sum_{k=0}^{2^n-1} |\rho(k)|.
$$

To conclude the proof, all that is remains to study the asymptotic behavior of $\sum_{k=1}^{2^n-1}|\rho(k)|$. It is readily checked that $\rho(k) \sim H(2H-1)k^{2H-2}$ as $k \to \infty$. Hence, if $H < \frac{1}{2}$ then $\sum_{k=0}^{2^n-1}|\rho(k)| \to \sum_{k=0}^{\infty}|\rho(k)| < \infty$, so that (D.3.1) holds. In contrast, if $H > \frac{1}{2}$ then $\sum_{k=0}^{2^n-1}|\rho(k)| \sim H(2H-1)\sum_{k=1}^{2^n-1}k^{2H-2} \sim H2^{2Hn-n}$ and (D.3.1) also holds, because $H < 1$. $\qquad\square$

As a direct application of the previous proposition, we deduce the following result.

Corollary D.3.3 *Let B^H be an fBm of Hurst index $H \in (0, 1)$, and let $p \in [1, +\infty)$. Then, in $L^2(\Omega)$ and as $n \to \infty$,*

$$
\sum_{k=1}^{2^n} \left|B_{k2^{-n}}^H - B_{(k-1)2^{-n}}^H\right|^p \to
\begin{cases}
0 & \text{if } p > \dfrac{1}{H} \\[2mm]
E[|N|^p] & \text{if } p = \dfrac{1}{H}, \text{ with } N \sim \mathcal{N}(0,1), \\[2mm]
+\infty & \text{if } p < \dfrac{1}{H}.
\end{cases}
$$

Proof Just apply (D.3.2) with $f(x) = |x|^p$. $\qquad\square$

We are now ready to prove that an fBm is never a semimartingale, except when its Hurst parameter is $\frac{1}{2}$. This explains why the study of fBm has often to be carried out without the use of semimartingale theory.

Theorem D.3.4 *Let B^H be an fBm with Hurst index $H \in (0, 1)$. If $H \neq \frac{1}{2}$ then B^H is not a semimartingale.*

Proof By the self-similarity property of B^H (Proposition D.1.4), it suffices to consider the time interval $[0, 1]$. Let us recall two main features of semimartingales on $[0, 1]$. If S denotes such a semimartingale, then:

(i) $\sum_{k=1}^{2^n}(S_{k2^{-n}} - S_{(k-1)2^{-n}})^2 \to \langle S\rangle_1 < \infty$ in probability as $n \to \infty$;

(ii) if, moreover, $\langle S \rangle_1 = 0$ then S has bounded variations; in particular, with probability one, $\sup_{n \geq 1} \sum_{k=1}^{2^n} \left| S_{k2^{-n}} - S_{(k-1)2^{-n}} \right| < \infty$.

The proof is divided into two parts, according to the value of H.

- If $H < \frac{1}{2}$, Corollary D.3.3 yields that $\sum_{k=1}^{2^n} (B_{k2^{-n}}^H - B_{(k-1)2^{-n}}^H)^2 \to \infty$, so (i) fails, implying that B^H cannot be a semimartingale.

- If $H > \frac{1}{2}$, we deduce from Corollary D.3.3 that $\sum_{k=1}^{2^n} (B_{k2^{-n}}^H - B_{(k-1)2^{-n}}^H)^2 \to 0$. Now, let p be such that $1 < p < \frac{1}{H}$. We then have, still by Corollary D.3.3, that $\sum_{k=1}^{2^n} \left| B_{k2^{-n}}^H - B_{(k-1)2^{-n}}^H \right|^p \to \infty$. Moreover, because of the (uniform) continuity of $t \mapsto B_t^H(\omega)$ on $[0, 1]$, we have that $\sup_{1 \leq k \leq 2^n} \left| B_{k2^{-n}}^H - B_{(k-1)2^{-n}}^H \right|^{p-1} \to 0$. Hence, from the inequality

$$\sum_{k=1}^{2^n} \left| B_{k2^{-n}}^H - B_{(k-1)2^{-n}}^H \right|^p$$

$$\leq \sum_{k=1}^{2^n} \left| B_{k2^{-n}}^H - B_{(k-1)2^{-n}}^H \right| \times \sup_{1 \leq k \leq 2^n} \left| B_{k2^{-n}}^H - B_{(k-1)2^{-n}}^H \right|^{p-1},$$

we deduce that $\sum_{k=1}^{2^n} \left| B_{k2^{-n}}^H - B_{(k-1)2^{-n}}^H \right| \to \infty$. These two facts being in contradiction to (ii), here again B^H cannot be a semimartingale. $\square$

D.4 Bibliographic comments

Standard references for fractional Brownian motion are the books by Nualart [98, chapter 5], Samorodnitzky [127] and Samorodnitsky and Taqqu [128], as well as the surveys by Taqqu [144] and Pipiras and Taqqu [115]. See also the forthcoming monograph by Nourdin [84].

Appendix E

Some results from functional analysis

E.1 Dense subsets of an L^q space

In what follows, we denote by $(S, \| \cdot \|)$ a Banach space over $K = \mathbb{R}$ or $\mathbb{C}$, and we write $(S', \| \cdot \|')$ to denote the dual of S (that is, S' is the collection of all bounded linear functionals from S into K, endowed with the usual dual norm). The following statement provides a necessary and sufficient condition for a given element of S to be in the span of a linear subspace.

Theorem E.1.1 *Let M be a linear subspace of S, let $s_0 \in S$, and denote by $\overline{M}$ the closure of M in S. Then, $s_0 \in \overline{M}$ if and only if there is no $T \in S'$ such that $T(s) = 0$ for every $s \in M$ and $T(s_0) \neq 0$.*

Proof This is a consequence of the Hahn–Banach theorem – for a complete proof, see [126, Theorem 5.19]. □

Now let $(A, \mathscr{A}, \mu)$ be a σ-finite measure space. The following result, known as the 'Riesz representation theorem', provides a description of $(S', \| \cdot \|')$ in the special case where $S = L^q(A, \mathscr{A}, \mu) := L^q(\mu)$ for some $q \geq 1$.

Theorem E.1.2 (Riesz representation theorem) *Let $q \in [1, \infty)$, and let $p \in (1, \infty]$ be such that $q^{-1} + p^{-1} = 1$. Then, the mapping $g \mapsto T(g)$, defined as*

$$T(g)(f) = \int_A f(x)g(x)d\mu(x), \quad f \in L^q(\mu),$$

is an isometry from $(L^p(\mu), \| \cdot \|_{L^p(\mu)})$ onto $(L^q(\mu)', \| \cdot \|'_{L^q(\mu)})$.

Proof See [32, Theorem 6.4.1]. □

Combining Theorems E.1.1 and E.1.2, we obtain a useful characterization of dense linear subspaces of L^q spaces. This result is used several times in this book.

Proposition E.1.3 *Let the above notation prevail, fix $q \geq 1$, and denote by M a linear subspace of $L^q(\mu)$. Let p be such that $q^{-1} + p^{-1} = 1$. Then M is dense in $L^q(\mu)$ if and only if the following implication holds for every $g \in L^p(\mu)$:*

$$\text{if } \int_A fg \, d\mu = 0 \text{ for every } f \in M, \text{ then } g = 0 \text{ a.e.-}d\mu. \tag{E.1.1}$$

Proof If condition (E.1.1) is satisfied, then Theorem E.1.2 implies that there is no $T \in L^q(\mu)'$ such that $T(f) = 0$ for every $f \in M$, and $T(f_0) \neq 0$ for some $f_0 \in L^q(\mu)$. We therefore deduce from Theorem E.1.1 that $f_0 \in \overline{M}$ for every $f_0 \in L^q(\mu)$, that is, M is dense in $L^q(\mu)$. For the opposite implication, assume that M is dense in $L^q(\mu)$. Then Theorem E.1.1 implies that every $T \in L^q(\mu)'$ satisfying $T(f) = 0$ for every $f \in M$ is necessarily such that $T(f_0) = 0$ for every $f_0 \in L^q(\mu)$. By virtue of Theorem E.1.2, this is equivalent to saying that if $g \in L^p(\mu)$ is such that $\int_A fg \, d\mu = 0$ for every $f \in M$, then $\int_A f_0 g \, d\mu = 0$ for every $f_0 \in L^q(\mu)$. This last relation implies that $\int_C g \, d\mu = 0$ for every measurable C such that $\mu(C) < \infty$. Using the fact that μ is σ-finite, together with a monotone class argument, we conclude that $g = 0$ a.e.-$d\mu$. $\square$

E.2 Rademacher's theorem

The following result, known as 'Rademacher's theorem', is used in this book when dealing with Lipschitz mappings.

Theorem E.2.1 (Rademacher's theorem) *Let $m, n \geq 1$, let $\Omega \subset \mathbb{R}^n$ be open, and let $f : \Omega \to \mathbb{R}^m$ be Lipschitz. Then, f is Lebesgue almost everywhere differentiable on Ω.*

Proof See [37, p. 81]. $\square$

E.3 Bibliographic comments

Two standard references for the content of Section E.1 are Dudley [32, chapter 6] and Rudin [126, chapter 5]. A discussion of Rademacher's theorem can be found, for example, in Evans and Gariepy [37, section 3.1.2].

References

[1] W.J. Adams (2009). *The Life and Times of the Central Limit Theorem* (2nd edition). Providence, RI: American Mathematical Society; London: London Mathematical Society.

[2] H. Airault, P. Malliavin and F.G. Viens (2010). Stokes formula on the Wiener space and n-dimensional Nourdin–Peccati analysis. *J. Funct. Anal.* **258**, 1763–1783.

[3] G. Anderson, A. Guionnet and O. Zeitouni (2010). *An Introduction to Random Matrices*. Cambridge Studies in Advanced Mathematics 118. Cambridge: Cambridge University Press.

[4] M.A. Arcones (1994). Limit theorems for nonlinear functionals of a stationary Gaussian sequence of vectors. *Ann. Probab.* **22**, 2242–2274.

[5] R. Arratia, A.D. Barbour and S. Tavaré (2003). *Logarithmic Combinatorial Structures: A Probabilistic Approach*. EMS Monographs in Mathematics. Zurich: European Mathematical Society.

[6] A.D. Barbour (1986). Asymptotic expansions based on smooth functions in the central limit theorem. *Probab. Theory Rel. Fields* **72**(2), 289–303.

[7] A.D. Barbour (1990). Stein's method for diffusion approximations. *Probab. Theory Rel. Fields* **84**(3), 297–322.

[8] O.E. Barndorff-Nielsen, J.M. Corcuera and M. Podolskij (2009). Power variation for Gaussian processes with stationary increments. *Stoch. Process. Appl.* **119**, 1845–1865.

[9] O.E. Barndorff-Nielsen, J.M. Corcuera and M. Podolskij (2011). Multipower variation for Brownian semi-stationary processes. *Bernoulli*, **17**(4), 1159–1194.

[10] O.E. Barndorff-Nielsen, J.M. Corcuera, M. Podolskij and J.H.C. Woerner (2009). Bipower variation for Gaussian processes with stationary increments. *J. Appl. Probab.* **46**, 132–150.

[11] B. Bercu, I. Nourdin and M.S. Taqqu (2010). Almost sure central limit theorems on the Wiener space. *Stoch. Process. Appl.* **120**, 1607–1628.

[12] H. Biermé, A. Bonami and J.R. Léon (2011). Central limit theorems and quadratic variations in terms of spectral density. *Electron. J. Probab.* **16**(13), 362–395.

[13] P. Billingsley (1995). *Probability and Measure* (3rd edition). New York: Wiley.

[14] P. Billingsley (1999). *Convergence of Probability Measures* (2nd edition). New York: Wiley.

[15] R. Blei (2001). *Analysis in Integer and Fractional Dimensions*. Cambridge Studies in Advanced Mathematics 71. Cambridge: Cambridge University Press.

[16] E. Bolthausen (1984). An estimate of the remainder in a combinatorial central limit theorem. *Z. Wahrscheinlichkeitstheorie Verw. Gebiete* **66**, 379–386.

[17] J.-C. Breton and I. Nourdin (2008). Error bounds on the non-normal approximation of Hermite power variations of fractional Brownian motion. *Electron. Comm. Probab.* **13**, 482–493 (electronic).

[18] P. Breuer and P. Major (1983). Central limit theorems for non-linear functionals of Gaussian fields. *J. Mult. Anal.* **13**, 425-441.

[19] D. Chambers and E. Slud (1989). Central limit theorems for nonlinear functionals of stationary Gaussian processes. *Probab. Theory Rel. Fields* **80**, 323–349.

[20] S. Chatterjee (2009). Fluctuation of eigenvalues and second order Poincaré inequalities. *Probab. Theory Rel. Fields* **143**, 1–40.

[21] S. Chatterjee and E. Meckes (2008). Multivariate normal approximations using exchangeable pairs. *ALEA* **4**, 257–283.

[22] L.H.Y. Chen, L. Goldstein and Q.-M. Shao (2011). *Normal Approximation by Stein's Method*. Berlin: Springer-Verlag.

[23] L.H.Y. Chen and Q.-M. Shao (2005). Stein's method for normal approximation. In: *An Introduction to Stein's Method* (A.D. Barbour and L.H.Y. Chen, eds). Lecture Notes Series, Institute for Mathematical Sciences, National University of Singapore, Vol. 4. Singapore: Singapore University Press and World Scientific, pp. 1–59.

[24] H. Chernoff (1981). A note on an inequality involving the normal distribution. *Ann. Probab.* **9**(3), 533–535.

[25] K.L. Chung (2001). *A Course in Probability Theory* (3rd edition). San Diego, CA: Academic Press.

[26] J.-F. Coeurjolly (2001). Estimating the parameters of a fractional Brownian motion by discrete variations of its sample paths. *Statist. Inference Stoch. Process.* **4**, 199–227.

[27] J.M. Corcuera, D. Nualart and J.H.C. Woerner (2006). Power variation of some integral fractional processes. *Bernoulli* **12**(4), 713–735.

[28] Y.A. Davydov and G.V. Martynova (1987). Limit behavior of multiple stochastic integral. In: *Statistics and Control of Random Process*. Moscow: Preila, Nauka, pp. 55–57 (in Russian).

[29] P. Diaconis and S. Zabell (1991). Closed form summation for classical distributions: variations on a theme of de Moivre. *Statist. Sci.* **6**(3), 284–302.

[30] R.L. Dobrushin and P. Major (1979). Non-central limit theorems for nonlinear functionals of Gaussian fields. *Z. Wahrscheinlichkeitstheorie Verw. Gebiete* **50**(1), 27–52.

[31] R.M. Dudley (1967). The sizes of compact subsets of Hilbert space and continuity of Gaussian processes. *J. Funct. Anal.* **1**, 290–330.

[32] R.M. Dudley (2003). *Real Analysis and Probability* (2nd edition). Cambridge: Cambridge University Press.

[33] D.D. Engel (1982). The multiple stochastic integral. *Memoirs of the AMS* **38**, 1–82.

[34] T. Erhardsson (2005). Poisson and compound Poisson approximations. In: *An Introduction to Stein's Method* (A.D. Barbour and L.H.Y. Chen, eds). Lecture Notes Series, Institute for Mathematical Sciences, National University of Singapore, Vol. 4. Singapore: Singapore University Press and World Scientific, pp. 61–115.

[35] C.G. Esseen (1956). A moment inequality with an application to the central limit theorem. *Skand. Aktuarietidskr.* **39**, 160–170.

[36] S. Ethier and T. Kurtz (1986). *Markov Processes. Characterization and Convergence.* New York: Wiley.

[37] L.C. Evans and R.F. Gariepy (1992), *Measure Theory and Fine Properties of Functions.* Studies in Advanced Mathematics. Boca Raton, FL: CRC Press.

[38] W. Feller (1971). *An Introduction to Probability Theory and Its Applications,* Volume II (2nd edition). New York: Wiley.

[39] R. Fox and M.S. Taqqu (1987). Multiple stochastic integrals with dependent integrators. *J. Mult. Anal.* **21**(1), 105–127.

[40] M.S. Ginovyan and A.A. Sahakyan (2007). Limit theorems for Toeplitz quadratic functionals of continuous-time stationary processes. *Probab. Theory Related Fields* **138**, 551–579.

[41] L. Giraitis and D. Surgailis (1985). CLT and other limit theorems for functionals of Gaussian processes. *Z. Wahrscheinlichkeitstheorie Verw. Gebiete* **70**, 191–212.

[42] L. Goldstein and G. Reinert (1997). Stein's method and the zero bias transformation with application to simple random sampling. *Ann. Appl. Probab.* **7**(4), 935–952.

[43] L. Goldstein and G. Reinert (2005). Distributional transformations, orthogonal polynomials, and Stein characterizations. *J. Theoret. Probab.* **18**, 237–260.

[44] F. Götze (1991). On the rate of convergence in the multivariate CLT. *Ann. Probab.* **19**, 724–739.

[45] P. Hall (1992). *Bootstrap and Edgeworth Expansions.* Berlin: Springer-Verlag.

[46] P. Hall and A.D. Barbour (1984). Reversing the Berry–Esseen inequality. *Proc. Amer. Math. Soc.* **90**(1), 107–110.

[47] D. Hirsch and G. Lacombe (1999). *Elements of Functional Analysis.* New York: Springer-Verlag.

[48] S.-T. Ho and L.H.Y. Chen (1978). An L_p bound for the remainder in a combinatorial central limit theorem. *Ann. Probab.* **6**(2), 231–249.

[49] R.A. Horn and C.R. Johnson (1996). *Matrix Analysis.* Cambridge: Cambridge University Press.

[50] C. Houdré and A. Kagan (1995). Variance inequalities for functions of Gaussian variables. *J. Theoret. Probab.* **8**, 23–30.

[51] C. Houdré and V. Pérez-Abreu (1995). Covariance identities and inequalities for functionals on Wiener and Poisson spaces. *Ann. Probab.* **23**, 400–419.

[52] Y. Hu and D. Nualart (2010). Parameter estimation for fractional Ornstein–Uhlenbeck processes. *Statist. Probab. Lett.* **80**, 1030–1038.

[53] H.E. Hurst (1951): Long-term storage capacity in reservoirs. *Trans. Amer. Soc. Civil Eng.* **116**, 400–410.

[54] I.A. Ibragimov and M.A. Lifshits (2000). On limit theorems of 'almost sure' type. *Theory Probab. Appl.* **44**(2), 254–272.

[55] K. Itô (1951). Multiple Wiener integral. *J. Math. Soc. Japan* **3**, 157–169.

[56] S.R. Jammalamadaka and S. Janson (1986). Limit theorems for a triangular scheme of U-statistics with applications to inter-point distances, *Ann. Probab.* **14**(4), 1347–1358.

[57] S. Janson (1997). *Gaussian Hilbert Spaces.* Cambridge: Cambridge University Press.

[58] P. de Jong (1990). A central limit theorem for generalized multilinear forms. *J. Mult. Anal.* **34**, 275–289.

[59] O. Kallenberg (1991). On an independence criterion for multiple Wiener integrals, *Ann. Probab.* **23**, 817–851.

[60] O. Kallenberg (2002). *Foundations of Modern Probability* (2nd edition). New York: Springer-Verlag.

[61] I. Karatzas and S.E. Shreve (1991). *Brownian Motion and Stochastic Calculus* (2nd edition). New York: Springer-Verlag.

[62] T. Kemp, I. Nourdin, G. Peccati and R. Speicher (2010). Wigner chaos and the fourth moment. *Ann. Probab.*, to appear.

[63] S. Kwapień and W.A. Woyczyński (1992). *Random Series and Stochastic Integrals: Single and Multiple.* Basel: Birkhäuser.

[64] L. Le Cam (1960). An approximation theorem for the Poisson binomial distribution. *Pacific J. Math.* **10**, 1181–1197.

[65] M. Ledoux (2011). Chaos of a Markov operator and the fourth moment condition. *Ann. Probab.*, to appear.

[66] O. Lieberman, J. Rousseau and D.M. Zucker (2001). Valid Edgeworth expansion for the sample autocorrelation function under long dependence. *Econometric Theory* **17**, 257–275.

[67] J.W. Lindeberg (1922). Eine neue Herleitung des exponential-Gesetzes in der Warscheinlichkeitsrechnung. *Math. Z.* **15**, 211–235.

[68] P. Major (1981). *Multiple Wiener–Itô Integrals.* Lecture Notes in Mathematics 849. Berlin: Springer-Verlag.

[69] P. Malliavin (1978). Stochastic calculus of variations and hypoelliptic operators. In: *Proceedings of the International Symposium on Stochastic Differential Equations, Kyoto, 1976* (K. Itô, ed.). New York: Wiley, pp. 195–263.

[70] P. Malliavin (1997). *Stochastic Analysis.* Berlin: Springer-Verlag.

[71] P. Malliavin and E. Nualart (2009). Density minoration of a strongly non-degenerated random variable. *J. Funct. Anal.* **256**, 4197–4214.

[72] P. Malliavin and A. Thalmaier (2005). *Stochastic Calculus of Variations in Mathematical Finance.* Berlin: Springer-Verlag.

[73] D. Marinucci (2007). A central limit theorem and higher order results for the angular bispectrum. *Probab. Theory Rel. Fields* **141**, 389–409.

[74] D. Marinucci and G. Peccati (2011). *Random Fields on the Sphere.* London Mathematical Society Lecture Note Series 389. Cambridge: Cambridge University Press.

[75] G. Maruyama (1982). Applications of the multiplication of the Itô-Wiener expansions to limit theorems. *Proc. Japan Acad.* **58**, 388–390.

[76] G. Maruyama (1985). Wiener functionals and probability limit theorems, I: The central limit theorem. *Osaka J. Math.* **22**, 697–732.

[77] P. McCullagh (1987). *Tensor Methods in Statistics*. London: Chapman & Hall. A partial version of this book can be downloaded from the webpage http://www.stat.uchicago.edu/~pmcc/tensorbook/

[78] N. Meyers and J. Serrin (1964). $H = W$. *Proc. Nat. Acad. Sci. USA* **51**, 1055–1056.

[79] E. Mossel, R. O'Donnell and K. Oleszkiewicz (2010). Noise stability of functions with low influences: invariance and optimality. *Ann. Math.* **171**(1), 295–341.

[80] J. Nash (1956). Continuity of solutions of parabolic and elliptic equations. *Amer. J. Math.* **80**, 931–954.

[81] E. Nelson (1973). The free Markoff field. *J. Funct. Anal.*, **12**, 211–227.

[82] S. Noreddine and I. Nourdin (2011). On the Gaussian approximation of vector-valued multiple integrals. *J. Mult. Anal.* **102**,1008–1017.

[83] I. Nourdin (2011). Yet another proof of the Nualart-Peccati criterion. *Electron. Comm. Probab.* **16**, 467–481.

[84] I. Nourdin (2012). *Selected Aspects of Fractional Brownian Motion*. New York: Springer-Verlag, to appear.

[85] I. Nourdin and D. Nualart (2010). Central limit theorems for multiple Skorohod integrals. *J. Theoret. Probab.* **23**(1), 39–64.

[86] I. Nourdin, D. Nualart and C.A. Tudor (2010). Central and non-central limit theorems for weighted power variations of fractional Brownian motion. *Ann. Inst. H. Poincaré (B) Probab. Statist.* **46**, 1055–1079.

[87] I. Nourdin and G. Peccati (2007). Non-central convergence of multiple integrals. *Ann. Probab.*, **37**(4), 1412–1426.

[88] I. Nourdin and G. Peccati (2009). Stein's method on Wiener chaos. *Probab. Theory Rel. Fields* **145**(1), 75–118.

[89] I. Nourdin and G. Peccati (2010). Stein's method and exact Berry–Esséen asymptotics for functionals of Gaussian fields. *Ann. Probab.* **37**(6), 2231–2261.

[90] I. Nourdin and G. Peccati (2010). Universal Gaussian fluctuations of non-Hermitian matrix ensembles: from weak convergence to almost sure CLTs. *ALEA* **7**, 341–375 (electronic).

[91] I. Nourdin and G. Peccati (2010). Cumulants on the Wiener space. *J. Funct. Anal.* **258**, 3775–3791.

[92] I. Nourdin, G. Peccati and M. Podolskij (2011). Quantitative Breuer–Major theorems. *Stoch. Process. Appl.* **121**, 793–812.

[93] I. Nourdin, G. Peccati and G. Reinert (2009). Second order Poincaré inequalities and CLTs on Wiener space. *J. Funct. Anal.* **257**, 593–609.

[94] I. Nourdin, G. Peccati and G. Reinert (2010). Invariance principles for homogeneous sums: universality of Gaussian Wiener chaos. *Ann. Probab.* **38**(5), 1947–1985.

[95] I. Nourdin, G. Peccati and A. Réveillac (2010). Multivariate normal approximation using Stein's method and Malliavin calculus. *Ann. Inst. H. Poincaré (B) Probab. Statist.* **46**(1), 45–58.

[96] I. Nourdin and F.G. Viens (2009). Density estimates and concentration inequalities with Malliavin calculus. *Electron. J. Probab.* **14**, 2287–2309 (electronic).

[97] D. Nualart (1998). Analysis on Wiener space and anticipating stochastic calculus. In: *Lectures on Probability Theory and Statistics. École de Probabilités de St. Flour XXV (1995)*. Lecture Notes in Mathematics 1690. Berlin: Springer-Verlag, pp. 123–227.

[98] D. Nualart (2006). *The Malliavin Calculus and Related Topics of Probability and Its Applications* (2nd edition). Berlin: Springer-Verlag.

[99] D. Nualart (2009). *Malliavin Calculus and Its Applications*. Providence, RI: American Mathematical Society.

[100] D. Nualart and S. Ortiz-Latorre (2008). Central limit theorems for multiple stochastic integrals and Malliavin calculus. *Stoch. Process. Appl.* **118**, 614–628.

[101] D. Nualart and G. Peccati (2005). Central limit theorems for sequences of multiple stochastic integrals. *Ann. Probab.* **33**(1), 177–193.

[102] D. Nualart and L. Quer-Sardanyons (2009). Gaussian density estimates for solutions to quasi-linear stochastic partial differential equations. *Stoch. Process. Appl.* **119**, 3914–3938.

[103] D. Nualart and L. Quer-Sardanyons (2011). Optimal Gaussian density estimates for a class of stochastic equations with additive noise. *Infinite Dimensional Anal. Quantum Probab. Related Topics* **14**(1), 25–34.

[104] E. Nualart (2004). Exponential divergence estimates and heat kernel tail. *C. R. Math. Acad. Sci. Paris* **338**(1), 77–80.

[105] G. Peccati (2007). Gaussian approximations of multiple integrals. *Electron. Comm. Probab.* **12**, 350–364 (electronic).

[106] G. Peccati, J.-L. Solé, M.S. Taqqu and F. Utzet (2010). Stein's method and normal approximation of Poisson functionals. *Ann. Probab.* **38**(2), 443–478.

[107] G. Peccati and M.S. Taqqu (2007). Stable convergence of generalized L^2 stochastic integrals and the principle of conditioning. *Electron. J. Probab.* **12**(15), 447–480 (electronic).

[108] G. Peccati and M.S. Taqqu (2008). Limit theorems for multiple stochastic integrals. *ALEA* **4**, 393–413.

[109] G. Peccati and M.S. Taqqu (2008). Stable convergence of multiple Wiener–Itô integrals. *J. Theoret. Probab.* **21**(3), 527–570.

[110] G. Peccati and M.S. Taqqu (2010). *Wiener Chaos: Moments, Cumulants and Diagrams*. New York: Springer-Verlag.

[111] G. Peccati and C.A. Tudor (2005). Gaussian limits for vector-valued multiple stochastic integrals. *Séminaire de Probabilités XXXVIII*. Lecture Notes in Mathematics 1857. Berlin: Springer-Verlag, pp. 247–262.

[112] G. Peccati and M. Yor (2004). Hardy's inequality in L^2 ([0, 1]) and principal values of Brownian local times. In: *Asymptotic Methods in Stochastics: Festschrift for Miklós Csörgő* (L. Horváth and B. Szyszkowicz, eds). Providence, RI: American Mathematical Society, pp. 49–74.

[113] G. Peccati and M. Yor (2004). Four limit theorems for quadratic functionals of Brownian motion and Brownian bridge. In: *Asymptotic Methods in Stochastics: Festschrift for Miklós Csörgő* (L. Horváth and B. Szyszkowicz, eds). Providence, RI: American Mathematical Society, pp. 75–87.

[114] G. Peccati and C. Zheng (2010). Multidimensional Gaussian fluctuations on the Poisson space. *Electron. J. Probab.* **15**, paper 48, 1487–1527 (electronic).

[115] V. Pipiras and M.S. Taqqu (2003). Fractional calculus and its connections to fractional Brownian motion. In: *Theory and Applications of Long-Range Dependence* (P. Doukhan, G, Oppenheim and M.S. Taqqu, eds.). Boston: Birkhäuser, pp. 165–201.

[116] M. Raič (2004). A multivariate CLT for decomposable random vectors with finite second moments. *J. Theoret. Probab.* **17**(3), 573–603.

[117] G. Reinert (2005). Three general approaches to Stein's method. In: *An Introduction to Stein's Method* (A.D. Barbour and L.H.Y. Chen, eds). Lecture Notes Series, Institute for Mathematical Sciences, National University of Singapore, Vol. 4. Singapore: Singapore University Press and World Scientific, pp. 183–221.

[118] G. Reinert and A. Röllin (2009). Multivariate normal approximation with Stein's method of exchangeable pairs under a general linearity condition. *Ann. Probab.* **37**(6), 2150–2173.

[119] D. Revuz and M. Yor (1999). *Continuous Martingales and Brownian Motion*. Berlin: Springer-Verlag.

[120] Y. Rinott and V. Rotar (1996). A multivariate CLT for local dependence with $n^{-1/2} \log n$ rate and applications to multivariate graph related statistics. *J. Mult. Anal.* **56**(2), 333–350.

[121] M. Rosenblatt (1961). Independence and dependence. In: *Proceedings of the Fourth Berkeley Symposium on Mathematical Statistics and Probability*, Vol. II (J. Neyman, ed.). Berkeley: University of California Press, pp. 431–443.

[122] S. Ross and E. Peköz (2007). *A Second Course in Probability*. Boston: Probability Bookstore.com.

[123] G.-C. Rota and C. Wallstrom (1997). Stochastic integrals: a combinatorial approach. *Ann. Probab.* **25**(3), 1257–1283.

[124] V.I. Rotar (1979). Limit theorems for polylinear forms. *J. Mult. Anal.* **9**, 511–530.

[125] V. Rotar (2005). Stein's method, Edgeworth's expansions and a formula of Barbour. In: *An Introduction to Stein's Method* (A.D. Barbour and L.H.Y. Chen, eds). Lecture Notes Series, Institute for Mathematical Sciences, National University of Singapore, Vol. 4. Singapore: Singapore University Press and World Scientific, pp. 59–84.

[126] W. Rudin (1987). *Real and Complex Analysis*. New York: McGraw-Hill.

[127] G. Samorodnitsky (2006). *Long Range Dependence*. Hanover, MA: Now Publishers.

[128] G. Samorodnitsky and M.S. Taqqu (1994). *Stable Non-Gaussian Random Processes*. New York: Chapman & Hall.

[129] W. Schoutens (2001). Orthogonal polynomials in Stein's method. *J. Math. Anal. Appl.* **253**(2), 515–531.

[130] M. Schreiber (1969). Fermeture en probabilité de certains sous-espaces d'un espace L^2. *Z. Warscheinlichkeitstheorie Verw. Gebiete* **14**, 36–48.

[131] S. Sheffield (2007). Gaussian free field for mathematicians. *Probab. Theory Rel. Fields* **139**(3-4), 521–541.

[132] I. Shigekawa (1980). Derivatives of Wiener functionals and absolute continuity of induced measures. *J. Math. Kyoto Univ.* **20**(2), 263–289.

[133] E.V. Slud (1993). The moment problem for polynomial forms in normal random variables. *Ann. Probab.* **21**(4), 2200–2214.

[134] M. Sodin and B. Tsirelson (2004). Random complex zeroes. I. Asymptotic normality. *Israel J. Math.* **144**, 125–149.

[135] Ch. Stein (1972). A bound for the error in the normal approximation to the distribution of a sum of dependent random variables. In: *Proceedings of the Sixth Berkeley Symposium on Mathematical Statistics and Probability*, Vol. II (L. Le Cam, J. Neyman and E.L. Scott, eds.). Berkeley: University of California Press, pp. 583–602.

[136] Ch. Stein (1986). *Approximate Computation of Expectations*. Institute of Mathematical Statistics Lecture Notes – Monograph Series, 7. Hayward, CA: Institute of Mathematical Statistics.

[137] D.W. Stroock (1987). Homogeneous chaos revisited. *Séminaire de Probabilités XXI*. Lecture Notes in Mathematics 1247. Berlin: Springer-Verlag, pp. 1–8.

[138] T.-C. Sun (1965). Some further results on central limit theorems for nonlinear functions of a normal stationary process. *J. Math. Mech.* **14**, 71–85.

[139] D. Surgailis (2003). CLTs for polynomials of linear sequences: Diagram formula with applications. In: *Theory and Applications of Long-Range Dependence* (P. Doukhan, G. Oppenheim and M.S. Taqqu, eds.). Boston: Birkhäuser, pp. 111–128.

[140] M. Talagrand (2003). *Spin Glasses: A Challenge for Mathematicians*. Berlin: Springer-Verlag.

[141] M. Taniguchi (1986). Berry–Esseen theorems for quadratic forms of Gaussian stationary processes. *Probab. Theory Related Fields* **72**, 185–194.

[142] M.S. Taqqu (1975). Weak convergence to fractional Brownian motion and to the Rosenblatt process. *Z. Wahrscheinlichkeitstheorie Verw. Geb.* **31**, 287–302.

[143] M.S. Taqqu (1979). Convergence of integrated processes of arbitrary Hermite rank. *Z. Wahrscheinlichkeitstheorie Verw. Geb.* **50**, 53–83.

[144] M.S. Taqqu (2003). Fractional Brownian motion and long-range dependence. In: *Theory and Applications of Long-Range Dependence* (P. Doukhan, G. Oppenheim and M.S. Taqqu, eds.). Boston: Birkhäuser, pp. 5–38.

[145] H.F. Trotter (1959). An elementary proof of the central limit theorem. *Arch. Math.* **10**, 226–234.

[146] C.A. Tudor (2008). Analysis of the Rosenblatt process. *ESAIM Probab. Statist.* **12**, 230–257.

[147] C.A. Tudor (2011). Asymptotic Cramér's theorem and analysis on Wiener space. In: *Séminaire de Probabilités XLIII* (C. Donati-Martin, A. Lejay and A. Rouault, eds.). Lecture Notes in Mathematics. Berlin: Springer-Verlag, pp. 309–325.

[148] I.S. Tyurin (2009). New estimates of the convergence rate in the Lyapunov theorem. http://arxiv.org/abs/0912.0726.

[149] A.S. Üstünel (1995). *An Introduction to Analysis on Wiener Space*. Lecture Notes in Mathematics 1610. Springer-Verlag.

[150] A.S. Üstünel and M. Zakai (1989). On the independence and conditioning on Wiener space, *Ann. Probab.* **17**, 1441–1453.

[151] F.G. Viens (2009). Stein's lemma, Malliavin calculus, and tail bounds, with application to polymer fluctuation exponent. *Stoch. Process. Appl.* **119**, 3671–3698.

[152] N. Wiener (1938). The homogeneous chaos. *Amer. J. Math.* **60**, 879–936.

Author index

Notation index

Subject index

Printed in the United States
By Bookmasters